SUSTAINING
THE WEST

ENVIRONMENTAL
HUMANITIES

SUSTAINING THE WEST
CULTURAL RESPONSES TO CANADIAN ENVIRONMENTS

Liza Piper & Lisa Szabo-Jones, editors

WILFRID LAURIER UNIVERSITY PRESS

This book has been published with the help of a grant from the Canadian Federation for the Humanities and Social Sciences, through the Awards to Scholarly Publications Program, using funds provided by the Social Sciences and Humanities Research Council of Canada. Wilfrid Laurier University Press acknowledges the financial support of the Government of Canada through the Canada Book Fund for our publishing activities.

Library and Archives Canada Cataloguing in Publication

Sustaining the West : cultural responses to Canadian environments. Liza Piper and Lisa Szabo-Jones, editors.

(Environmental humanities)
Includes bibliographical references and index.
Issued in print and electronic formats.
ISBN 978-1-55458-923-4 (pbk.).—ISBN 978-1-55458-924-1 (pdf).—
ISBN 978-1-55458-925-8 (epub)

1. Human ecology—Canada, Western. 2. Environmental sciences—Social aspects—Canada, Western. 3. Environmentalism—Social aspects—Canada, Western. 4. Canada, Western—Environmental conditions. I. Piper, Liza, 1978–, editor II. Szabo-Jones, Lisa, 1969–, editor III. Series: Environmental humanities series

GF512.P7S88 2015   304.209712   C2014-905564-1
                                                 C2014-905565-X

Cover design by Daiva Villa, Chris Rowat Designs. Front-cover image *Archipelago*, 2008 (detail); photo by Mark Freeman. Text design by Angela Booth Malleau.

© 2015 Wilfrid Laurier University Press
Waterloo, Ontario, Canada
www.wlupress.wlu.ca

Excerpts in Chapter 8 ("Poetry, Science, and Knowledge of Place") from "Apostrophe," "Astonished –" and "First Philosophies," from *Strike/Slip* by Don McKay, copyright © 2006 Don McKay, reprinted by permission of McClelland & Stewart, a division of Random House of Canada Limited, a Penguin Random House Company. Excerpt in same chapter from "Oh Lovely Rock," from *The Collected Poetry of Robinson Jeffers, Volume 2, 1928–1938*, edited by Tim Hunt, copyright © 1938, Garth and Donnan Jeffers, renewed 1966, used with permission of Stanford University Press, www.sup.org. All rights reserved. Excerpts in same chapter from "Mapmaking," "On a Mountainside," and "The Words" from *Traveling Light: Collected and New Poems*, copyright © 1999 David Wagoner, used with permission of the University of Illinois Press.

This book is printed on FSC® certified paper and is certified Ecologo. It contains post-consumer fibre, is processed chlorine free, and is manufactured using biogas energy.

Printed in Canada

Every reasonable effort has been made to acquire permission for copyright material used in this text, and to acknowledge all such indebtedness accurately. Any errors and omissions called to the publisher's attention will be corrected in future printings.

No part of this publication may be reproduced, stored in a retrieval system, or transmitted, in any form or by any means, without the prior written consent of the publisher or a licence from the Canadian Copyright Licensing Agency (Access Copyright). For an Access Copyright licence, visit http://www.accesscopyright.ca or call toll free to 1-800-893-5777.

CONTENTS

List of Illustrations — viii
Acknowledgements — xi

INTRODUCTION
What if the Problem Is People? — 1
Liza Piper

PART 1: ACTING ON BEHALF OF

CHAPTER 1
Grass Futures: Possibilities for a Re-engagement with Prairie — 15
Trevor Herriot

CHAPTER 2
Wastewest: A State of Mind — 23
Warren Cariou

CHAPTER 3
Sustaining Collaboration: The Woodhaven Eco Art Project — 33
Nancy Holmes

CHAPTER 4
A Natural History and Dioramic Performance: — 43
Restoring Camosun Bog in Vancouver, British Columbia
Lisa Szabo-Jones & David Brownstein

CHAPTER 5
A Subtle Activism of the Heart — 65
Beth Carruthers

CHAPTER 6
Sublime Animal   79
Maria Whiteman

CHAPTER 7
The Becoming-Animal of *Being Caribou*: Art, Ethics, Politics   87
Dianne Chisholm

INTERLUDE
Creating Metaphors for Change   109
Lyndal Osborne

PART 2: CONSTRUCTING KNOWLEDGE

CHAPTER 8
Poetry, Science, and Knowledge of Place: 117
A Dispatch from the Coast
Nicholas Bradley

CHAPTER 9
Deception in High Places: 139
The Making and Unmaking of Mounts Brown and Hooker
Zac Robinson & Stephen Slemon

CHAPTER 10
Escarpments, Agriculture, and the Historical Experience   159
of Certainty in Manitoba and Ontario
Shannon Stunden Bower & Sean Gouglas

CHAPTER 11
Whatever Else Climate Change Is Freedom: Frontier Mythologies,   175
the Carbon Imaginary, and British Columbia Coastal Forestry Novels
Richard Pickard

CHAPTER 12
Endangered Species, Endangered Spaces: 193
Exploring the Grasslands of Trevor Herriot's *Grass, Sky, Song*
and the Wetlands of Terry Tempest Williams's *Refuge*
Angela Waldie

CHAPTER 13
What Should We Sacrifice for Bitumen?   211
Literature Interrupts Oil Capital's Utopian Imaginings
Jon Gordon

INTERLUDE
Symphony for a Head of Wheat Burning in the Dark   233
Harold Rhenisch

PART 3: MATERIAL EXPRESSIONS

CHAPTER 14
Propositions from Under Mill Creek Bridge: A Practice of Reading   241
Christine Stewart

CHAPTER 15
Understory Enduring the Sixth Mass Extinction, ca. 2009–11   259
Rita Wong

CHAPTER 16
Seeding Coordinates, Planting Memories: 273
Here, There, & Elsewhere in W.H. New's *Underwood Log*
Travis V. Mason

CHAPTER 17
Re-Envisioning Epic in Jon Whyte's Rocky Mountain Poem   289
The fells of brightness
Harry Vandervlist

CHAPTER 18
Ware's Waldo: Hydroelectric Development and the Creation   303
of the Other in British Columbia
Daniel Sims

AFTERWORD
Humming Along with the Bees: A Few Words on Cross-Pollination   325
Pamela Banting

Bibliography   331
Contributors   349
Index   355

LIST OF ILLUSTRATIONS

4.1	Camosun Bog	43
4.2	Camosun Bog sign – A Community Working Together	46
6.1	Embryonic horse, vertical head and legs in jar	79
6.2	Curled fawn with spots in jar	81
6.3	Frog hand and body in jar	83
6.4	Embryonic fawns wrapped together in jar	84
Interlude 1	*Archipelago* (2008), detail	111
Interlude 2	*ab ovo* (2008), detail	112
Interlude 3	*ab ovo* (2008), detail	113
Interlude 4	*Endless Forms Most Beautiful* (2006–11)	113
9.1	David Douglas (1798–1834)	140
9.2	Douglas's 1828 manuscript, *A Sketch of a Journey …*	142
9.3	The first map showing Douglas's mountain giants	143
9.4	Map showing Mount Brown and Mount Hooker, 1901	145
9.5	The summit of Mount Brown	151
10.1	Map of Manitoba and Ontario scarp landscapes	160
13.1	Memorial for killed and injured workers in Waterways, Alberta	217
13.2	Syncrude's Wood Bison Gateway	224
13.3	Bison grazing on reclaimed land at Syncrude's Beaver Creek Wood Bison Ranch	224
18.1	Mike Halleran on the shore of the Williston Lake Reservoir, *CBC Hourglass*	307
18.2	Waldo, BC, burning, *The Reckoning*	309
18.3	Village of Finlay Forks from the air, *CBC Hourglass*	310
18.4	Finlay Forks, *CBC Hourglass*	310
18.5	Finlay Forks, from McKay, *Crooked River Rats*	311
18.6	"Tse Keh Nay" rivermen, *CBC Hourglass*	312

18.7	SS *Minto*, *The Reckoning*	313
18.8	Houses in the Columbia River Valley, BC, *The Reckoning*	314
18.9	Houses in the Columbia River Valley, BC, *The Reckoning*	315
18.10	Muskeg in the Peace River Country, BC, *CBC Hourglass*	316
18.11	Cabin on the banks of the Williston Lake Reservoir, *CBC Hourglass*	316
18.12	Debris at low water of a reservoir in the Columbia River Valley, BC, *The Reckoning*	318

ACKNOWLEDGEMENTS

This volume emerged from the workshop "Cross-Pollination: Seeding New Ground for Environmental Thought and Activism across the Arts and Humanities," held in Edmonton, Alberta, in March 2011. We thank Melanie Marvin and Cheryl Williams for making the workshop happen—it would not have been possible without their hard work and commitment. We thank Martha Campiou for helping us welcome participants to Treaty Eight lands. We thank all of the participants at that original event for their contributions and involvement which shaped the direction of this volume. We would also like to thank Cate Sandilands and Alan MacEachern for their enthusiasm for this collaboration.

Financial and logistical support from the Social Sciences and Humanities Research Council of Canada; NiCHE: the Network in Canadian History and Environment; the Association for Literature, Environment, and Culture in Canada; the Pierre Elliott Trudeau Foundation; the Edmonton Nature Club; and the Faculty of Arts, the Department of History and Classics, and the Department of English and Film Studies at the University of Alberta made both the workshop and this volume possible.

Thanks to Lisa Quinn, Rob Kohlmeier, and Blaire Comacchio at Wilfrid Laurier University Press for ensuring this volume saw the light of day. We thank the anonymous reviewers for their comments and suggestions which strengthened the cohesion of this collection. Heather Green and Denny Brett provided valuable assistance on various aspects of the publication.

Conceptualizing, bringing people together, working with contributors, and ultimately producing this volume has been an inspiring, enjoyable, and truly collective effort for which we are grateful.

INTRODUCTION

What if the Problem Is People?

Liza Piper

In his *The Future of Environmental Criticism*, Lawrence Buell emphasizes that "issues of vision, value, culture, and imagination are keys to today's environmental crises at least as fundamental as scientific research, technological know-how, and legislative regulation."[1] He makes this point in order to demonstrate the essential contributions from humanists to solving our environmental crises. Buell goes on to note that obstacles faced by humanists are of their own making, arising more out of internal disputes and uncertainties than out of claims to irrelevance by other parties.[2] Frankly, given where we, as a species, find ourselves with regard to the continued health of our home, the earth, any constructive approach that addresses our current circumstance must be encouraged. Buell's latter point thus goes right to the heart of why a volume such as *Sustaining the West* is needed. Through conversation and interaction, the artists, writers, and arts and humanities scholars present here speak clearly and emphatically about how we find ourselves at this crossroads, and how we might imagine ways forward.

My co-editor, Lisa Szabo-Jones, and I initiated this project from a desire to promote rigorous interdisciplinary engagement within arts and humanities disciplines and practices. For those interested in interdisciplinarity, greater emphasis is placed upon the value and need for work between the sciences and arts rather than among disciplines within the arts and humanities. This arises from the perceived gulf between these forms of knowledge (to the extent that either the "sciences" or "arts" can be considered in a unitary fashion); the importance of social studies of science (pursued by arts and humanities scholars, among others) to better understanding the form and function of scientific discovery and knowledge; and even the fact that the greater value typically accorded to scientific knowledge in contemporary Western society gives a practical impetus to non-science researchers to ally themselves with their science counterparts. Interdisciplinary encounters between the sciences and arts aim, primarily through contrast and juxtaposition, to stimulate new insights into the subject(s) of study.[3] The other more problematic form of such interdisciplinary encounters occurs where culturally rooted knowledge is transformed into data that can then be applied to science-based models. The challenge in these instances, highlighted by, but not exclusive to, the conflict between traditional ecological knowledge and Western science, is that the process strips cultural knowledge of much of its value in order to be useful to an abstract modelling process.[4] In an age when Western researchers generate highly specialized knowledge and speak primarily to small groups of experts, there is readily apparent value to interdisciplinary encounters in general—and between sciences and arts in particular—to deepen and more effectively disseminate understanding of the problems we face as a species.

In *Sustaining the West*, we proceed from the recognition that individuals working in the arts and humanities already share much common ground and produce work that either speaks to a broad public, or to narrow disciplinary audiences, yet rarely to each other. The challenges, though different from those that confront scientists and artists seeking to work together, are nevertheless significant and warrant focused attention. To bring these artists and scholars together in conversation, we organized an interdisciplinary workshop called "Cross-Pollination: Seeding New Ground for Environmental Thought and Activism across the Arts and Humanities" in Edmonton, Alberta, in the spring of 2011. We promoted cross-pollination in a number of different ways at this workshop and, subsequently, in the preparation of this volume. Beyond bringing selected artists, poets, historians, and ecocritics (among others) into a room together for three days and seeing what would ensue, we focused participants' attention on the opportunities for interdisciplinarity—and its specific application to environmental concerns—in a number of key ways. First, we asked participants to provide their contributions, which as this volume attests took a variety of forms, in advance of our meeting, and we selected commentators for each session from a different discipline or creative practice from those who were presenting. This forced the commentators and presenters to engage formally and with generosity across disciplines, as well as encouraging a range of responses from the remaining participants in our open discussions. Second, we organized public plenary sessions with two or three presenters bringing different disciplinary or creative perspectives to the fore. Again, this aimed to provide a range of voices on a given topic but also to highlight the overlaps we share in engaging with issues of environmental concern. By making these plenaries public, we reached out to the wider community in our conversation, highlighting the fact that this workshop was very much about pressing, public concerns. Lastly, we organized a field trip to Edmonton's world-renowned waste management facility, and ensured ample opportunities for casual interactions where participants could initiate and continue conversations provoked by the workshop presentations. In the aftermath of the workshop, we asked participants interested in contributing to this volume to revise their work with the view to incorporating what they learned (formally and informally) in our time together. One of the chapters (Szabo-Jones and Brownstein) was a product of a collaboration that emerged from the workshop itself. Pamela Banting's "Afterword" was authored after the volume had come together to re-engage with our goal of interdisciplinarity.

We understand collaboration as learning from one another to create something better—more communicative, more compelling, more insightful. This process, however, does not mean dulling distinct disciplinary voices or asking contributors to compromise the methodologies, rigour, and artistic licence that underpin their work. Readers will find, therefore, that this volume includes works that are more personal and intimate alongside others that appear more dispassionate or philosophical; the different voices are harmonized by their shared concern for the environment. Ultimately, this volume brings together a select group from the workshop, as well as some additional invited contributions that reinforced our central aims of interdisciplinary environmental engagement. Our hope, ambitious though it might be, is that this volume will continue to work in the world by engaging new audiences and demonstrating how we can, as artists and scholars in the arts and humanities, work together more effectively to push for change in dealing with pressing and ubiquitous environmental concerns.

Given the importance of place to the disciplines (environmental history, ecocriticism, eco-art, ecopoetics) and artists presented here, we felt the most effective interdisciplinary engagements would come from having our place in common. We chose the West for two reasons: for its physical and biological diversity and because it is for us, in the present, home. As the contributors illustrate, the West is not a *particular* place; rather, it is a range of different environments: urban, coastal, grassland, parkland, mountain, agricultural, and industrial. Each of these poses distinct environmental challenges: Where is nature in the city? How can native grasslands thrive in the twenty-first century? What happens to wildlife when confronted with oil exploitation? What happens to people when their homes and lands are submerged, or burned, or contaminated in the interests of energy, security, and profit? Each place and each question casts light on a different aspect of human relations with the rest of nature. We adopted an expansive interpretation of the West, relying on contributors to self-identify their work as situated here, rather than insisting upon hewing to particular boundaries. Readers will thus find intimate explorations of many western places in each of the chapters. These range from the Yukon and Alaska, to Utah and Nevada, over to western Ontario and right across the Prairie provinces and British Columbia. These different environments are also united at certain moments in culture and history by more than their orientation on the compass. We see this in the unanticipated connections that appear between historical and poetic considerations of mountains (in chapters by Robinson & Slemon and Vandervlist), for instance. There are likewise important cross-border comparisons in Bradley and Waldie, which connect the American and Canadian wests. Some of the contributors to this volume offer insight to larger questions of region, and the West as a region, in Canadian culture and history.[5] Beth Carruthers argues that the more recent settlement of western Canada relative to elsewhere in the country, and indeed the continent, has served to highlight the "clash of differing world views" between Aboriginal peoples and colonizers.[6] Several essays collected here examine, compare, and contrast Aboriginal relationships to place, and express the importance of Indigenous ways of knowing to understanding social and cultural relationships to nature in this part of the world. Situating ourselves in the West has meant that certain issues have figured more prominently than others, and in this way, the collection implicitly refines our understanding of the West as a region. Yet we did not intend for discussions of the West as a region to act as a significant contribution of this volume. To the contrary, to quote Harry Vandervlist in his essay on Jon Whyte, we "aim[ed] not to reify what is observed." Hence our second reason for choosing the West: that it is our home. The idea of home as an organizing locus for critical consideration speaks to one of the core themes of this volume, namely, the importance of sustainable and ethical relationships to place. Dwelling in a particular place produces intimacy; it encourages concern for long-term environmental and cultural health and brings the consequences of inaction into sharp focus. To live in the West in the present is to be confronted by change: many of us continue (like waves of immigrants before us) to come here from elsewhere or are only here in passing; even when we establish roots here, western places are changing dramatically in the face of ecological and economic forces—the deracinating power of neoliberalism in particular—that are beyond individual or community control. From organizing our 2011 workshop to shaping this volume, we have asked contributors and readers to "tarry a while" so we can stop and look for shared experiences across the larger West and in this way build a community of understanding and a network of scholars and artists to engage with the challenges of the present.

Sustaining the West serves as a positive example of and reflection upon how the diversity apparent within the arts and humanities can turn focused attention to a particular issue, in this instance, the environments, past and present, of western North America. It also acts as an example of the critical importance of the environmental humanities to addressing contemporary environmental crises. Increasingly, humanists with "environmentally oriented perspectives" are organizing themselves under the rubric of the environmental humanities, a more focused gathering of souls than the larger environmental studies groupings that have, to date, offered shelter for disparate approaches to understanding human relations to the rest of nature.[7] Although the category is relatively recent, there are works such as Hessing, Raglon, and Sandilands' *This Elusive Land: Women and the Canadian Environment* (2005) that can be seen as seminal in the Canadian environmental humanities literature. Following *This Elusive Land*, *Sustaining the West* takes an expansive, interdisciplinary approach to considering Canadian environment and culture, drawing upon different perspectives—here from artists and scholars—to enable creative and critical readings of place. *Sustaining the West* also shares with Davidson, Park, and Shields' *Ecologies of Affect* (2011) the goal to move beyond materialist readings of places to focus instead upon the role of imagination and sentiment in knowing nature. *Sustaining the West* signals an important departure in the way that it has brought visual and written art, theory, and analysis together. This volume builds upon recent trends in environmental history and ecocriticism where it focuses specifically on deconstructing knowledge of western nature and identifying the goals (both sustainable and non-sustainable) that such constructed ways of knowing serve. By bringing historians and ecocritics together in this manner, this volume demonstrates the significant similarities in origin and approach shared by these two divergent disciplines, with each giving particular attention to interdisciplinarity, the recognition of material realities, and activism.

The momentum driving the environmental humanities forward derives from a fundamental dilemma: if we are going to be honest about the environmental crises we face, the problems before us lie with people. This isn't a reference to the bogey of overpopulation and Malthusian or neo-Malthusian anxieties about the scarcity of resources; even in our degraded environmental present, there are still ample resources to sustain human life—it is distribution and equity that remain the issues.[8] But that question—of how to equitably support ever-growing populations—is just one part of a much larger canvas of catastrophe with fingerprints strewn across. There is widespread acceptance of human responsibility for contemporary environmental change. This is evidenced by the initiative to formally identify the period since about 1800 CE as a new geological epoch, the Anthropocene: a time interval "in which many geologically significant conditions and processes are profoundly altered by human activities."[9] It is also evidenced in the backlash from climate change "skeptics" who target their criticisms at evidence of the human role in climate and widely broadcast the red herrings of non-anthropogenic climate variability.[10] Notwithstanding such fringe ideologues, the role of humans in twenty-first-century environmental change is clear. Framed as such, who better to grapple with the cultural issues at the core of our environmental crises than artists, writers, and scholars in the humanities? As Warren Cariou notes in his chapter, "Wastewest," even with a scientific consensus regarding anthropogenic climate change, informed public policy does not necessarily ensue. Clearly, our current troubled relationship with the natural world in western Canada and elsewhere arises not simply from insufficient information or understanding but from a cultural disconnect: from an inability at times to see, at other times to act, upon the dramatic changes under way all around us. Some of the contributors to this volume engage explicitly with the possibilities for artists and humanities scholars to effect meaningful change. On the whole, this volume seeks new cultural avenues to reconnect with the natural world and bring its concerns to the fore, and in doing so, to perpetuate—to sustain—healthy and ethical relationships with the nature we call home.

The Poet and the Critic

Nicholas Bradley writes in his chapter on the poetry of Don McKay and David Wagoner, "Poets, like alpinists and *flâneurs* alike, are explorers. They report on what they have found." Reading, writing, speaking, listening, visualizing, and imagining are, like the work of the poet, essential environmental practices. As is evidenced in Harold Rhenisch's five poems from "Symphony for a Head of Wheat Burning in the Dark," such work connects us to the world around us and forges relationships between people and place as profound as the physical links established by working in nature. That said, the works in this volume aim not to reinforce a dichotomy opposing material and cultural engagements with nature. To the contrary, ecocriticism and environmental history have each acted as important contributions to their larger disciplines (literary criticism and history, respectively) in part through their attention to material circumstance. As Christine Stewart asserts in her chapter on the Mill Creek Bridge, "Reading/listening under the bridge, we are compelled by the material world, to see and hear its surfaces, to reconcile ourselves to the specific world within which we move ..." Imagination, in Lyndal Osborne's installations, is provoked through objects found and grown at home; imagination is also called upon to visualize past landscapes, such as unaltered grasslands, that are no longer present. Cultural practices are called upon in this volume to create relationships with the rest of nature, particularly where other kinds of interactions might not suffice.

Maria Whiteman and Harry Vandervlist each draw attention to the divide between sentiment and criticism. The practices of reading, imagining, visualizing, listening, and creating characterized above often rely significantly upon sentimental attachments to place or beings; however, such connections are not to the exclusion of critical engagement. As Whiteman notes, "Affect need not be the other of critical theory or philosophy." The place of criticism cannot be understated. The ability of the scholars whose work is presented here to demonstrate, in the case of Zac Robinson and Stephen Slemon, for instance, the cultural politics at work in entrenching the archetypal mountain landscapes of the Rockies, furthers our understanding of western landscapes as cultural products with social consequences. By ascribing character—chaos, for example, or threat—to the land, it becomes fundamentally "cultured" and is valued accordingly. Seen through a cultured gaze, physical land is transformed into landscapes, a term, not coincidentally, associated with the genre of artistic pictorial representations of nature. For British historian Simon Schama, "landscape is the work of the mind. Its scenery is built up as much from strata of memory as from layers of rock."[11] Mountains, after all, are merely contingencies of geology. They do not deliberately please, nor do they intentionally kill: any meaningful properties they hold are vested in them by human imagination, experience, power, and language. And, in turn, these values are used to justify protecting certain landscapes over others: mountains over grasslands, for instance, as Angela Waldie notes. Eliciting the cultural work of landscapes is core to the critic's project.

So, too, is identifying the values that are inscribed in places or in practices that connect and disconnect people and the rest of nature. Richard Pickard draws a thread from the logging industry's discourse of independence, evident in early-twentieth-century logging novels, to contemporary resistance to addressing the imminent realities of climate change. It is a focus upon this independence that frees the logging industry from facing the future. Warren Cariou characterizes a similar desire to free ourselves from contemplation of our waste as the "containment and sequestration ... fantasies of modern industrial culture." Such values undermine sustainable relationships to western ecosystems in the present and emphasize that we need to identify where such values are expressed and what consequences they will have, if we hope to circumvent the worst of these.

How do we know? How do we experience? And to what ends? The production of certainty, in contrast to the production of accurate knowledge, is the focus of the essay by Shannon Stunden Bower and Sean Gouglas. They interrogate how a sense of knowing nature comes into being, an essential form of analysis given that "accurate knowledge" can be ephemeral and contested in the past and the present. Certainty is nevertheless what drives competing interpretations of what is going on in nature and how those in the West who have inherited the legacies of the Enlightenment assert influence in their world and drive change. Certainty has long been produced not only by the verifiability of science, but also by other ways of knowing, as Dianne Chisholm demonstrates with wildlife biologist Karsten Heuer's powerfully unscientific experience of "becoming-animal of man."

Dwelling in the West

As noted above, the West in this volume is, for either the contributors themselves or for their subjects of study, a home and valued place. This is part of Heuer's experience, for instance, as he travels with the caribou herd sharing the caribou's nomadic experience of home in the Yukon and Alaska. Daniel Sims offers a close analysis of the images and text presented in two documentaries to show how the transformation of the Peace and Columbia rivers by hydroelectric development has created new representations of the Tse Keh Nay and people in the Kootenays who had called the region home. In this respect his chapter demonstrates how values external to a given place can transform understandings of home. Standing on home ground offers a powerful vantage point from which to consider environmental relations and consequences. Our contributors emphasize not necessarily the construction of "home" per se, but rather the significance of sustaining connections with place through remaining, belonging, and careful attention. Carruthers asserts that the history of colonization keeps many "from being fully engaged with and committed to a place," and several other contributors highlight Aboriginal relationships with western places to provide a crucial connection with a pre-colonial (or extra-colonial) experience and time. It is Nancy Holmes who writes, "One of the great problems of ethical knowledge and sustainable action is the difficulty of remaining." Her emphasis upon the need to stay is in turn elaborated upon by other contributors, who signal for us what we should do while we remain. Rita Wong calls for us to be "present" and "alert," attentive to, not fearful of the overwhelming environmental changes under way; Trevor Herriot calls this "a radical form of patience." Collectively, contributors speak to the need to establish and nurture our relationships to these places we care for, not just in the West, but beyond—to understand how cultural forces alternatively mask and reveal particular aspects of a place's history and nature, and to thus participate in a commitment to sustaining them well into the future.

Change

If remaining and sustaining seem like passive cultural acts, the contributors to this volume call them up in service of bringing about real, significant, and immediate change. Activism was one of the core themes explored at our Cross-Pollination workshop, and one that is particularly important to those concerned with the health of the environment in Canada in the twenty-first century. In 2012, in public statements and through the passage of Bill C-38, Stephen Harper and his Conservative government not only gutted essential environmental protections found in, for instance, the Environmental Assessment, Fisheries, and Species at Risk acts, but also characterized environmental activists and organizations as "radical groups" with a "radical ideological agenda."[12] Such aggressive and hostile actions and words are not isolated manifestations of the particular position of Harper's Conservatives, but rather indicative of wider global trends manifest in North America and the United Kingdom that seek to undermine concern for environmental health by promoting the notion that there is a simple choice to be made between "the environment" and "jobs," and by characterizing those who persist in choosing the environment as marginal, radical others.[13] Such tactics have a long history in North America, perhaps most evident in agribusiness's characterization of Rachel Carson as a Communist, in a time and place when such a charge could not be taken lightly, just as charges of terrorism (the unstated extension of Canada's Minister of Natural Resources Joe Oliver's "radicals") cannot in ours.[14] And, in spite of the name calling, it remains the case that, as I have written elsewhere, "the environmentally unconstrained character of capitalism ensures its continued liability for unprecedented ecological degradation."[15] If identifying the root causes of our current environmental crises makes one a radical, then so be it. Radical change is needed.

The contributors to this volume suggest creative and productive ways of moving forward, ways that respond to the needs of communities comprised of people and other animals, of the culture of forests as well as the culture of artists, and are attentive to both the ecological consequences of cultural inattention and the cultural consequences of ecological destruction.[16] Jon Gordon, for example, turns his attention to the "forms of loss caused by oil capital in Alberta" and the hopeful possibility for literature to defy attempts by government and industry "to put those losses in the service of liberal capital." Lisa Szabo-Jones and David Brownstein interrogate the value and significance of restorationist activity in Camosun Bog and conclude, "Revitalizing natural history and remnant ecological restoration through design that promotes an ecological theatre provides compelling and engaging material to inspire motivation to protect and acquire interest in the environment." These works endeavour to help us see differently and think differently about relations between people and the rest of nature in the West. Deeper understanding of these relationships will empower people to act not only on their own behalf, but also on behalf of the species and places explored in this collection. By encouraging cross-pollination, we have enlarged our community of understanding. These two contributions: (1) to empowering those who contributed to and who will engage with this volume and (2) to building a larger, more connected community, will—we anticipate—lead to acting differently on behalf of a sustainable future. We are optimists about what lies ahead.

The Collection

What follows has been organized into three parts, joined together by visual and poetic interludes from Osborne and Rhenisch, respectively, and closing with Banting's reflections on interdisciplinarity. The first part brings together essays and art to explore different ways of acting on behalf of western nature and the significance of such activism in effecting change. The perspective is variously that of the contributors themselves as artists or curators (Holmes, Carruthers, Whiteman); of others whose actions are reflected upon or theorized by contributors (Szabo-Jones & Brownstein, Chisholm); or the essays are themselves explicit calls to action (Herriot, Cariou). In the second part, the chapters examine how scientists, farmers, naturalists, poets, industrialists, and others arrived at their knowledge and understanding of western environments. Influences such as length of residence, class conflict and aspirations, and the authority given to science figure significantly in these examinations. In considering the construction of knowledge, many of the chapters explore (Bradley, Pickard, Waldie) or are themselves examples of (Robinson & Slemon) interdisciplinary approaches to knowing western nature, demonstrating the importance of intersecting alternative ways of knowing. The third and final part reflects upon sense and place. Poetry informs four of the five chapters brought together here, which is otherwise united by the contributors' concern for the sentiments, personal and projected by others, that connect people to place. The categorizations of both people and place figure importantly in such considerations (Mason, Vandervlist, Sims). Stewart, Wong, and Sims, moreover, offer sustained consideration of the particular relations between Aboriginal peoples and their historical landscapes in the West. Organizing the chapters in such a fashion is, of course, a somewhat arbitrary task: here we have chosen the optimal organization to meet with the necessarily linear structure of a book. Most significantly, the works in this volume speak clearly and passionately about matters that are global in connections and consequences, and it is to this broadest audience that we present our considerations of our corner of Earth. It is in this context that we exhort readers to approach this volume as an exploration. Though we encourage you to read it from front to back, to engage with its writings, analyses, and art, foremost we invite you to take advantage of the connections that it advances from a range of cultural responses, past and present, to the West and beyond.

Notes

Many thanks to Lisa Szabo-Jones, Zac Robinson, and Sarah Wylie Krotz for their comments and feedback on this chapter.

1. Lawrence Buell, *The Future of Environmental Criticism: Environmental Crisis and Literary Imagination* (Malden, MA: Blackwell, 2005), 5.
2. For an example of such uncertainty, see Derrick Jensen, "Loaded Words: Writing as a Combat Discipline," *Orion Magazine*, Mar./Apr. 2012, http://www.orionmagazine.org/index.php/articles/article/6698/.
3. See, e.g., Robert Crawford, ed., *Contemporary Poetry and Contemporary Science* (Oxford: Oxford University Press, 2006) or Beatriz Calvo-Merino and Patrick Haggard, "Neuroaesthetics of Performing Arts," in *Art and the Senses*, ed. Francesca Bacci and David Melcher (Oxford: Oxford University Press, 2011), 529–41. Tipping Point Canada and Tipping Point UK are organizations that practise and promote interdisciplinarity between artists and scientists. Jen Rae, one of the directors of the Canadian organization, participated in the Cross-Pollination workshop.
4. For discussions of the implications of the encounter between Indigenous knowledge and Western science see Julie Cruikshank, "Glaciers and Climate Change: Perspectives from Oral Tradition," *Arctic* 54, no. 4 (2001), 377–93; Zacharias Kunuk and Ian Mauro, *Inuit Knowledge and Climate Change* (2010), http://www.isuma.tv/hi/en/inuit-knowledge-and-climate-change; Jill Oakes and Rick Riewe, eds., *Climate Change: Linking Traditional and Scientific Knowledge* (Winnipeg: Aboriginal Issues Press, 2006).
5. For this wider literature on the West as region, see, e.g., Sarah Carter, Alvin Finkel, and Peter Fortna, eds., *The West and Beyond: New Perspectives on an Imagined "Region"* (Edmonton: University of Alberta Press, 2010); Roger Epp and Dave Whitson, eds., *Writing Off the Rural West: Globalization, Governments, and the Transformation of Rural Community* (Edmonton: University of Alberta Press, 2001); Gerald Friesen, "The Evolving Meanings of Region in Canada," *Canadian Historical Review* 82, no. 3 (2001), 530–45.
6. This characterization would not apply, however, to the part of Ontario that is included in this volume in Stunden Bower and Gouglas' chapter on historical constructions of certainty in agricultural settings.
7. Quotation from Environmental History Project, "About," http://ehp.stanford.edu/about.htm. To name a few of the many examples, there is now a journal, *Environmental Humanities*, which published its inaugural issue in November 2012, http://environmentalhumanities.org/; a transatlantic research network, http://environmental-humanities-network.org/; and the Rachel Carson Center's Environment and Society Portal includes a Multimedia Library for environmental humanities, http://www.environmentandsociety.org/mml; see also Maya Lin's *What Is Missing?* Memorial, http://whatismissing.net/#/home.
8. Amartya Sen's classic *Poverty and Famines: An Essay on Entitlement and Deprivation* (Oxford: Clarendon, 1981) remains critical in this regard. See the writings of George Monbiot for more recent examinations of these issues, http://www.monbiot.com. For a very interesting analysis of the evolution of debates over environmental crisis and resource scarcity in particular in relation to political ideology, see Paul Sabin, *The Bet: Paul Ehrlich, Julian Simon, and Our Gamble over Earth's Future* (New Haven, CT: Yale University Press, 2013).
9. For details about the use of the term "anthropocene" and some of the debate surrounding it, see "Welcome to the Anthropocene: A Planet Transformed by Humanity," http://www.anthropocene.info/en/home, and "Subcommission on Quaternary Stratigraphy: Working Group on the 'Anthropocene,'" http://quaternary.stratigraphy.org/workinggroups/anthropocene. Quotation is from the latter webpage. Most members of this

working group are scientists; however, specialists from other fields (law, political science) are also involved including environmental historian John McNeill.

10 In the spring of 2014, the so-called Friends of Science paid for a billboard in Calgary declaring: "The sun is the main driver of climate change. Not you. Not CO_2." For more on the "Friends of Science" and this controversy, see Chris Turner's post, "Why it's not enough to be right about climate change," 28 Jan. 2014, http://desmog.ca/2014/01/27/why-it-s-not-enough-be-right-about-climate-change.

11 Simon Schama, *Landscape and Memory* (Toronto: Random House, 1995), 7.

12 Bill C-38, An Act to Implement Certain Provisions of the Budget Tabled in Parliament on March 29, 2012 and Other Measures, 1st sess., 41st Parliament, 2012 (first reading, 26 Apr. 2012), http://www.parl.gc.ca/HousePublications/Publication.aspx?DocId=5524772. Joe Oliver, "An open letter from the Honourable Joe Oliver, Minister of Natural Resources, on Canada's commitment to diversify our energy markets and the need to further streamline the regulatory process in order to advance Canada's national economic interest," 9 Jan. 2012. http://www.joeoliver.ca/news/an-open-letter-from-the-honourable-joe-oliver-minister-of-natural-resources-on-canada%E2%80%99s-commitment-to-diversify-our-energy-markets-and-the-need-to-further-streamline-the-regulatory-process/.

13 For sustained discussions of this trend in Canada, see Chris Turner, *The War on Science: Muzzled Scientists and Wilful Blindness in Stephen Harper's Canada* (Vancouver: Greystone Books, 2013). For North America more broadly, see Will Potter, *Green Is the New Red: An Insider's Account of a Social Movement under Siege* (San Francisco: City Lights, 2011) and the author's blog, http://www.greenisthenewred.com/blog/; and the film, *If A Tree Falls: A Story of the Earth Liberation Front*, DVD, directed by Marshall Curry and Sam Cullman (New York: Oscilloscope Pictures, 2010).

14 Mark Stoll, "Rachel Carson's *Silent Spring*: A Book that Changed the World," Environment and Society Portal, http://www.environmentandsociety.org/exhibitions/silent-spring/personal-attacks-rachel-carson. Sun News explicitly connected "radicals" and "eco-terrorists," in an article, "One in Two Worried about Eco-terrorist Threats: Poll," 20 Aug. 2012, http://www.sunnewsnetwork.ca/sunnews/politics/archives/2012/08/20120820-072651.html. The article opens, "The Conservative government's verbal attacks on 'environmental and other radical groups' have sparked a fear, most prevalent among Conservative voters, of an eco-terrorist attack on Canada's energy infrastructure, a new poll has found."

15 Liza Piper, "Nature, History, and Marx," *Left History* 11, no. 1 (2006), 45.

16 Activism, particularly on the environmental front, is an important theme for academics in the twenty-first century. See, e.g., Susan A. Crate and Mark Nuttall, eds., *Anthropology and Climate Change: From Encounters to Actions* (Walnut Creek, CA: Left Coast Press, 2009).

PART 1
ACTING ON BEHALF OF

CHAPTER 1

Grass Futures
Possibilities for a Re-engagement with Prairie

Trevor Herriot

I want to start by telling you a story. I learned it from a friend, Margaret Hryniuk, who was part of the trio that produced a wonderful book a couple of years ago, *Legacy of Stone*, about the stone buildings of Saskatchewan. The story though is about a woman named Mary Ann McNabb. Mrs McNabb came from Scotland with her husband to homestead on the prairie at the foot of Moose Mountain in Saskatchewan in 1882. To hold on to your claim you had to improve the land, which meant raising a building and ploughing at least thirty acres. But things did not go well for the McNabbs. Within a couple of years Mr McNabb and several of their children died, and Mary Ann was left alone with two remaining children. A covetous neighbour who wanted her land erected a building on her unimproved quarter. Then he reported that it was abandoned and claimed it for himself.

A Presbyterian minister wrote a long letter appealing to the Department of the Interior on Mrs McNabb's behalf, citing the biblical injunction to plea for the widow. Ultimately the government gave the land to the neighbour because Mrs McNabb had done nothing to break the native prairie. Taking it over, the neighbour, whose name, significantly, was "Philander," turned around and sold the land. And like the rest of the land in the area it was quickly broken.

This kind of thing happened again and again in prairie places because at the foundation of our law and social contract was this principle that to possess land you must break it; to civilize a place and settle in the landscape, it must first be legally alienated and then broken.

I will come back to that idea of breaking land later, but for my purposes here I am not going to go into too much of the nasty details of what we have done to our grasslands up to now. I will give just enough to provide the baseline from which we can look at possible futures—so we understand where we are starting from.

Instead, I will focus on possible directions for the future of prairie. Which is a bit more fun—partly because it is speculative and there are so many possibilities and factors and so we have some freedom to imagine how things might be—and no one can say with any certainty that we are wrong or right.

The main reason I want to speak about possibilities is that prairie conservation can be an utterly depressing and hopeless endeavour. Hope is something you have to search for in the dim light of the present moment and in prairie conservation we don't do near enough groping about for ways ahead, we don't spend enough time looking for the means that are worthy of our ends.

First though we do have to name the present moment. Where are things at for our prairie ecology right now? Ecologists tell us that we have four kinds of grassland in the 241,000 square kilometres that make up Canada's Prairie Ecozone, four "ecoregions": Aspen Parkland, Moist Mixed Grassland, Mixed Grassland, and the Cypress Hills Uplands. Of course, we have lost a lot of our native grass since we settled the prairie. How much is gone?

The figures are bad, but it really struck home last year when I read a *National Geographic* article on the loss of the rainforest in the Amazon basin. They reported that 20 percent of that rainforest is gone, mostly to agriculture, 80 percent remains.[1] That is terrible and we must all do what we can to stop it, but here on the Canadian plains those numbers are flipped. We have lost 80 percent and have only 20 percent left. Using satellite imagery, researchers estimate that we have somewhere between 17 and 21 percent of our native grassland remaining.[2] Alberta has preserved more of its native grass—around 30 percent, Manitoba has only 18 percent of mixed-grass prairie remaining, and Saskatchewan perhaps 20 percent.[3] In some landscapes where the soil was particularly fertile and good for growing crops, the numbers are much worse. Where I live on the Regina clay district—excellent soil—we have 0.03 percent, less than one percent of the native grassland is left. Only 10 percent of the grassland original to the Aspen Parkland, where we are here in Edmonton, remains and very little is protected.

It would be better perhaps if most of the 20 percent of native grass remaining were all in one big chunk, but of course what we have is a prairie that is cut up into a thousand pieces, varying in size and for the most part surrounded by cropland. What this means, of course, is that the native biodiversity of the prairies is in rapid retreat. Grassland birds, for example, are generally recognized as having experienced the greatest declines of all bird groups, more than forest birds, wetland birds, Arctic birds, and so on. The Canadian Wildlife Service estimates that twenty-one out of twenty-four grassland birds are in decline, and in the last forty years their overall populations have dropped by half. What will happen in the next forty years? The Native Plant Society of Saskatchewan estimates that more than 24 percent of our remaining native grassland is at medium-to-high risk of being broken. Things like subsidies for biofuels and advances in crop development, including crop varieties that can grow under drier conditions and on infertile soil, will continue to threaten the prairie.

Before I try to describe alternative futures, it's important to name the monster, the thing that is behind all of the damage. And, of course, that monster is us.

It's our entire model of progress and prosperity and wealth, which is founded on making energy and food easy to acquire so that very few people actually have to get their hands dirty. The land is suffering under the effects of the technologies and trade policy we apply to make sure that food and energy are relatively cheap and abundant, freeing the majority of people to dedicate their labour to other pursuits: selling real estate, designing video games, running liposuction clinics, whatever. But the last thing we want to do is grow most of our own food or use our own bodily energy to get some work done.

The highest goal in conventional agriculture has been to find a method, breed, machine, or chemical that will increase overall yield at minimal cost. And that pursuit as much as anything is destroying the prairie and destroying other biomes all over the planet. Changing that reality and overcoming human desire for comfort, ease, and progress at all costs is, of course, a daunting task and has wider implications for much more than prairie conservation.

The good news is that we are living in a time when the work of making that change is under way. A lot of people are finally calling the established norms of high-yield, industrialized agriculture into question and asking if it might be saner of us to have another goal for our agriculture: to produce food that is healthier and to do it in ways that will keep the land's communities and human communities healthier as well.

So, now to look at five ways toward a better engagement with prairie.

Better Prairie through Better Eating

For each of these possibilities I think we need a champion or symbol, and for this one our champion is food writer Michael Pollan. Pollan's motto—"Eat food. Not too much. Mostly plants."—is good advice for keeping ourselves healthy, but in his writing he also looks at the virtues of grass-based agriculture.[4] The "mostly plants" part recognizes that we are going to keep eating animal products but shouldn't eat too much. But Pollan and many others have shown that eating animal products from animals that eat only or mostly grass is much better for us and for the ecology of agricultural landscapes. Millions of acres of former grassland are currently used to grow feed grains for industrialized livestock operations responsible for cheap dairy products and beef, pork, and poultry. Using the prairie as feedstock for this kind of unsustainable agriculture pollutes watersheds, contributes significantly to climate change, breeds *E. coli*, and produces meat that contains unhealthy fats, antibiotics, and other chemicals, at low prices that encourage us to eat far too much animal product. If we switched to much healthier and more sustainable grass-fed livestock, we'd have a lot more land that could be put back into grass, we'd sequester more carbon, and we would all be healthier and slimmer. And if we fostered a market for beef finished on ecologically ranched native grass, we'd have a stronger economic incentive protecting native grassland from those who would like to convert it to cropland.

Better Prairie through Better Science

The person who symbolizes this model best is Wes Jackson. Jackson is a plant geneticist who uses conventional plant breeding as opposed to messing around with DNA. He has been patiently working with native prairie grasses for several decades, trying to create what he calls a "perennial polyculture."[5] What he is imagining is a major revolution in agriculture, which for thousands of years has been based on annual monocultures. Dependency on annuals rather than perennials and growing them in monocultures has put our agriculture in conflict with nature and natural processes. The use of annuals means that every year you have to cultivate and seed the soil, which erodes it, releases greenhouse gases, and sends soil and nutrients into local waterways. By growing crops in monocultures, getting plants to produce maximum yield usually requires pesticides and other harmful practices. Jackson's Land Institute in Kansas' Tallgrass Prairie Region is working on a new paradigm, which is to mimic the processes, efficiency, and productivity in natural prairie. I think this is one of the more promising paths toward a renewed engagement with the prairie, but it will not happen overnight. We should be putting more of our university dollars into this kind of agricultural and genetic research, and much less into supporting agribusiness and its destructive technologies.

Better Prairie through Better Use of Wealth

The symbolic figure here is Ted Turner, the media mogul and billionaire who is the largest landowner in the United States. He now has more than two million acres, a lot of it grassland, protecting 45,000 bison and 250,000 prairie dogs. You can do worse things with money than create a land trust foundation to protect prairie. Others are following his lead. Wealthy benefactors and smaller donors are now donating millions of dollars to land trust organizations that are buying up grassland especially in the United States. Here I am thinking, of course, of groups like the Nature Conservancy in the United States and

Nature Conservancy of Canada, but a lesser-known organization I want to mention is the American Prairie Foundation. This fast-growing land trust currently protects more than 180,000 acres of native grassland in a large reserve in north-central Montana. It has a growing herd of two hundred bison and its goals are "to accumulate and wisely manage, based on sound science, enough private land to create and maintain a fully-functioning prairie-based wildlife reserve."[6] This kind of big-thinking philanthropy will not solve all of the prairie's problems, but there is definitely a role for private money to play in helping to conserve our grassland.

Better Prairie through Better Community and Governance

For a figurehead, I could not think of a suitable model other than to use nature herself. The land and the water are wonderfully self-governing and are the ultimate models for the governance and distribution of resources. By governance I mean the way we steer our course in all of our organizations and institutions that have a say over how land and resources are used, so not just in official government but in NGOs and in small community organizations. This look toward a possible future asks, What if we had better leadership, management, policies, guidance processes, and decision making? And what would that look like?

The defining crises of our time, climate change, the global water crisis, the economic crisis, and general environmental degradation, have been brought on by human governance systems that are structured to serve the rights of autonomous individuals or corporations to pursue their private interests no matter what it does to the commonwealth. We all support this self-destructive model of governance by voting with our pocketbooks, as consumers. That in turn gives our influence over to the economists and corporations, which now seem to represent the public interest by proxy. The other elite our policy-makers and governance systems consult is composed of scientists and technologists. However, science and technology are only listened to if their council serves the same agenda that the business and economic world exists to serve, which is, again, the unlimited right of the individual and the corporation to act autonomously in pursuing private wealth and happiness.

Better governance would be driven by a healthy tension between the rights and freedom of the individual and the need to sustain the commonwealth we all depend on: healthy air, water, soil, biodiversity. If our institutions and decision-making bodies were aligned to the limits of and opportunities offered by the land and local ecology, they would consult a different set of oracles before acting. Science would still be there, but guided by the larger values of community, ecology, and health, instead of the right to become as rich as possible. But in the place of business leaders and economists, there would be the community's commonwealth of wisdom and traditional knowledge—Indigenous and non-Indigenous—arising from both humanist and cultural/spiritual traditions based in a respect for nature and natural processes. A kind of ecoliteracy in the general public would ultimately have to replace our consumer literacy.

This kind of talk sounds dangerously utopian, but history has a few encouraging examples for us to consider. Prairie people have in the past made decisions to act together in co-operatives and small community-based groups to find ways to create a more equitable and just society. The farmer co-op movement that began in Saskatchewan, for example, at one time helped defend farmers from the predatory practices of the grain cartels. The original prairie farmer impulse to cooperate with neighbours, sharing equipment, trading labour, helping out during a crisis has gone into remission, but it is not dead. It could rise again, if we had governance that fostered community cooperation, the public good, and social capital instead of only fostering competitiveness, individual ambition, private wealth, and financial capital.

On other continents, this kind of Community Ecological Governance is already happening. In African, Asian, and Latin American communities, where corporate interests have destroyed entire watersheds and privatized aquifers, the people are organizing community-based resistance and turning things around, saving their soil and water and replacing bad governance with good. Here we might at least begin to take some steps in that direction, and make smaller changes that would begin to conform our human governance systems to the design and processes within nature.

One of those steps might be to overhaul the spurious "stakeholder consultation" processes that we currently pay lip service to and replace them with a council of respected elders who represent communal and not private or corporate interests. Here are some other things we could do right now:

We could find ways for government policy to reward the use of human energy, in general but particularly in growing food. As Wendell Berry likes to say when people brag of "labour-saving devices": "there is no such thing as a reservoir of bodily energy ... by saving it we simply waste it."[7]

We could begin to use the tools of government policy to put some restraint on the accumulation of property. Our tax systems and farm policy could favour small family farms with diversified operations that are feeding local communities rather than the mega-farms producing grain for international markets and feedlots.

We should make low-interest loans available for people who want to buy family-size farms to grow food in ways that work with and conserve natural systems.

We could implement programs to promote local food self-sufficiency, fostering direct marketing from farmers to consumers and encouraging producer and consumer co-ops.

We could disengage biotech and agribusiness interests from our universities. If we could turn our agriscience resources away from the corporate world that has made food into a commodity for profit and toward the growing of food that is healthy for people, communities, and the land, we'd receive a big boost in the transformational work we are facing as a civilization.

Finally, we could begin to make more grassland. In as little as ten years you can take cropland and grow a facsimile of native grassland—not as diverse but a net gain over what was on the land before. These places sequester carbon, build the soil, protect watersheds, and almost always show improved biodiversity over tame grass fields. Returning large tracts of prairie to native cover is a vital step in halting the decline of grassland ecologies but in the Prairie provinces we have scarcely begun to talk about it.

Our policy-makers must begin to see that every decision they make about technology, agriculture, or economic development has to be measured against the absolute good of health and wholeness. What will this tool, this incentive, this project do to the health of individual people, to the health of our families and communities, to the health of the land and its ecosystems? Of course, it is easy to scoff and say that all of this is far too idealistic, that it simply won't happen because no one wants to give up the easy access to cheap food that we currently enjoy. If that were true there would be no local food, fair trade coffee, and organic food movements, no outcry against genetically modified food, intensive livestock operations, and pesticide use. Yes, these are relatively small movements today but they are making a difference and growing fast. Imagine what we could achieve if we could divert even one-tenth of the energy and ingenuity we currently devote to the accumulation and protection of wealth and put it to work finding ways to grow and distribute food that arise out of respect for the land and the health and wholeness such a respect fosters.

I would like you to think again of the story of the widow Mrs McNabb losing her land because she did not break it soon enough. We founded our prairie culture and economy on a principle that said to possess land you have to break it. The sum of what I have been pondering here is a new social contract with prairie, a new covenant as it were. Instead of to possess land you must break it, we would have incentives and disincentives urging the opposite: to possess land you would want to keep it whole, or if it has been destroyed you would want to truly improve it, restore it, heal it. In conservation circles it seems we are always starving for hope, but our only real hope is to align ourselves with the healing and recovery nature offers. Think of it as a radical form of patience where we invest in the distant future by choosing the right way to act today. In deforested lands that means planting trees even though you may never see them become a true forest. In prairie damaged by years of bad agricultural policy, hope is having the patience to plant grass and trust that, with the right care and attention, it will come to good.

Notes

1 Scott Wallace, "Last of the Amazon," *National Geographic*, Jan. 2007, 44–71.
2 A.M. Hammermeister, D. Gauthier, and K. McGovern, *Saskatchewan's Native Prairie: Statistics of a Vanishing Ecosystem and Dwindling Resource* (Saskatoon: Native Plant Society of Saskatchewan, 2001), 5.
3 Ibid., 6–8; E. Saunders, R. Quinlan, P. Jones, B. Adams, and K. Pearson, *At Home on the Range: Living with Alberta's Prairie Species at Risk* (Lethbridge: Alberta Conservation Association and Alberta Sustainable Resource Development, 2006), 2; C.J. Lindgren and K. De Smet, *Community Conservation Plan for the Southwestern Manitoba Mixed-Grass Prairie Important Bird Area*, prepared for the Canadian Nature Federation, Bird Studies Canada, BirdLife International, and the Manitoba Naturalists Society, Winnipeg, Manitoba (2001), 12.
4 Michael Pollan, *In Defense of Food: An Eater's Manifesto* (New York: Penguin, 2008).
5 See, among other of his publications, Wes Jackson, *New Roots for Agriculture* (San Francisco: Friends of the Earth, published in cooperation with the Land Institute, 1980).
6 American Prairie Reserve, "About: Mission," http://www.americanprairie.org/aboutapf/.
7 Wendell Berry, *The Unsettling of America: Culture and Agriculture* (San Francisco: Sierra Club Books, 1986).

CHAPTER 2

Wastewest
A State of Mind

Warren Cariou

We humans are destroying ourselves with our waste. It floats in our air, it seeps into our water, it penetrates into every corner of our world and our lives. Much as we try to move it away from us and make it disappear, our waste always finds its way back into our ecosystems, our neighbourhoods, our bodies. I am thinking of waste here not in the narrow sense of excrement, but rather in a more general sense that is encapsulated by Georges Bataille's idea of excess. What Bataille calls "*la part maudite*" ("the accursed share"), in his three-volume work of that title can be defined as whatever material is left over, expended, unaccounted for, or repressed in our attempts to create value, or simply to live our lives. By calling it "our" waste, I am reasserting our intimate connection to that waste, our responsibility for it, even though we would often prefer to believe that it is not ours at all because we have jettisoned it, expunged it from our consciousnesses. But as anyone with the most basic understanding of natural systems (or Freudian economies of repression) will know, what is left over does not conveniently vanish, much as we might want it to. It persists. It builds up. And eventually we have to come to terms with it whether we want to or not.

Our waste is altering the earth's climate, decimating the natural world, destabilizing our economies, and making us sick. In many ways, our relationship to our waste will determine the course of our future on this planet. And yet in the current state of environmental crisis that has been brought on by our inability to contain our waste, not many of us have stepped back and asked ourselves: what is this waste, anyway? Could we choose to relate to it differently? If we changed the way we think about waste, might that enable us to make the practical changes that are going to be necessary in order to leave our grandchildren with an inhabitable world?

I confess I am not an optimist about human nature. Some people believe that if the scientists can only get their stories perfectly straight about what is going on in our environment, we will all listen to them and act accordingly. Unfortunately, I see very little evidence to suggest this will happen any time soon. In the case of climate change, it is already quite clear that the scientific community overwhelmingly agrees that the waste products of human activities are causing disruptions of climate patterns and that this process will become much worse unless we do something drastic very soon. We know what we should do: change our lives so that we produce far less carbon-based waste material than we currently do. That is the only solution that the scientific data suggest. So yes, we know what we *should* do, but the big question is: *will we do it?* And how can we increase the odds that enough of us will do it?

The answer to those questions is for the most part not scientific but cultural. A huge part of our current problem is that we are clinging to a set of cultural values that has got us into this mess in the first place, and that will certainly make things worse if we don't disengage from it. I call this set of cultural values "the wastewest." The wastewest is a state of mind as well as a history and a set of practices. It is also a series of relationships—economic, ethical, and theological—that has produced a great deal of wealth but has also brought with it a legacy of increasingly negative consequences, most pertinent for my present argument being the consequences for the environment (though I could focus instead on human rights consequences, or consequences for community identity). The wastewest is a particular attitude toward waste that is embedded in Western culture and that is also implanted within the West's most successful export product: the ideology of globalized capitalism. Thus it has now become, I would argue, a worldwide phenomenon, belonging to ideologies of development and modernity wherever they are found. This attitude can be described quite neatly in psychoanalytic/Marxist terms: waste as the unconscious, essentially, or as the Kristevan abject; waste as the repressed term of modernity.

To elaborate briefly: in the wastewest, humans' relationship to their waste products, be they sewage or industrial waste or environmental devastation, is characterized by a movement of separation or repression. One might even see this movement as a defining feature of modernity, this need to put the waste out of sight, to keep it away from what we consider to be ourselves. We can see this very clearly in the implementation of modern sanitation systems, which of course serve a practical and important health function by keeping our excrement at a safe distance from our living quarters. When Western countries sponsor "development" in less wealthy nations (or in their own Indigenous communities), one of the things they tend to focus on is providing sanitation.

And while I agree that there is much to be said for the health benefits of having good sanitation, I also believe that the naturalization of such systems can reify a dangerous misapprehension: they make it all too easy for us to believe that our waste is truly flushed away into a magical zone where we will never have to interact with it again. We think we can forget about our waste because we have the benefits of "modern" sanitation, which save us from having to deal with it in a more intimate way. What it means to be modern, in a sense, is to be insulated from your excrement.

And hallelujah for that, you might say. Fair enough. I am not arguing that we should do away with our sanitation systems, which of course also hugely mitigate the environmental damage that "raw" sewage would otherwise do. But I am arguing that we need to be aware of the psychological and even ontological side effects of these systems. They teach us that we can be separate from our waste. In fact I would go so far as to say that they indoctrinate us into a belief that this process of separation is in some way a sign of our modern humanity. And one of the problems of this situation is that if you believe you can truly and forever separate your waste products from yourself, then you cease to care about how toxic or virulent that waste becomes.

It seems clear that this kind of self-deception about waste is occurring on a massive scale in contemporary industrialized societies. The leftover materials of industrial production are subject to a gigantic act of collective repression. We don't want to see them, even though they are sometimes hiding in plain sight. And when they do become visible, people generally want to get them as far away as possible, back out of mind. Thus we see the prevalence of the Not In My Back Yard phenomenon. The ideology of the wastewest also explains the popularity of terms like "containment" and "sequestration" in industrial language about waste. We often hear catchphrases like "carbon capture and storage," "waste containment facility," and "carbon sequestration." The enticing thing about these terms is that they tie into the hope for a separation between self and waste. They replicate the ideology of sanitation, in a way. But the notions of containment and sequestration are really more psychological strategies than they are valid ways of dealing with waste. Dams break. Nuclear containment facilities fail. Tailings ponds leach. This is the nature of our physical environment. As far as I am aware, there are no known cases in which a part of our environment has been completely and permanently sequestered away from everything else. Yet we persist in believing that we can create an exception to this phenomenon. So, rather than devoting ourselves to stopping the production of dangerous waste materials, instead we buy into the convenient notion that we can simply separate them from our living space and that we will then be freed from dealing with the consequences.

Of course, many activists and artists have been trying to point out the wastewest's ideologies of self-deception for a long time by drawing public attention to the aspects of our way of life that people don't like to see. Edward Burtynsky is one of the most accomplished contemporary artists to do this. His work is about simultaneously revealing and reshaping the meaning of contemporary industrial waste, and he definitely walks a line between aestheticizing and condemning the industrial practices that create such waste. But to me the most important function of his art is that it makes people see what they wouldn't normally want to see. If that vision must be sugar-coated with aestheticism, so be it. Burtynsky's work at its best is not simply landscape art but is instead a view into the unconscious of modernity. The hellish landscapes of his recent *Alberta Oil Sands* series reveal something that may not seem entirely "real" because we have trouble conceiving of the nightmarish reality depicted there. We are being presented with an extraordinary spectacle of un-containment, of waste become sublime.

Burtynsky is participating in a photographic tradition alongside artists like David T. Hanson, whose 1997 book *Waste Land* presents aerial photographs of many industrial waste sites as well as military installations such as missile silos and landing strips. Hanson's work functions similarly to Burtynsky's in its focus on revealing what is normally not visible, and he relies upon the aerial perspective to show things that are not accessible from the ground because the companies generally do not want photographers documenting their waste practices. (Burtynsky is an exception to this, since he is able to gain access to these places, probably because of the way he aestheticizes his images of waste in such a way that they can be seen as gorgeous abstracts, and indeed they can be misinterpreted as validations of industrial processes as a kind of art production.) Where Hanson differs from Burtynksy is in the geographical specificity his work provides: on the facing page of each image in the *Waste Land* series, he includes a detailed map with coordinates to show exactly where each part of the "waste land" is located in the real world, and he also includes a brief narrative outlining some of the documented violations of the U.S. Environmental Protection Act (EPA) that have occurred there. This mapping foregrounds the political as well as the psychological dimensions of Hanson's work: he is very concerned with anchoring these images in the real, so they cannot easily be dismissed as merely aesthetic fancies.

Artists like Hanson and Burtynsky illustrate my point that contemporary ideologies of containment and sequestration are the overriding fantasies of modern industrial culture, and I could use the rest of my space here to give further documentation of that point with examples from contemporary art. But I would like to shift the investigation to look at the wastewest from a different perspective. We certainly do need to look critically at the ideas and cultural beliefs that have led us to this place, as Burtynsky and Hanson and many activists continue to do, but I think we also need to try to find new ideas, new cultural norms and values that might enable us to take the necessary action to avert the environmental catastrophe that most scientists predict will happen within the next two hundred years. And I hold out a tenuous hope that we can accomplish this by trying to change the culture of the wastewest: by altering collective beliefs about the human relationship to waste.

Of course, for a very long time, many non-Western cultures have been embodying alternative ways in which humans can relate to the waste products they produce. These cultures are sometimes described today as "traditional cultures," or "hunter-gatherers," or even "pre-capitalist societies," and they are often the groups that are most threatened by the juggernaut of modernity. But I think they are, in fact, incredibly valuable precisely because of their differences from modern mass culture. I believe they can teach us something about how to negotiate the contemporary crisis of waste, because they can present alternatives to the wastewest's untenable ideas of containment, sequestration, and Not In My Back Yard. Traditional cultures contain the knowledge and the ethical sensibility that can help the global community to regain a sense of proximity to our waste, and thus a responsibility for it.

My experience with traditional cultures is mostly limited to the Métis and Cree traditions, so that's what I will focus on here. I believe both of those traditions often suggest a relationship to waste that is very different from what we see in the wastewest. In my own family, for example, there seems to be little stigmatization of waste spaces such as garbage dumps. Going out to the dump—and staying there all day—is not viewed as a morally questionable activity but rather a perfectly reasonable thing to do, because you can find good stuff in there! My Uncle Eli recently retired from the only real job he's had in the last twenty years, as the manager of the Ituna (Saskatchewan) Landfill. He doesn't call it the landfill, though. He calls it the Métis Mall. For him, the garbage dump is a cornucopia, a source of all kinds of items that he can make use of. He is happy to be out there in the dump, and in fact, he goes there regularly even now that he's retired. For him the dump is not a place to be repressed or avoided, but rather a place to be examined closely because of the many valuable things it holds. He doesn't send his junk there to put it out of his sight; instead he goes there to see what new things other people have jettisoned. I have heard many similar stories about other Indigenous communities that treat their garbage dumps not as wastelands but rather as places of exchange. My colleague Peter Kulchyski tells the story of his students in Pangnirtung, Nunavut, doing a study of the town dump as a space utilized for socializing and mutual exchange of items. Something like a mall, perhaps. Some of the students emerged from the study with new clothes gleaned from the dump.

You might say: well, that's simply recycling. However, I think it's not exactly recycling, or not only that. My Uncle Eli's joyful attention to garbage bespeaks a different general attitude toward it than what we see in recycling, which is essentially about a different kind of containment: a capturing of value, or a minimization of one's "waste footprint." For Eli, being at the dump is an activity of gathering, very similar to what we in the family do when we go berry picking or go hunting. The personal contact with the waste materials is important to him, as is the sense that newly deposited materials represent an opportunity for new finds. He enjoys his time at the Métis Mall.

I fear that I am perpetuating an unfortunate stereotype here, in my description of Métis people frequenting garbage dumps, and I have to state categorically that *not all Métis people do this!* My point, though, is that hanging out in the garbage dump need not, and should not, be thought of as a sign of depravity or cultural insufficiency. To me, Uncle Eli's activities at the Métis Mall in some ways represent a persistence of a traditional way of life, one that is signalled again and again when I listen to Cree and Métis Elders talking about their traditional practices. My idea for this essay can be traced back to my observation, some years ago, that when Elders want to describe their people's traditional ways of life, they often use the phrase "Nothing was wasted." It seems that for these Elders, the attitude to waste is something that clearly distinguishes their traditional ways from contemporary Westernized values. For example, Granny Mary Fletcher, a Cree Elder from Norway House, says of her grandchildren, "If only they could see how the elders used every part of the animals. Nothing was wasted. Every part of the animal was used in one way or another."[1] The renowned Omushkego Cree storyteller and Elder Louis Bird expresses a similar sentiment when he says, "According to our ancestors, everything works in order, systematically. Nothing was overused, there was nothing that overextended its usefulness or its benefit to humans."[2] I have heard similar statements from many other Elders over the years, and I think these comments signal something very important about the relationship between humans and the natural world. They indicate that there is a moral imperative to make use of everything given to us by nature, and that taking too much, or not using what you have taken, are serious transgressions. Uncle Eli's corollary to this idea would be that using what someone else has wasted is also part of one's responsibility toward the natural world.

Louis Bird expands upon this notion of waste as transgression in his teachings about the concept of *pastahowin*, which he translates as "a blasphemous act"[3] or as "a sin against nature."[4] He explains that a hunter is taught "never to kill an animal for nothing, never to kill an animal and leave it there to rot and waste. If he does that, he has committed a sin against nature, a *paastaho*, and he will not be able to kill the animal until he has declared that he has done so and why he did it."[5] Bird also explains that the idea of *pastahowin* is closely connected to the importance of respect for the natural world: "The way it was then, before the appearance of the European, the teachings were about how to respect animals and all nature. There were rules about respecting nature and the environment—the animals and the birds. If one of these were broken by a member of the family, a kid maybe, the punishment was a retraction of the benefits from nature."[6] The only way to make up for this kind of transgression, this lack of respect for nature, is to speak about it. In the example of the hunter, it is not until he talks in public about his transgression, makes his wasteful actions known to everyone, that nature will allow him to have some of its bounty again. This making-public of the waste is very different from what we see in the wastewest, where waste is kept hidden and the entire culture in a way colludes to keep it from being brought into plain sight. In the Cree conception, nature itself creates the punishment for a *paastaho*, by withholding its gifts, its bounty. And in a sense the transgressor's relationship to nature—and his or her ability to support himself or herself—is interrupted until that transgression is made public, and is thereby atoned for.

If only the modern world of the wastewest operated this way. In a sense, though, it does; it just takes a somewhat longer time for nature to respond to the transgressions when the perpetrators refuse to acknowledge what they have done. As Louis Bird himself points out, Cree spirituality and cultural rules are very much derived from close examination of the way nature operates in Cree territory. What we are seeing in the contemporary climate change crisis could be seen as a large-scale response of nature to the global acts of *pastahowin* perpetrated by corporations, individuals, and governments in the wastewest. And until the true nature of these transgressions is made public, and is admitted to by the perpetrators, these damaging activities will continue to happen, and the disastrous consequences will continue to build up.

A recurring theme in traditional Cree stories is the idea that you can't hide what you have done to nature. *Pastahowin* always rebounds back against the transgressor eventually. Several of Louis Bird's favourite Wisakaychak stories illustrate this theme by showing us Wisakaychak's greed, and the ways in which that greed gets him into trouble. As a cultural hero and trickster, Wisakaychak is a character who embodies important traditional teachings, but he often does so by providing an example of what *not* to do. His greed and his disregard for the natural world can be seen as illustrations of *pastahowin*, and very often these stories provide fascinating allegories about the Cree philosophy of waste. For example, in one of Louis Bird's stories, Wisakaychak eventually manages to kill a bear, and he decides that he wants to eat the whole animal by himself. But when he gets full, he realizes that there is still a great deal more meat to be eaten, so he decides, "I should squeeze myself between some trees so I can digest fast and eat more!"[7] The tamarack trees seem to oblige him by allowing him to squeeze his body between two trunks, but then he discovers he is held fast, and the trees refuse to let him go. He is forced to watch a parade of other animals coming to eat the bear that he wanted to have all to himself. Nature prevents him from having access to its bounties, and the trees only release him when there is no more meat left. Louis Bird points out at the end of the story that Wisakaychak "teaches that you should live moderately and that you should not kill any animal that you can not put away or preserve for use. Most of all, you should not be too greedy because you will always lose out in the end."[8]

Another more complex Wisakaychak story about greed and waste is found in a version of "Wisakaychak and the Geese" that Louis Bird told to a group of us at the University of Manitoba in August 2010. This version is quite different from the one published in his book *The Spirit Lives in the Mind*. The story begins with a narrative of Wisakaychak's clever use of songs to capture and kill a huge number of geese. After this excessive killing, he is faced with the dilemma of what to do with all these geese, but instead of sharing them with anyone else, he decides to roast them in a pit, which he does by burying them almost completely, with only their feet sticking up out of the sand that he has covered his bonfire with. But while he is waiting for the geese to cook, Wisakaychak gets tired, so he decides to go to sleep. He appoints a sentinel to prevent anyone from stealing his geese: this sentinel is his own anus, which he exposes to the sky, telling it to make a noise if anyone comes near. Needless to say, some creatures see him sleeping there with his ass sticking out, and they wonder what he is up to. Then they see all the feet of the geese, and they decide they will play a trick on the trickster and take all this food that he's hoarding for himself. So these beings manage to lull Wisakaychak's anus by saying "Shhhhhhh!" to it every time it is about to sound the alarm, and they take all the geese out of the sand and then just stick the feet back in. When Wisakaychak awakes, he is ready for his feast and he grabs the first goose to pull it out and begin eating. But there's nothing on the other end of the feet! So he tries another, and another. And then he realizes what has happened. He says, "Asshole, why didn't you warn me?" and he punishes it by walking over to the hot coals and sitting right down on them. But after that, every time Wisakaychak gets close to an animal that he's hunting, his ass makes that warning sound. This goes on for days and days, and eventually Wisakaychak is so weak and hungry, so desperate for food, that he breaks off a piece of the scab on his ass and eats that.

That was an abridged version of the story, and it lacks so much of Louis Bird's own personality and presence which he was able to impart when he told it, but I wanted to recount it here, even so imperfectly, because I see it as an important allegory about waste and nature. Wisakaychak's greed is what causes the problem in the first place: his belief that he should kill as many animals as he can, and that he should then hoard them for himself. The fact that he chooses his anus as the sentinel to watch over this overabundance of food is interesting. As the origin point or at least the conduit of human waste, the anus is perhaps a symbol of Wisakaychak's fixation on destroying more than he can possibly consume. In addition, his ass is not going to have any vested interest in watching over this collection of food because it doesn't feel hunger. So it is no wonder that the interlopers are able to keep the ass quiet and steal all the geese. But what I find most fascinating about the story is Wisakaychak's response to his anus after the geese are stolen, and then its response back to him. By punishing that part of himself, he seems to be creating a kind of division between the rational and the bodily, between the valued intellect, which is also associated with nearly inexhaustible hunger, and the devalued entrails, which are associated with filth and disobedience. And if we step back from this scenario, it's clear that such a division is quite similar to what we see in the wastewest, in which waste is sublimated at the same time as it is targeted as something to be controlled, tamed, or mastered. I think Wisakaychak's actions in this story can be read as advice to the so-called developed world, with its fixation on consumption and its obliviousness to the consequences of waste. In this case, Wisakaychak's punishment of his ass results in it becoming actively disobedient, sounding the very alarm that he expected it to sound when he was asleep and the geese were being stolen—but now the alarm is sounded at precisely the wrong time, when he is trying to stalk his prey. It is as if his ass is no longer a part of him, and it has instead become his enemy. And finally the culmination of this allegory comes when he is so hungry that he resorts to eating the scab from his own cauterized anus. Because he has broken the rules of nature and blamed the results of that transgression on his failure to master his nether regions, he now has nothing left to feed on except his own waste. After separating himself from his ass, he ends up being reconnected to it in the most intimate and degraded way.

Let's try to avoid that fate, shall we? Maybe there is still time to reverse the trajectory that the wastewest has propelled us toward. If contemporary mass culture can start to absorb some of the lessons about waste from Cree and Métis and other Indigenous traditions, I think we will have a better chance of changing cultural expectations and perceptions in the future so that humans can arrive at a more sensible relationship with the rapidly wasting world we inhabit.

Notes

1 Byron Apetagon, ed., *Norway House Anthology: Stories of the Elders*, vol. 2 (Winnipeg: Frontier School Division #48, 1992), 33.
2 Louis Bird, *The Spirit Lives in the Mind: Omushkego Stories, Lives, and Dreams*, ed. Susan Elaine Gray (Montreal and Kingston: McGill-Queen's University Press, 2007), 75.
3 Louis Bird, *Telling Our Stories: Omushkego Legends and Histories from Hudson Bay* (Toronto: University of Toronto Press, 2005), 164.
4 Bird, *Spirit Lives in the Mind*, 77.
5 Ibid.

6 Ibid., 75.
7 Ibid., 193.
8 Ibid.

CHAPTER 3

Sustaining Collaboration
The Woodhaven Eco Art Project

Nancy Holmes

Near where I live in Kelowna, British Columbia, is a place that I and many others love and regard as worth protecting. The Woodhaven Nature Conservancy is a small corner of land that contains within it an intersection of four major bioclimatic zones of the Central Okanagan. The Conservancy's striking features come from this unusual overlapping of forest zones, where the hot, dry ponderosa pine landscape joins a humid western red cedar grove, both of which weave into the regionally prevalent (though disappearing) cottonwood riparian ecosystem and a more typical British Columbia mountainous forest dominated by Douglas fir. Woodhaven is also home to several rare and one nearly extirpated species, the western screech owl. The Conservancy is also special because of its importance to Kelowna's environmental activist history: these unique twenty-two acres were saved from development in 1973 by a couple, Joan and Jim Burbridge, who at the time were living on the land in a rented cabin. Through their intervention, Woodhaven was purchased by a coalition of donors, the Nature Conservancy, and the BC Government, and it became a nature preserve, now often called the "jewel" of the regional park system. In fact, the main boardroom of the offices of the Regional District of the Central Okanagan is called the Woodhaven Room in honour of the Conservancy's importance to the district's park-creation history.

 This story of preservation was of its time; such conservation activities were happening all across the country in the 1970s when "saving the environment" was newly popular. DDT had just been banned, and people were beginning to worry about issues many of us now find our daily fare: pollution, oil consumption, habitat and species preservation. The Burbridges were local environmental activists who went on to do much more environmental work, chairing the local naturalist club, and guiding children and biology students through the Conservancy. In 1989, Joan wrote and photographed a compact guidebook to flowers of the southern interior of British Columbia and adjacent American states, which is still widely used by amateur naturalists in the region although it is out of print.[1] The couple continued to collaborate with scientists and naturalists in a variety of ways until their deaths, Jim's in 1990 and Joan's in 2001.

An interesting feature of the Woodhaven story is that the Burbridges moved back into the Conservancy once it had been saved from destruction. The Regional District of the Central Okanagan supports several live-in caretakers in its parks, and in Woodhaven the caretaker lives year-round in the original 1920s heritage cabin on the property. The Burbridges were the first official live-in caretakers of Woodhaven. For nearly thirty years, they took care of the place and lobbied for it when necessary, particularly in the 1980s, when housing developments went up around the Conservancy; to prevent the new suburban basements from flooding, the city diverted the water so that the rare (for the dry Okanagan) groves of western red cedar in the park began to die. The Burbridges were on the ground and advocated for the place and its species, with no personal financial investment in it other than the fact that the Regional District charges an extremely reasonable rent in return for caretaking duties. The water never returned to its original flow, but after a vociferous campaign, the Burbridges managed to convince the city and the Regional District to consider the needs of the Conservancy as well as those of the adjacent homeowners. Now, each spring, a heavily managed creek nearby is diverted into Woodhaven and it douses the trees for a few months. The Burbridges created connections and community in good activist fashion, but beyond that, they committed to staying in one place and accepted the responsibility of lifelong stewardship. The Cross-Pollination workshop asked us to examine how the production and interpretation of text and visual imagery enable more sustainable and ethical knowledge and action on behalf of places. I keep the Burbridges in mind for they were effective activists who worked on behalf of a beloved place. They did it through teaming up with scientists and community, educating themselves about the natural world, and communicating their knowledge to others—in other words, through *collaborative* community building—but they also did this by ensuring that their work lasted as long as their lives did through *sustaining* their relationship with the place and with the community. As a person concerned about environmental health, I am thinking a great deal about collaboration and continuity, as these seem to be keys to the difficult challenge of sustainable dwelling. Also, as a poet and now a coordinator of various eco-art projects in the Okanagan, I have been thinking about the actions of artists and environmentalists in the light of the possible need to sustain connection; such demands for commitment and collaboration complicate ideas about art and our relation to places we value. My thinking has also been challenged by the too-brief contact I've had with Okanagan Indigenous philosophy, particularly as explained by the Okanagan Elder and artist Jeannette Armstrong, and that informs the stories of Okanagan storyteller Harry Robinson. Their influences hover over my musings in this chapter, though I do not pretend to be able to do credit to the depth and range of their ideas.

In 2010, I and my collaborator Lori Mairs undertook a year-long eco-art project in Woodhaven Nature Conservancy and thus began a new phase in the human relationship to the Conservancy. How this art project came about and how the project affected my thinking about place is explored in the rest of this essay. In 2002, Lori Mairs was hired as the new caretaker for Woodhaven. She is a long-time environmental activist (once involved in the *Rainbow Warrior* activities of Greenpeace when she was a young woman in New Zealand), and she is a sculptor who works in bone, shed antler, metal, and beeswax, creating large-scale structures and small-scale jewellery. In 2005, I moved a block away from Woodhaven. When I moved into the area, no one (neither neighbours nor real estate agents) mentioned the nature conservancy nearby. Once I discovered it, I went for a few walks in the park and was struck by the unusual cedar groves and wildness of the place, but after only a few visits the park was shut down for three years as assessments about tree health were done and negotiations with the province were conducted over the screech owl habitat. From 2006 and 2009, Mairs lived alone in the locked-in, fenced-up park; in 2008, I met her at an art event in town. When we began to talk, we realized we had a Joan Burbridge connection: she lived in the Burbridges' old home in Woodhaven, and I had owned Joan's wildflower book for some time. I was delighted to find out that this book had been written right around the corner from where I now lived.

When the park reopened in June 2009, Mairs and I decided we should reintroduce the community to Woodhaven and to the legacy of the Burbridges; with a welcome grant from the University of British Columbia's Hampton Fund, we began the nearly year-long Woodhaven Eco Art Project to do just that.

From the beginning, this was an intensely collaborative effort, not only between myself and Mairs, but also among the many artists who worked with us (a total of eighty-one artists eventually created over sixty works of art for the park). Some of these artists collaborated with each other and with me, and Mairs helped nearly every artist situate his or her art in the park, since she knows the place so well. The collaboration was also with the Regional District of the Central Okanagan who provided support and naturalists and park workers and several practical aids in terms of administration. The collaboration eventually included the public, and from the beginning Mairs and I felt we were collaborating through time with Jim and Joan, furthering the work of stewardship that they had begun over thirty-five years ago. We also felt we were collaborating with the park itself. Our collaborators—or as dancer Elizabeth Langley says, *co-labourers*[2]—were multiple: friends and artists, students, co-workers, neighbours, people long dead touching us across time, and the physical objects of the forest and other species.

The collaborative nature of much eco-art is often remarked. In a research report commissioned by the Canadian Commission for UNESCO, Beth Carruthers lists, as examples, several collaborative eco-art projects in western Canada, most of them in British Columbia.[3] Suzi Gablik, as many eco-art critics do, argues for collaborative processes and results in eco-art.[4] Eco-art's collaboration with the natural world itself has also been talked about often, especially in relation to remediating projects where natural places or systems are restored; restoration projects are often "spurred on by new technology and an increased spirit of collaboration among artists, scientists, and engineers."[5] Such collaborative, multi-created eco-art is part of the community-based and cultural activism movements in art, variously described as "new genre public art," "littoral art," "relational aesthetics," "conversational art," "dialogue-based public art," or "interventionist practice."[6] Eco-artists, more and more, are inviting public participation, activist inspiration, and environmental remediation in the process of making; such art is becoming more participatory and less specular. Eco-art seems particularly suited to this kind of artmaking as the art usually happens outside conventional art spaces such as journals or art galleries, distancing art from art's standard economic systems. Ephemeral and community-driven eco-art seems to bypass some of the capitalist values of property rights, ownership, and exploitation. Additionally, eco-art's collaborative and dialogical relations can mimic or reflect ecological processes as collaborative communities may function analogously to ecological communities. Ecophilosopher Andrew Brennan notes that ecological modelling is based on an "individualistic approach to communities."[7] This phrase encapsulates the specific yet relational thinking that environmental work promotes—the specific must be married to the whole but this whole, this ecocommunity, is as unique as an individual. The individual is embedded in a larger but also individually unique structure—an ecosystem or an environmentally rooted cultural community—and thus distinctive collaborations and even idiosyncratic groupings respond more completely to a particular place than universal theories, individual actions, or single visions. Ideally, through recognition of and response to places' *and* cultures' simultaneous uniqueness and complex interdependence, collaborative art can respond to both environmental problems and impoverished cultural and historical knowledge.

In all these ways, the Woodhaven Eco Art Project was typical. It was participatory: the public were not only spectators who attended in droves, especially on the four open house days held throughout the year, but they were also creators contributing to various processions, interventions, drawing projects, and community poem writing. Much of the art was ephemeral: especially music and performances, but also popular installations such as the Log Poem where walkers could create their own ephemeral poems from words painted on stones and laid upon a fallen log. Some of the artwork is currently decomposing in the park, and as of July 2012, the Log Poem is not only still attracting participants, but some people have begun painting their own words on stones and bringing these into the park to add to the ones made by the original artists. Other sculpture has been "used" as habitat (within a few days, every scrap of buffalo wool on four separate sculptures by Mairs had been confiscated by squirrels for bedding), and some of the artwork seems to linger in the park still as we walk about—some literally and some in memory, as if some of the most vivid performances have inscribed themselves into the park's landscape. At times we still find scraps or traces of performances or installations that seem to drift out of crevices or off branches and onto paths (for example, paper "blessings" from the Lois Huey-Heck's "Blessing Tree" installation or tiny silk puppets tied to twigs and branches from Denise Kilshaw's children's puppet theatre).[8]

The artistic skills of the artists, and the public relations effects, were put at the disposal of our goal to raise awareness about the site, and this project seems to have been a success. The eco-art project increased visits to the park fourfold and the neighbourhood's sense of stewardship also increased substantially; the project instilled a sense of pride in the special qualities of the park that drew such prolonged and loving attention from a large number of artists. The project opened up the park in a fresh way, so that neighbours and community members who had never been in the park, or had used it merely as part of a jogging route, or who had not used it for years, began to see it as something special. It gathered an aura traditionally generated by gallery space or a published book; it was the platform for works of art and this quality made it more than "just" a park, yet it was a gallery that could be "used" for completely other purposes and even rearranged by visitors, destabilizing both the idea of a park and of a gallery or other art world space. The tendency of collaborative art or eco-art to bypass traditions of individual authorship and possession was also evident. In the local art and naturalist communities of Kelowna, the word "Woodhaven" has become a sort of code for a new nexus of nature and art, but one attached to no particular artist or name. Thus, the project seems to have also subverted the highly competitive art "economy of recognition," as critic Stephen Wright says.[9] I don't wish to over-romanticize collaboration. Wright provides several cautionary notes in his fine essay, "The

Delicate Essence of Artistic Collaboration," in particular warning of "intellectually and aesthetically impoverished practices ... [where] artists make forays into the outside world, 'propose' (as artworlders like to say) usually very contrived services for people who never asked for them ... then expropriate as the material for the work ... from these participants."[10] This sort of appropriation and exploitation is a moral morass in any project such as the Woodhaven one or similar eco-art community-based projects. These tendencies raise serious concerns that undermine the much-lauded "alternative" economy of eco-art. Certainly, in the Woodhaven project, we experienced several challenges. The difficulty of managing consent forms, the threat of hurt feelings, the dodginess of proprietary fences around the documentation of work, and the contribution of various works of art that really made no attempt at integrating with the other work or the place were all a part of this project, which should come as no surprise to anyone who is attempting a process that mimics ecological processes. Any complex system is always in flux, imbalances occur, and violence and destruction are as much a part of such systems as nourishment and creation. Among people, issues of copyright and ownership, issues around boundaries and public space, issues of tenure and publication, issues of control and organically evolving forms are all areas that can be problematic especially for artists who value the individual gesture and voice, as they should for these are poignantly human. There needs to be room in collaborative projects for the specific, unique, and individual. As Wright notes, "the question is how to channel [artistic] competences and perceptions beneficially into collaborative endeavours."[11] In the end, I think we were largely successful, as most of the artists approached their participation in the project as a problem (What kind of art is right for this park? What kind of artistic tools can contribute to the overall aims of awareness and stewardship?) rather than as an opportunity for reputation building or showcasing. The art became primarily a gift to a place and a larger purpose (though we did provide honoraria for some artists, these were token amounts. Some students were funded through small research assistantships). The excitement of being part of such a large project provided a great deal of the motivational "fuel" for many of these artists.

While collaboration feels like the preferred method of eco-artistic production, especially collaboration focused on specific places or sites, the collaboration is often a contingent coming together—artists leave. These kinds of community-based art projects are, in fact, often called "contingent communities."[12] The Woodhaven Eco Art Project is now over. When I think of the Burbridges' example of continuous stewardship and collaboration, this contingency troubles me. What's troublesome comes from how I am coming to understand relationship to place. Gary Snyder quotes the Zen philosopher Dōgen: "When you investigate mountains thoroughly, this is the work of the mountains. Such mountains and waters of themselves become wise persons and sages."[13] Similarly, David Abram speaks of how in oral literature:

[story] envelopes its protagonists much as we ourselves are enveloped by the terrain.... For a deeply oral culture this relation may be experienced as something more than a mere analogy: along with the other animals, the stones, the trees, and the clouds, we ourselves are characters within a huge story that is visibly unfolding all around us, participants within the vast imagination, or Dreaming, of the world.[14]

This ancient view of our relation to knowledge, art, and nature reconfigures the sages (scholar, artist, naturalist) so *they* are the instruments and even the elements of nature knowing itself. To "know" a mountain is to evolve and enhance the mountain's knowledge, to be an instrument or bearer or piece of the knowledge. However, what happens to the mountain's or forest's wisdom if the human instrument, highly honed and trained over time, drifts away? Unless the "sages" stay in place, remain "enfolded," does the forest lose its knowledge? If we no longer are participants in places that we know, does the world's dreaming shrink and become impoverished, thus leaving the world vulnerable to human carelessness and ruthless exploitation? It seems that one of the great problems of ethical knowledge and sustainable action is the difficulty of remaining. Humans are both nomads and home builders; many cultures have found ways to accommodate these seemingly contradictory qualities. Our civilization, however, has untied this complex knot, turned most of us into unattached nomads with nostalgic dreams and hearts, building huge, shrinelike houses, gated housing developments, fenced-in parks like Woodhaven, and neighbourhoods with "views" of place but no connection to it but carbon-costly roads. We long for rootedness in a home that we cannot really understand or relate to for we cannot create livable cities and healthy economies that recognize or are integrated into their place. Rootedness is a great anathema to our mobile culture, in spite of the nostalgia, and nearly impossible in our current political, social, technological, and economic systems. Similarly, although we have made our libraries portable with our laptops and e-tablets, is a sage, a generator of the knowledge of a mountain or a forest or a seashore, seen now as a portable knowledge device? If so, does this call into question any knowledge about place that sages are capable of? I think it may.

Knowledge of the mountain or forest is more than mere information, and the human response to place is more than an equivalent to a portable storage system like an external hard drive. My experience over the many months of creating in Woodhaven is that the artist or knower needs to be "booted up" by physical contact with the place, through feet and hands and eyes and ears and scents. The knowledge exists fully only in relationship, as Abram notes in his exploration of Indigenous thinking. In Woodhaven, many of the artists felt they were literally creating the work for the place itself to absorb and to enjoy. We have numerous photographs of animals watching us. Over and over again, the artists came into the park not really knowing what they would do for the project; part of the process was to spend time in the place and, as filmmaker Michael V. Smith said, "I'll have to let Woodhaven tell me what sort of film to make." (In the end, he made a little hand-held video meditation on the ants in the park.) While it is possible Michael's film and some of the other artwork will leave Woodhaven, Woodhaven's art-generation engine has stopped. Art does have the power to cast an aura of special attention around a place, which means its local worth is increased. The aura around Woodhaven will linger for some time. Poems and films and art catalogues will be published, as our art-world systems demand, although, wrongly, in my view, artist-scholars will have to show their films abroad or in large cities if they wish to add "reputable" credits to their CVs, even if the film shown in Kelowna attracts more people and is of more significance to people here than anywhere else. Nevertheless, in the end, will all this collaborative energy and care be fossilized into libraries and archives and a website a few years from now? If care and knowledge are not passed down over time to new artists and new collaborators, will this living knowledge fade and be forgotten as Woodhaven was for some years in the community before we began the project? And what of all the other special, local places that are also crying out to be learned, storied, and lavished with art?

Okanagan Indigenous thinker and writer Jeannette Armstrong, at her June 2009 Association for the Study of Literature and the Environment plenary talk at the University of Victoria, spoke eloquently of the importance of re-indigenating (as have others such as Wes Jackson). Part of such projects like the Woodhaven Eco Art Project is to further feelings of indigenation: belonging to a place. We talk about "raising awareness," but really what we mean is "pay attention to this place which supports you. Look at it and know it. Be in it and stay." This message may be tainted with forms of nostalgia—I feel that nostalgia for home in myself—but it can also be a radical message. Most Western forms of knowing, whether art or science, have a spotty record of effectively delivering a message of connection to specific places. While collaboration can be a strategy for generating short-term, specific knowledge about natural places, sustaining collaborative work over time and through generations is a larger matter and may be a more difficult long-term goal in our society, since it requires a commitment to place most of us are unable to give. Such work requires, to use that Heideggerian term, "dwelling." This is what the Burbridges mean to me; they were place-based thinkers and dwellers. If Joan and Jim brought scientists and naturalists into Woodhaven to learn about the place, Mairs has brought art into the project of caring for Woodhaven, and the two of us have collaborated together on a major eco-art project in Woodhaven that has reawakened local knowledge and awareness of the place. But now what? Will I and the other artists involved begin to think through place and work from within relationship to place? It is hard to say. Certainly, Lori Mairs is continuing to explore this possibility in her thinking and her artwork. Some hope for the possibility of continuity comes from another comment by Wright: "the management of incompleteness is indeed an artistic competence."[15] Those of us involved in the eco-art project are stepping into incompleteness now that the Woodhaven Eco Art Project is "complete." A consciousness of incompleteness can urge us to continue to do more.

Complex systems are unknowable; even small parks like Woodhaven are ecosystems vast and full of chance, so much so that one has to wonder if we merely impose a belief that there *is* a system rather than simply rich chaos out of which new learning can come. My experience of Woodhaven and coming to know it over the years, and in Mairs' nine years and the Burbridge's much longer tenancy is that even as we untangle a pattern or skein of understanding from the park, chance or the will of wild things messes it up even as we observe it, and we need to search again for new structures and patterns within this familiar place. If a wild place—or any place—is always changing and ever incomplete this explains why the instrument of the place's knowing has to remain. I do not intend moving anytime soon, but my record is not good. I seem to move house every eight to ten years. Still, I hope I will continue to walk in the park as I get older, continue to learn from it and enact its learning, protect it, and, at times, give in to ever-changing, volatile, and wild Woodhaven's pressure to leave a small artistic offering. I think Jim and Joan would like their incomplete legacy to live on.

Notes

1. See Joan Burbridge, *A Field Guide to Wildflowers of the Southern Interior of British Columbia and Adjacent Parts of Washington, Idaho, and Montana* (Vancouver: University of British Columbia Press, 1989). The biographical note on the back of the book reads: "Joan Burbridge is a guide and warden of Woodhaven Nature Conservancy in Kelowna, BC, and has lectured and written on the flora and fauna of the region."
2. Quoted in Megan Andrews, "On Collaboration," in *Across Oceans: Writings on Collaboration*, ed. Maxine Heppner (Toronto: Across Oceans, 2008), 43.
3. Beth Carruthers, "Mapping the Terrain of Contemporary Ecoart Practice and Collaboration," report commissioned by Canadian Commission for UNESCO, presented at "Art in Ecology: A Think Tank of Arts and Sustainability," Vancouver, BC, 27 Apr. 2006, 9–17.
4. Suzi Gablik, "Alternative Aesthetics," *Landviews: Online Journal of LAND: Landscape, Art and Design* (2003), http://www.landviews.org/la2003/alternative-sg.html.
5. Robin Cembalest, "Turning Up the Heat," *ARTnews* 107, no. 6 (2008), 102.
6. See summary in Jack Richardson, "Interventionist Art Education: Contingent Communities, Social Dialogue, and Public Collaboration," *Studies in Art Education* 52, no. 1 (2010), 19.
7. Andrew Brennan, *Thinking about Nature* (Athens: University of Georgia Press, 1988), 54.

8 A record of all the artworks can be found online at www.woodhaven.ok.ubc.ca or in the catalogue by Nancy Holmes and Lori Mairs et al., *The Woodhaven Eco Art Project* (Kelowna, BC), 2011.
9 Stephen Wright, "The Delicate Essence of Artistic Collaboration," *Third Text* 18, no. 6 (2004), 534.
10 Ibid., 534–35.
11 Ibid., 535.
12 Richardson, "Interventionist Art Education," 18.
13 Gary Snyder, *The Practice of the Wild* (Berkeley, CA: Counterpoint, 1990), 123.
14 David Abram, *The Spell of the Sensuous: Perception and Language in a More-Than-Human World* (New York: Vintage, 1997), 163.
15 Wright, "Delicate Essence," 544.

CHAPTER 4

A Natural History and Dioramic Performance
Restoring Camosun Bog in Vancouver, British Columbia

Lisa Szabo-Jones & David Brownstein

Figure 4.1 Camosun Bog. Photo: Lisa Szabo-Jones

In the summer of 2008, during a Vancouver field trip to the restored Camosun Bog in Pacific Spirit Park, as we walked the circumference of the boardwalk, the two of us fell into conversation.[1] We remarked how the design by the Camosun Bog Restoration Group (CBRG) deliberately set up the feel of an outdoor natural history museum with tableau views that gave the sense of viewing living dioramas, open space articulated with rare flora so at odds with the dramatic backdrop of towering western hemlock, tangled salmonberry, and sprawling understory of salal. We experienced a sense of artifice, the uncanny feeling that the palpably moist, spongy bog before us was no more than simulated nature—real, but not the real thing. In this reaction, we were not alone. To illustrate, writing about Camosun Bog in a 2007 *Globe and Mail* article, "It's Not about Ecology—It's about Gardening," author Timothy Taylor came to a similar conclusion.[2] Positing precisely what his title suggests, he claimed that the present, restored incarnation of the bog has more to do with culture than nature. While just an observational piece intended for popular consumption, Taylor's complaint is indicative of one of several criticisms in the ecological restoration literature, exemplified in the writings of Eric Katz and Robert Elliot, for instance.[3] That is, while restoration projects may appear to revert nature to its original, "unspoiled" state, they are, in actuality, fake, or a deception tantamount to forgery.

But was this the full extent of the bog's meaning? Was it just a cultivated garden masquerading as a 3,000-year-old ecosystem? No, we thought. The Camosun restoration did not attempt to obliterate the history of the bog; it did not hide the series of events that necessitated its restoration. It did not seek to pass as the original. Along the boardwalk, there were signs that issued thanks to the CBRG and recognized corporate financial support. Even more importantly, a panel entitled, "A Community Working Together: Restoring and Protecting the Bog," outlined the near disappearance of the bog, and what community members were now doing to make a difference. Dissimulation was, thus, not the issue. The restored bog, instead, signalled its departure from the original via a structural and informational design that employed boardwalks, viewing points, and signage to call attention to its historical demise, its present revival, and what may lie in store for it in the future. The bog assumed the form, we suggest, of the habitat diorama.

In this chapter, we contend that the restoration efforts of the CBRG at Camosun Bog, resemble the aesthetic of, and can be seen as an organic, dynamic version of the habitat diorama, a staple of the natural history museum's educational arsenal. Following art historian Karen Wonders, we put forth an expanded view of the habitat diorama as ecological theatre, which sees the model as more than a didactic tool, but as a form that records and illuminates humanity's relationship with nature, contests a clear demarcation between humans and the biophysical world, and intentionally encodes and communicates a position regarding the natural space to which it specifically, or more broadly points.[4] We extrapolate Wonders' view that the habitat diorama is a representation, embedded with an ideological position. That is, we argue that its message can be a catalyst for action. The restored bog, and its continued maintenance under the CBRG, reflect, participate in, and *promote* environmental action, stewardship, and a preventative ethics. To support our position, we will first consider the historical narrative of Camosun Bog as put forth by the CBRG, and the way in which this story makes manifest a point of view through on-site text panels and the group's Internet blog page. We will then examine how this story and the inextricably related restoration, when considered under the framework of a habitat diorama, relate to and defy the accusations of ecologists, scientists, and literary critics brought forth in historical nature faker controversies. Finally, we will employ the habitat diorama concept as a means to expand and complicate the idea of nature and reconsider the aims of restoration, emphasizing the educational, activist, and ethical environmental message presented by the restored object of the Camosun Bog.

A Brief History of Camosun Bog

While Karen Wonders delimits her inquiry of the habitat diorama to its physical content, the accompanying text panel plays a crucial role in communicating meaning. It orients the viewer and provides context regarding the facsimile of the natural environment in question. It works in tandem with the display to denote its content and to connote a point of view regarding the represented object. The panels at Camosun Bog are no exception, and are, indeed, integral to the articulation and understanding of the bog as a habitat diorama space. Here, a dozen panels that recount a history of the bog are arrayed about the boardwalk. They variously describe a past in which the bog thrived, the events leading to its decline, its restored state, and the relationship between the bog and a heterogeneous set of partners within the community at large.

According to the educational panels in Camosun Bog's self-guided nature walk, "Although Camosun Bog/məqʷe:m was here for 3,000 years, in less than 50 years, it almost disappeared. Draining and filling from nearby developments lowered the water table, encouraging non-bog plants to establish and nearly eliminate specialized bog plants. Community members, the people of the Musqueam, scientists, park staff, students and park visitors are working to protect and restore the bog. Here is how they help." The narrative tersely outlines the problem as the rise and decline of the bog, loosely assigns blame to human development, and indicates how different community members have remedied, and continue to remedy the problem. The sign further depicts young and old, novices and experts, the general community, and more particularly members of the Musqueam First Nation, as involved in the restoration process, and tacitly invites further participation by the viewer—a point that is made explicit by additional handmade signage soliciting volunteers.

Yet, like the habitat diorama, the format of the panel dictates that only a cursory amount of information may be conveyed. The panel acts as a signpost to a richer, more intricate history, some of which is shared on the website and blog of the CBRG. This wider history tells us that before it was a bog, the site was a post-glacial lake. At its historical maximum, 12,000 years ago, the lake that preceded the bog probably extended 1,500 metres by 300 metres, and was likely about 6 to 8 metres deep, catching water from a basin of 250 hectares. Ecological succession resulted in deposition of material on the lake bed and edges, 7,000 years ago, allowing a marsh to form. The marsh, in turn, was replaced by a bog dominated by sphagnum in the first millennium BCE—the site we know today as Camosun Bog.[5]

The informational panels also point to the long-standing human engagement with the bog, first with the Musqueam First Nation, and later with the newcomers. In the panel, "Musqueam People's Source of Supplies and Stories," visitors learn that the Musqueam, the Indigenous people of the area, have used the bog as a "source of food, medicine, raw materials and trade commodities" for 4,000 years and have inhabited the coastal region for 9,000 years.[6] Musqueam resource management and harvesting activities are presented as in harmony with the bog, prior to the later degradation. This is in contrast to the more recent activities of the newcomers, who the signs represent in two ways. First, they are obliquely referenced but not named as responsible for lowering the water tables through nearby developments. Second, they figure prominently as restorationists engaged in reversing the bog's decline. The narrators of the signage opt to elide overwhelmingly the details of how competing visions for this space by the newcomers led to its compromised status. Rather, by underplaying blame, they choose to focus on

Figure 4.2 Camosun Bog Sign ~ A Community Working Together. Photo: David Brownstein.

the restoration as a positive and productive activity, shaping the present and future. The seemingly neutral signage, describing what is objectively a real space, presents a position through the story it has selectively chosen to tell, and omit. Like the habitat diorama, it conveys an educational message, which in this case is a plea for environmental action.

Omitted from the signs are the details recounting the various forms of disturbance that the site has experienced since the mid-nineteenth century when the newcomers first encountered the bog. There is no indication that the surrounding forest was logged, sometimes several times, or that the region experienced dramatic fires, all of which helped transform the bog into an area of drier, woodier vegetation. Nor is there detailed reference to a series of events that ensured that, by the 1970s, the bog was on the verge of disappearance. This includes both 1929 and 1950s drainage attempts to, respectively, protect adjacent housing from flooding and to assuage fears of a polio epidemic, as well as the addition of fill from a University of British Columbia (UBC) construction site between 1971 and 1973, which lowered the water table.[7]

Moreover, the bog's signage, which commemorates the inauguration of the Camosun Bog Nature Walk in 2006, the moment when the boardwalk and informational panels were opened to the public, does little to hint at the long-standing restoration efforts that preceded it. Indeed, contemporary bog advocacy began in the late 1980s, when members of the Vancouver Natural History Society, UBC faculty, and neighbourhood residents, appealed to the University Endowment Lands Technical Committee for permission to intervene in favour of the bog's preservation.[8] In 1988, with the help of UBC Professor Emeritus of Plant Science Bert Brink (1912–2007), the Vancouver Natural History Society received a federal Environmental Partners Fund grant. Work to remove hemlock from the bog was suggested by a technical committee and supervised by the Greater Vancouver Regional District, which assumed control of the park in 1989. Once removed, coppice-germinated birches colonized the opened space. While initially discouraged by this unforeseen consequence, bog advocates reorganized. In 1995, Laurence Brown, Brian Woodcock, and Mitch Sokalski met and agreed to form the Camosun Bog Restoration Group, also known as the "Crazy Boggers," with a mission statement of "A Successfully Restored Camosun Bog."[9] From 1997 to 2000, because there was meagre information available regarding restoration techniques, the group experimented with a test patch of the bog 8 metres by 10 metres in size. Between 2000 and 2005, they applied their findings to the larger bog, although it still required maintenance through weeding.[10] This experimental, evolving history is not obvious in the "completed" bog restoration, which we call the habitat diorama. Rather, the boardwalk panels' primary focus on bog species, and the present and future state of the bog, evokes momentum and hope for the landscape's future.

A Long Natural History Tradition

The restoration work by the Crazy Boggers can be located in a well-defined tradition of amateur natural history, that branch of knowledge dealing with natural objects including plants, animals, and minerals. Indeed, criticism of the Boggers' work regarding Camosun Bog can also be situated in a parallel context that probes the proper role of art and science, in both nature study and science education. This context includes historical controversies such as the "nature faker" debates that occurred in early-twentieth-century North American nature writing circles, and controversies regarding the introduction of habitat dioramas to European natural history museums. A brief review of these conversations will reveal that Taylor's contemporary critique of Camosun Bog's restoration is the most recent, isolated, episode of this long-standing dialogue. We focus on these two intersections primarily because of the nature of Camosun Bog's design: part outdoor natural history museum and part biology field lab, situated in one of Vancouver's large-scale forested recreational parks.

The scientific lineage of the Camosun Bog Restoration Group would not normally place it at the leading edge of conservation debates or contemporary knowledge generation. The Boggers self-identify as naturalists, and so draw upon a lengthy amateur tradition now viewed as antiquated, or at the very least, pursued by laypeople of small significance (birders, "rock hounds," and botany enthusiasts, for instance). Popularly understood in the nineteenth century to have interests limited to outdoor fieldwork and collecting, describing, and classifying natural objects, such close study historically intended to reveal patterns of uniformity and interrelatedness in nature, which would in turn reveal nature's basic laws.[11]

During the nineteenth century, naturalist clubs formed across North America, and because of the province's comparatively late settlement by colonists, during the early twentieth century across British Columbia.[12] In this later period, natural history societies were moving away from what the American Aldo Leopold referred to as their "dickey-bird" predecessors.[13] These clubs were not just collecting and classifying regional flora and fauna, but were also marking habitat changes, endangerment, migratory patterns, and impacts of invasives on native species, all with an eye to preserving the natural world. Over the course of the twentieth century, nature's existence for itself (biocentricism) became the focus, and Nature Vancouver (formerly the Vancouver Natural History Society) is a good example of this trend.

Coincident with these changes internal to the natural history clubs were the more general trends of urbanization and industrialization. City dwellers could not so easily access natural specimens anymore, a service that nature stories and natural history museums soon filled, and which then became a site of authentic nature/science controversy. Whereas contemporary restoration attempts face accusations of forgery, nature writers of a century ago and natural historians confronted similar issues as to how to bring their work on nature into popular literature and the museum in a dynamic and informative way. As Szabo-Jones contends elsewhere, the impetus for this move, particularly in North America, indicates a desire to familiarize others with new surroundings. These writings, often circulated through daily or local weekly newspapers and society bulletins, offered short explanatory poetic prose sketches that animated the local landscape through a syntax that evoked a particular region, which in turn expanded the linguistic boundaries of, in North America, a formative settler community to include (and legitimize) sensory knowledge of place. Szabo-Jones proposes that "natural history with its colloquial and non-technical language, its focus on subjects within immediate surroundings, its sentimentalized and empathic imagery, and its didactic function made the natural world accessible—a place where everyone had equal footing on common ground."[14]

With such a large audience willing to pay a great deal for the stories, authors vigilant regarding the factual details of natural history such as American John Burroughs felt that some put "too much sentiment, too much literature" in their depictions of animal heroes.[15] Burroughs complained, "In many of the narrations only a real woodsman can separate the true from the false."[16] Of course, Burroughs' criticism overlooked that the boundary between fiction and fact in nature writing was often a matter of interpretation. The nature faker controversy was, on one level, a literary expression of a larger conflict within the nature study movement. It was a debate over whether the goal of nature study was to educate students in the sciences or to teach them appreciation and a sense of harmony with nature.[17] The North American nature faker controversies were concurrent with other debates among European museum professionals. Indeed, there was a strong connection between natural history dioramas and writing of the nature movement. For instance, illustrations that accompanied many of Ernest Thompson Seton's stories were drawn from dioramas.[18] Nature study informed by ecology—and key to this model is promotion of imaginative perception—provides "mechanisms" for appreciating nature, allows the appreciator to see beyond the aesthetics of nature to the function of and the interrelations between human, non-human, and ecosystems. Thus perceiving the importance of the parts of the whole, the ecological model enables the perceiver to recognize the whole that the parts unite.

Habitat Dioramas

The contemporary reader will know of dioramas as educational, three-dimensional landscape models behind glass, showcasing mounted zoological specimens in the foreground, which merge imperceptibly into background landscape painting. The term "diorama" was coined around 1821 by J.L.M.

Daguerre and Charles M. Bouton, from Greek *dia*, through, and *horama*, to see.[19] The first diorama was a special-effects theatre—in which a light show took the audience from dawn to dusk twice in about thirty minutes—that so convincingly simulated nature that it was a visual surrogate for the original.[20] As mass entertainment, dioramas were very popular, until they experienced a decline in the mid-nineteenth century.[21] In the late nineteenth century, natural history museum curators adopted similar visual techniques in their construction of zoological displays meant to replicate field conditions for museum visitors unable to visit field sites. Natural history group displays, or habitat dioramas, underwent an evolution in which more and more elaborate recreations of foreground sculptural modelling and background paintings surrounded taxidermy specimens. Exhibition teams visited the represented site, where they collected specimens and measurements of all kinds, took photographs, and made field sketches. Such preparatory work then informed the recreation of the real world back in the glass case of the museum. As a result, dioramas became powerful tools for conveying the multiple and complex layers of information that comprise reality within an economy of contained space.

Foreshadowing the contemporary accusations that the ecological restoration of Camosun Bog is a forgery, the historical museum exhibits described here were similarly disparaged in their time. Among those who saw the natural history museum as holding a research mandate, dioramas were a threat, leading to conflict with museum staff who championed the museum's role as that of popular public education for the lay masses. As Lynn Nyhart has noted, diorama displays carried risks for museum professionals, who wanted to demarcate their institutions of science and education, from what curator Otto Lehman called in 1906 "the imbecilic panopticon that only satisfies sensual pleasures."[22] The habitat diorama's introduction to museums created tensions regarding competing views of authenticity's authorship. Did authenticity lie properly in the museum visitor, who experienced the diorama illusion and thus learned to read the relationships among organisms in the field without having to travel; or was it located in the authority of the scientist who vouched for the diorama's veracity based on his or her own empirical observations of the natural objects so displayed?[23] One of the diorama's eventual roles in environmental education was that of bringing the goals of nature preservation into public view. In time, though, the costly and laborious process of diorama construction came to be replaced by nature films, television programs, and omni-max films, all of which could immerse the viewer in a field site.[24]

In her examination of the history of habitat dioramas, an ecological theatre, as Wonders defines it in relation to habitat dioramas, comprises an exhibit that "go[es] beyond the reconstruction of a 'pretty picture' transferred behind glass from the outer world. Its purpose ought also to be ecologic, that is, it should elucidate natural interrelationships between organism and organism, and between all and the physical environment."[25] Wonders conceives of habitat dioramas as ecological theatrical scenes where "the animal actors star in an evolutionary play."[26] The artistry of the diorama, the three-dimensional "actors" integrated with the animal's habitat, and realistic sculpted and painted back- and foregrounds create a momentary suspension of disbelief that serves "as a specialized form of visual communication that both expresses and influences environmental thought."[27] The illusion brings the outdoors indoors and, Wonders argues, collapses the divide between humans and the natural world, represents a "living community," and "makes concrete the concept of an ecosystem."[28]

Where we envision remnant restoration ecology as practised by the Crazy Boggers as diverging from Wonders' notion of ecological theatre is how she fails to convey the dynamism that such a concept as "ecological theatre" proposes when applied to a living dioramic aesthetic. Her articulation does not detract from her concept as much as it prompts a quest for a new form of ecological theatre, one that is not reminiscent of the stuffed dead but celebratory of growing, evolving lives. And, though Camosun Bog, on the surface, seems to fit Wonders' ascription of the bog's aesthetic akin to habitat dioramas as housed in mausoleums, or Paul Gobster's assignation of "museumification"[29] to urban park restoration, a greater complex interplay emerges in the CBRG's own interpretation of the bog. As the CBRG website and blog attest, they designate the bog as, in part, "a memento of the last ice age, a rare ecosystem,"[30] which hints at Gobster's observation, but also they demonstrate through ongoing community endeavours and a routinely updated online presence that the bog is something much more than a "memento": the CBRG stages the bog as a dynamic space of ongoing, ritual human interactions with each other and with the biophysical world.[31]

Performing Restoration Ecology

A restorationist approach, which involves community participation, turns "from personal experience into the dimension of shared experience, performance."[32] The dynamics of Boggers, visitors, bog species, surrounding rainforest—the repeated performance of the shared, interactive experience manifests in this ecological "theatre" of Camosun Bog. Camosun, however, and its designed "natural" dioramic presentation set a stage of "true" "ecological theatre in which all the actors participate,"[33] as well as the "audience."

As naturalists and hunters created dioramas in the early twentieth century as a way to capture and preserve wilderness central to the North American consciousness, but that had been in decline, the bog, as living diorama, attempts similarly to reverse a decline. The beauty of the Camosun Bog "diorama" is that the viewers (and neighbourhood critics) take dis-/pleasure in its contained aesthetic design and either overlook or ignore that they are viewing what, arguably, appears as remnant restoration's equivalent of a habitat diorama. The artifice goes largely undetected because unlike a museum diorama, the visitor, as another living being moving into and throughout a live and *un*contained environment, becomes immersed in the display, participates in the makeup of the "diorama": an actor in the site's ecological theatre. Here the illusion lies in a misapprehension that human constraints (boardwalk, weeding programs, institutional ideologies) control entirely the bog. The bog retains an element of fakery through human design and, though arguably not a sense of its wilderness, but certainly a claim to its own wildness. This wildness interacts with human intervention and interaction to create a more dynamic interpretation of Wonders' conception of ecological theatre.

So as scientists and researchers during the nineteenth and early to mid-twentieth centuries sought to bring the natural world indoors with habitat displays, the late twentieth and twenty-first centuries invert that pursuit and appear to bring a revised museum-like display aesthetic to the outdoors.[34] Often this ecological design aesthetic, as Gobster claims, particularly in urban parks, rather than promoting intimate nature experience effects the opposite result and reinforces the nature/culture divide.[35] Gobster argues that such design practices, while beneficial in reducing undesirable impacts, also restrict movement and limit "the spectrum of otherwise acceptable behaviors down to those passive appreciative activities that are deemed appropriate for this revised context to ensure minimal degradation of the now fragile environment."[36] This helps to ensure, he further suggests, that the sites remain experientially limiting and dictates who has access to certain nature experiences.[37] Though Gobster raises valid points, his conception of how a site conveys nature experience suggests that there is only a particular way of experiencing nature. However, though he laments a loss of more unstructured play in these ecologically sensitive spaces, he acknowledges that restoration groups' volunteer initiatives offer new ways of accessing and experiencing nature. He argues that when considering "restoration, authenticity should be conceived as having both ecological and experiential dimensions, and management that considers both of these needs can help strengthen the role of urban parks as a bridge between nature and culture."[38] We agree with Gobster, yet, unpronounced in his conception is how these restoration sites create not just a bridge between nature and culture (as this suggests that these are two exclusive camps to be connected), but rather how these are performance sites that enact processes that constitute one another.

The remnant restoration project, one that crosses a design of outdoor natural history museum with field laboratory and recreational site, creates a "specialized sort of aesthetic" that offers a change in the way we approach the biophysical world. As Laurence Brown notes, the design is the Boggers' "vision of what the bog might have looked like,"[39] as they can only guess what the bog looked like 3,000 years ago. In reconfiguring a dioramic aesthetic into a living exhibit or memento from a past ice age, the Crazy Boggers reconstruct both natural and human histories that constitute one another. But more importantly, the Boggers' botanical and boardwalk storytelling construct a future narrative that offers, we propose, an alternative model to Western assumptions of a nature/culture divide. The design does not detract from intimately experiencing nature, as Gobster contends some restoration projects do, "leading to a form of detached observation not unlike what one might experience in a museum."[40] Rather, the design, with its multisensory biotic and abiotic engagement pushes visitors to experience nature in a way that resists passive observation. The ability to not just see, but to touch, taste, smell, and hear the living "actors" of Camosun Bog and the surrounding forest refutes the notion of isolation. Rather, the capacity to interact with the live habitat situates the spectator as participant, as actor. Jogging, walking, kneeling, digging, weeding, breathing in the bog and forest reinforces that this area, despite the boardwalks, is a live (and thus responsive to and a creator of change) ecological theatre in that human and non-human alike are both participant and spectator. We can therefore imagine how remnant ecological restoration projects like Camosun Bog offer an expanded ecological model "by broadening the interpretative pedagogy of the habitat dioramas, and by preserving, adapting and diversifying the exhibition tradition that it represents."[41] The small-scale restoration projects, such as Camosun, are an expansion of this exhibition tradition as pedagogical model, as these outdoor sites also become in part an outdoor laboratory, such as plant growth experimentation in small test areas referred to as "boglets" or "test bogs."[42] The distinction, of course, is that these planting experiments are focused back at the level of the plants' exterior, their functional relationships with other plants, and the environment at large (rather than internal function).

Like habitat dioramas, which allow for imaginary travel and escape for the urban dweller,[43] Camosun, situated within an urban setting, also provides respite for urban dwellers. The Camosun site diverges from the natural history diorama in that the "travels" are temporal as well as geographical. The indexing of exoticism (rare native species: arctic starflower and cloudberry) among the banal (everyday native species: salal and salmonberry): the rare plants seem out of place and time. The bog remnant gives a simultaneous view into both extant ecosystem and larger extinct or further ranging ecosystem that once constituted the area. The geographical travel, without some botanical knowledge is not so obvious, as some of the rare native species are found only in the northern tundra. This apprehension results in a cognitive "hiccup," as visitors not only make connections between two different bioregions, but also "traverse" deep time. Tied into this literal and imaginative spatial-temporal travel is an ethical imperative, whereby the Crazy Boggers conceive the site as a cultural and historical resource,[44] which they enact through human–material interaction, not passive reception as a museumification aesthetic would suggest. The Camosun restoration project, thus, tends to illustrate Wonders' assertion that "preservationists found their principal source of inspiration for nature conservation in the traditional field approach to natural history [and] were *dedicated to re-creating a visual impression of the beauty and complexity of nature*,"[45] which they argued would lead to "new appreciation of nature [and lead] to conservation rather than exploitation."[46] By combining natural history principles with restoration's philosophies and practices that promote community-based affiliation and preservationist ethics, we suggest that the Camosun Bog Restoration Group takes the original notion of dioramic conception beyond the visual and incorporates aspects that cue other sensory apperception and potentially changes ethical and ecological perceptions.

Static dioramic thinking stymies ecological thinking. Thus sentiments such as Henning's—"In the attempt to preserve vanishing worlds, they turn to reconstruction. The copy becomes a means to bring the original closer"[47]—reveals how critics of ecological restoration can fall back on Taylor's forgery story: they see restoration not as an evolving process but as a static end goal. They get caught up in vocabulary such as "vanishing," "reconstruction," "copy," and "original," rather than searching for new words that would represent a changed, ecological thinking, one that moves beyond binaries and static representation. The restoration outdoor "diorama" defies passive observation. As mentioned, a "natural dioramic habitat," like Camosun, incorporates all the senses so that observation remains only one aspect of the experience.[48] Camosun expands sensory perceptions, reinforcing the experience of being in nature literally, but with an emphasis that leads the visitor to appreciate or perceive through design, the aesthetic richness of the place in other ways besides sight. This fuller embodied access enhances aesthetic appreciation while also educating the visitor of the biological and ecological processes at work. The natural history that resonates in Camosun Bog goes beyond visual education, or a mode of visualization—rather, natural history combined with restoration ecology ends up promoting a more embodied education—or a full sensory education, a mode of not just of seeing, but of interacting, engaging, and promoting potential learning models for ethical environmental responsibility.

Wonders observes, "If one remembers that natural history is fundamentally a science of observation (as opposed to experimentation), then the use of illusionism to re-create the effect of scenery from the natural world is hardly a contradiction."[49] In the case of restoration ecology, however, we need to recontextualize this correlation between natural history as a science of observation and art's capacity to facilitate a re-creation of those observations. We need to rethink Wonders' notion of art as a specialized scientific function and acknowledge art and design's own agency in promoting ecological thinking and participation as it applies to restoration projects such as the Camosun Bog. In effect, restoration design can collapse the barrier between spectator and participant, and show how the interaction between nature and culture (volunteers, visitors, and the biophysical world) sets up a performative space for the complex interplay of science, human and natural design, and natural ecological and human cultural processes to unfold. What does that do to change the way we view the bog, and by extension nature, and further, global environmental issues? If we think of art as not having, as Wonders claims, "a specialized scientific function," but rather as complementing scientific function, then aesthetic design contributes along with the scientific practice to inform, entertain, and change perception.

Bronislaw Szerszynski, Wallace Heim, and Claire Waterton challenge forgery criticism through their definition of performance as "the manifestation of agency and action through which agency and creativity emerge. Performance is thus ephemeral, unpredictable, improvisatory, always contingent on its context."[50] And, they contend, key to performance is "iteration [...] the way that variation and difference emerge in the spontaneous, creative moments."[51] As they further note, linking the two terms "nature" and "culture" is not to shed new light on either term, but to initiate a different thinking. Because humans tend to view performance as cultural coupled with nature, they argue, different agencies emerge:[52] "Out of this mutual improvisation one loses a sense of nature as pre-figured and merely being 'played-out'; instead, the performance of nature appears as a process open to improvisation, creativity and emergence, embracing the human and the non-human."[53] The emergent interactions between nature and humans appear, to a certain degree, as unstaged and highlight the inter-agential dynamics involved in an ecosystem.

Camosun Bog is, arguably, at first encounter a site with more emphasis on human authorship, with the prevalence of interpretive signs, boardwalk, fencing, weeding, and selected species allowed to colonize the surrounding space (those that are edible, for instance). But, as Brown points out in an interview, after initial design and implementation of maintenance programs (e.g., a regular "Saturday Work Party"), the bog's natural processes do the majority of the work.[54] Then the space as a co-created process requires greater environmental imaginary (and observational acuity); coming to see the natural processes is where the natural history initiatives play a role (and even the human design, i.e., signs) with the community partnerships and place-based educational initiatives (e.g., university science students' mentorships of elementary school students). Such practice aligns with Eric Higgs' claim that

> [the] restorationist's aim is usually not to arrest change or recreate a plant-by-plant replica of a historic landscape—what restorationists disparage as "snapshot" objectives or a "diorama"—but rather to redirect it, getting the

system back on track, setting it in motion again, not only in an ecological, but ideally even in an evolutionary sense. The restored system should not only look (more or less) like the "original" or model system, *it should act like it*. It should undergo change and respond to disturbances in the same way. In the long run it should even support evolutionary processes, acting as a source of new species.⁵⁵

Arguably, our assertion that Camosun as analogous to a natural history diorama potentially equates it to a "taxidermy" of nature, a futile attempt to reanimate a sense of the entities' former essence. Yet, the difference, of course, is that the restoration site is a living entity, one that humans help shape or shift, but inevitably do not control. We view the diorama comparison as productive when thought of as performance—that the natural design and the unpredictability (to a degree) of the different agents/actors involved in the bog restoration inhabit an ecological theatre or habitat space, in which all the actors participate. The bog is a stage for both its animal/plant "stars" and its "audience," "stage managers," and "co-stars," humans. In fact, the ongoing maintenance activities and regular interactions of community members play up the "dynamic features of the system being restored."⁵⁶ The bog as a natural space contained by, but also free of, human cultural constraints challenges assumptions of remnant restoration projects as static and thus easily manipulated. Coupled with performance, restoration

> is an event and process that sets loose further actions, those actions eluding confinement [...] Performance here is mutual creation, an adaptation with variation where organisms, ideas, activities and memories move into new contexts, transforming both themselves and these contexts in the process. It is a conversing—between human and human, human and nature, or organism and environment—and in all cases is an exchange with unpredictable outcomes as its effects continue into the future. Those unanticipated consequences are, like the event itself, dependent on their contexts.⁵⁷

Like the diorama and early nature stories, restoration cannot hide the artifice, but efforts such as Camosun Bog, rather than attempt to hide it, we contend, accentuate that artifice by creating a dioramic experience. In Camosun there are still "staged" events that the visitor encounters (for instance, the way the boardwalk controls where someone can walk, the way signs direct a visitor's gaze), and despite the emergent processes that are evident in the bog, there remains in its constrained form, an element of the diorama or paludarium. The Camosun Bog Restoration Group promotes initiatives such as the Camosun Bog Buddies, which has local university biology students providing interpretive tours for elementary school children. The biology students coordinate interactive educational activities, such as games and storytelling. The CBRG also hosts an annual professional development workshop in the fall, inviting participants from various backgrounds—"youth leaders, urbanists, teachers, informal educators, *and* student teachers." Their aims are to "give educators the tools to bring their classrooms and small groups to the bog with confidence; to familiarize educators with the bog narrative, bog ecology and share connections to the elementary and secondary curriculum; to encourage educators to visit our bog. Camosun Bog is the most transit-accessible bog in Vancouver and we encourage everyone to experience this treasure in the city."[58] Further, what is excluded and included in the bog also indicates how artifice (for those who know their native plant ecologies) plays out in more subtle ways. As Brown pointed to a skunk cabbage among the site's sphagnum, he noted this plant exemplifies how this is not a natural bog, illustrates how its deliberate exclusion from weeding signals it is a managed site. The Boggers leave three skunk cabbages as a compromise to meet people's perception of what a bog should be.[59] We argue that the starting point or emphasis of critique should fall on the interactive processes—the performances that accentuate the indivisibility between authenticity and forgery. Restoration, like most wilderness areas—if not all spaces in this globalized world—have been touched to some degree by some form or another through human cultivation or industrial reclamation.

Jordan contends that, from a performative perspective, "it is not nature as an autonomous 'other' that is the ground and touchstone of being and authenticity, but rather nature in reflexive interaction with all its elements, including ourselves."[60] Thus, following Jordan's suggestion, species (even in restoration projects) have their own strategies and characteristics that constrain and limit to what degree they interact with other species. This, then, suggests restoration as a co-performed design between cultural and natural processes. The species can, to some degree, hold in check the human action often by subverting the "plot" through unpredictable growth behaviour, climate anomalies, or browsing deer, for instance. Subsequently, such an approach allows for appreciation of restoration not as a good forgery, but rather in its design enables recognition of nature's "integrity as a self-organizing subject"[61] that participates in an ongoing "work in progress." Jordan notes, "While restoration necessarily entails the manipulation of the landscape and other species, this is, at best, not so much the imposition of a plan on nature, as it is the basis for a conversation with it. The best restoration work does not so much impose on the landscape as *propose* to it, opening a dialogue of which the practice of restoration represents only the technical and restoration ecology only the formally scientific component."[62]

But Wonders' theatre analogy is perhaps closer to the ecological processes emergent in Camosun Bog in that as with each visit, each performance is never entirely the same enactment. Within each change there emerges from the established meanings and interpretations new meanings and interpretations of the restored site. And, as Jordan rightly points out, perhaps the focus should not be put so much on the "authenticity" as on the story(ies) being told, and "it is story we need."[63] Or the stories not being told. Which returns us again to ask, is this the right question—authenticity versus real—or should we instead ask how these processes or events or performances are inhibiting or changing ways of environmentalist thinking? How do these old debates hold us back from creating new forms of dynamic, community-based environmental ethics and preventative acts?

Conclusion: Giving Back

In the preceding pages, we refute accusations of fakery or forgery levelled at the restored Camosun Bog, based on our rejection of a nature/culture binary. Such debates, we argue, perpetuate static perceptions of the biophysical world; they either evoke views of unchanging ecosystems or set ideas about how relationships between nature and humans must evolve. As a means to disassemble that static representation, we reconceptualize ecological restoration within a diorama framework as posited by Karen Wonders, and extend her understanding of ecological theatre to comprise both human and non-human, as both participant and spectator within a dynamic, ever-evolving ecosystem. Constant intervention on the part of the Crazy Boggers is a theatrical performance that sustains non-human members as part of the community identity. There is much to learn here, for those in many quarters.

Camosun Bog's restoration was begun by a mixed group of specialists and dedicated amateur laypeople (many who have become, through their dedication to broadening their knowledge of the bog's natural history, specialists in their own right), which points to the potential power of natural history clubs in educating the wider public and effecting change. Museums everywhere should take careful note. As now neglected nineteenth-century institutions, repositories of natural history collections are struggling to find relevance in contemporary society, and they would do well to reconnect with amateur natural historians and bring their collective expertise to bear on similar projects.

Good interpretation aims to provoke. In the Camosun story that the Boggers share with visitors to the site, casual interlopers could be made more aware of the connection between this site and their own daily activities. Peter Davis suggests that combining progressive museum activities with other interpretive facilities in the urban area "provides the opportunity to emphasize the threads which link natural resources with cultural identity, nature with human-kind."[64] Understanding that the signage at Bog entrances and around the boardwalk was the result of a protracted negotiation by many parties, perhaps it is asking too much that a preventative ethic be made more explicit. Experience of the restored bog will have to carry the burden of communicating the urgency of ecological theatrical performance unfolding there.

"We are in a restorative, as opposed to say, a conservationist mode."[65] A metaphor, Higgs contends, represents "a larger cultural shift to restoration."[66] However, what is the underlying consequence of this thinking—or even of these two approaches? We cultivate an environmental ethos that is not about prevention but about restoring health to an ailing ecosystem. This said, however, we do not want to discredit a restorative ethic. Instead, we advocate that what is missing to begin with is a social ecological (active) thinking that prevents us from allowing an environment to reach a degraded state in the first place. What is needed is an ecological thinking that puts onus on individuals, communities, corporations, governments, and policy-makers to act collaboratively and set up controls and regulations that prohibit or diminish harmful environmental impacts. Of course, at a large-scale project such as a resource extraction operation, like mining, such a thinking model meets with much resistance as it contravenes the potential for gross financial gain.

Richard Evanoff contends, "We must learn not how to better manage *nature* but rather how to better manage *ourselves* by creating forms of culture which go with, rather than against, the flows/directionalities of natural, evolutionary processes."[67] Evanoff's approach promotes a thinking that prioritizes the biophysical world's health. The bog as mini-ecosystem surrounded by the forest can signal or point to other ecosystems, such as the surrounding park, potentially inciting citizens to mobilize against further harmful development. The Camosun blog's contact page provides links to other place-based educational groups, which indicate their own perception as a place-based educational model and site, but also the links reveal the extent of their regional partnerships and, we would argue, the stable infrastructures they have in place for mediating and negotiating with other parties.

Is the bog diorama an improvement on the traditional diorama? What is its significance? We believe that nature faker critics raise valid criticisms of Camosun Bog. They draw attention to important assumptions of ecology, assumptions that emphasize contemporary beliefs of human relationships with the environment. Bringing the diorama aesthetic into the outdoors or bringing a field site to urban dwellers—is this only possible because of its location in Dunbar, an affluent Vancouver neighbourhood? Perhaps, and such critique aligns with traditional criticism against natural history societies and their composition of citizens who can afford the leisure time to restore a 3,000-year-old bog in an urban recreational forest. We cannot overlook the historical context of the bog's restoration and history, but a brief survey of other urban bog restoration projects in BC's Lower Mainland reveals how these efforts do not locate specifically in affluent neighbourhoods.[68] As issues of environmental justice, for instance, overlap with naturalist endeavours, arguably, more remnant restoration projects will occur in economically suppressed neighbourhoods and regions. In fact, as Brown indicates, the Boggers in various capacities have begun to work with other Lower Mainland restoration and natural history education efforts, such as at Stanley Park, Trout Lake, and Killarney. As well, that the Boggers coordinate outreach educational programs to schoolchildren suggests, in keeping with naturalist tradition in British Columbia, a level of inclusiveness that welcomes diversity.

So, maybe what is a successful component of Camosun is its dioramic illusion—an aesthetic display that does not hide its artifice, its artistry; rather, because it is integrated with other ecosystems it tends to de-emphasize—or at least like the conjurer's tricks (and as some critics argue regarding the diorama's effect) makes visitors feel a part of the natural scene (though really they are, literally, unlike a diorama in a museum, immersed in, a part of/participant/actor in the diorama). As ecological crisis becomes more prevalent in the public eye, revitalizing natural history and remnant ecological restoration through design that promotes an ecological theatre provides compelling and engaging material to inspire motivation to protect and acquire interest in the environment. For central to projects like Camosun Bog, and framing these motivations, is the cultivation of belonging—belonging to place, belonging to community. Gobster raises some unsettling truths about the loss of unstructured nature experience in urban ecological parks, but as we suggest, expanding and applying Wonders' conception of ecological theatre to remnant restoration projects reveals different forms of unscripted play, play that evokes human and non-human ethical improvisations. Inherent in a performativity of belonging, as Vikki Bell contends, are iterations of "the norms that constitute or make present the 'community' or group as such. The repetition, sometimes ritualistic repetition, of these normalized codes makes material the belongings they purport to simply describe."[69] And, thus we propose, a sense of belonging that manifests in a reading of the bog and dedicated group of Crazy Boggers as a preventative form of stewardship. Camosun's contained model of community, ecology, and natural history provides an arena in which to teach that restoration and conservation should be the last resort, that environmental education and advocacy should be calling for a change of social behaviour, ethics, and habit of mind that respects and maintains an environment so that it never gets to the point of being a near-extinct ecosystem.

Notes

1. Canadian History and Environment Summer School, UBC, Vancouver, BC, 30 May to 1 June 2008.
2. Timothy Taylor, "It's Not about Ecology—It's about Gardening," *Globe and Mail*, 27 Aug. 2007, http://www.theglobeandmail.com/life/its-not-about-ecology---its-about-gardening/article4107735/.
3. Eric Katz, "Another Look at Restoration: Technology and Artificial Nature," in *Restoring Nature: Perspectives from the Social Sciences and Humanities*, ed. Paul H. Gobster

and R. Bruce Hull (Washington, DC: Island Press, 2000), 37–48; Robert Elliot, *Faking Nature: The Ethics of Environmental Restoration* (London: Routledge, 1997).
4 Karen Wonders, *Habitat Dioramas: Illusions of Wilderness in Museums of Natural History* (Uppsala: Acta Universitas Upsaliensis, 1993); Karen Wonders, "Habitat Dioramas as Ecological Theatre," *European Review* 1, no. 3 (1993), 285–300.
5 Sally Hermansen and Graeme Wynn, "Reflections on the Nature of an Urban Bog," *Urban History Review* 34, no. 1 (2005), 11; Nadia Baker, Patrick Lilley, Toshiko Sasaki, and Heather Williamson, "Investigation of Options for the Restoration of Camosun Bog, Pacific Spirit Regional Park," Environmental Studies 400 thesis, University of British Columbia (2000), 15, http://www.ensc.ubc.ca/about/pdfs/theses/baker_et_al.pdf.
6 Camosun Bog Restoration Group signage, 2012.
7 Baker et al., "Investigation of Options," 53; Hermansen and Wynn, "Reflections," 14–19.
8 Baker et al., "Investigation of Options," 11.
9 Laurence Brown, interview by Lisa Szabo-Jones and David Brownstein, transcribed by Sandi Kingston, Camosun Bog, Vancouver, BC, 2 Sept. 2011.
10 Camosun Bog Restoration Group, website, http://www.camosunbog.org/.
11 *Correspondence of John Bartram*, quoted in Richard W. Judd "George Perkins Marsh: The Times and Their Man," *Environment and History* 10, no. 2 (2004), 176.
12 David Brownstein, "Sunday Walks and Seed Traps: The Many Natural Histories of British Columbia Forest Conservation, 1890–1925," PhD dissertation, Institute for Resources, Environment and Sustainability, University of British Columbia, 2006.
13 Aldo Leopold, *Round River: From the Journals of Aldo Leopold*, ed. Luna B. Leopold (Toronto: Oxford University Press, 1993), 58.
14 Lisa Szabo, "Wildwood Notes: Nature Writing, Music, and Newspapers," MA thesis, University of British Columbia, 2007, 11.
15 John Burroughs, "Real and Sham Natural History," *Atlantic Monthly*, Feb. 1903, 299.
16 Ibid., 303.
17 Ralph H. Lutts, *The Nature Fakers: Wildlife, Science, and Sentiment* (Charlottesville: University of Virginia Press, 1990), 167.
18 Wonders, *Habitat Dioramas*, 162.
19 Toby Kamps and Ralph Rugoff, *Small World: Dioramas in Contemporary Art* (San Diego, CA: Museum of Contemporary Art, 2000).
20 Frances Terpak, "Diorama," in *Devices of Wonder: From the World in a Box to Images on a Screen*, ed. Barbara Maria Stafford and Frances Terpak (Los Angeles: Getty Research Institute, 2001), 325–26.
21 Wonders, *Habitat Dioramas*, 13.
22 Lynn Nyhart, "Science, Art, and Authenticity in Natural History Displays," in *Models: The Third Dimension of Science*, ed. Soraya De Chadarevian and Nick Hopwood (Stanford, CA: Stanford University Press, 2004), 313.
23 Ibid., 319.
24 Wonders, *Habitat Dioramas*, 222.
25 Ibid., 287.
26 Ibid., 285.
27 Ibid., 285–86.
28 Ibid., 287.
29 Gobster explains that "'museumification' is a process in which places or subjects of the everyday world are transformed in ways that can lead people to think and act toward them as if they had been placed in a museum. Museumification can be accidental or intentional and its aim might be to conserve or commodify, but the end result is a shift in the meanings, behaviors, and experiences people have in relation to a place

or subject." Paul H. Gobster, "Urban Park Restoration and the 'Museumification' of Nature," *Nature and Culture* 2, no. 2 (2007), 100.
30 Brown, interview; CBRG, http://www.camosunbog.org/main_intro.htm.
31 In our definition of "ritual," we refer to secular repeated activities that have social and ecological import on the Camosun Bog's continuance: maintenance routines, planned community endeavours, educational outreach programs, and visitors' strolls or jogs through the area. A spiritual aspect that does apply to ritual and Camosun Bog as unceded traditional territory relates to the site's spiritual significance to people of the Musqueum First Nation and their use of the area and some of its plants for ceremonial purposes.
32 William R. Jordan III, *The Sunflower Forest: Ecological Restoration and the New Communion With Nature* (Berkeley: University of California Press, 2003), 20.
33 Wonders, *Habitat Dioramas*, 228.
34 See esp. chapter 5, "From Interpretive Centre to Ecomuseum—Museums Beyond the Walls," in Peter Davis, *Museums and the Natural Environment: The Role of Natural History Museums in Biological Conservation* (London: Leicester University Press, 1996).
35 Gobster, "Urban Park Restoration," 96.
36 Ibid., 105.
37 Ibid., 107.
38 Ibid., 111.
39 Brown, interview.
40 Gobster, "Urban Park Restoration," 96.
41 Wonders, *Habitat Dioramas*, 228.
42 Baker et al. "Investigation of Options," 12.
43 Wonders, *Habitat Dioramas*, 160.
44 Brown, interview.
45 Wonders, *Habitat Dioramas*, 163 (emphasis added).
46 Ibid.
47 Michelle Henning, *Museums, Media and Cultural Theory* (New York: Open University Press, 2006), 59.
48 The Boggers are also developing a "hands-on" area for children where they can use the sense of touch so that observation remains only one aspect of the experience.
49 Wonders, *Habitat Dioramas*, 192.
50 Bronislaw Szerszynski, Wallace Heim, and Claire Waterton, "Introduction," in *Nature Performed: Environment, Culture, and Performance*, ed. Bronislaw Szerszynksi, Wallace Heim, and Claire Waterton (Malden, MA: Blackwell, 2003), 3.
51 Ibid.
52 Ibid., 4.
53 Ibid.
54 Brown, interview.
55 Eric S. Higgs, *Nature by Design: People, Natural Process, and Ecological Design* (Cambridge, MA: MIT Press, 2003), 21–22 (emphasis added). In his accompanying footnote that comes after *diorama*, Higgs observes, "There are exceptions—at historic sites, for example, where an attempt may be made to restore and maintain specific features of the landscape such as the peach orchard at Gettysburg or the historic vegetation at Abraham Lincoln's boyhood farm in Kentucky. Though most restorationists regard 'postcard' projects of this kind as incidental or irrelevant to the practice of *ecological* restoration, they are nevertheless restoration projects in the sense that I am using the term here and worth noting because they represent the principles of faithfulness to the model in an extreme and especially clear way. Their weakness is that they downplay the dynamic features of the system being restored" (206n6; original emphasis).

56 Ibid., 206.
57 Szerszynski et al., "Introduction," 5.
58 Camosun Bog Restoration Group Blog, http://camosunblog.blogspot.ca/p/where-wild-things-are-pro-d.html.
59 Brown, interview.
60 Jordan, *Sunflower Forest*, 123.
61 Ibid., 125.
62 Ibid., 124; original emphasis.
63 Ibid., 131.
64 Davis, *Museums*, 124.
65 Higgs, *Nature by Design*, 11.
66 Ibid.
67 Richard Evanoff, *Bioregionalism and Global Ethics: A Transactional Approach to Achieving Ecological Sustainability, Social Justice, and Human Well-Being* (New York: Routledge, 2011), 58; original emphases.
68 See Rachel Wiersma's 2003 map for the bogs in BC's Lower Mainland at the Natural History of Richmond's website page "The Biodiversity of Richmond, British Columbia: Bogs of the Lower Mainland," June 2009, http://www.geog.ubc.ca/richmond/city/bogslowermainland.html.
69 Vikki Bell, "Performativity and Belonging: An Introduction," in *Performativity and Belonging* (Thousand Oaks, CA: Sage, 1999), 3.

CHAPTER 5

A Subtle Activism of the Heart

Beth Carruthers

This chapter is rooted in hybridity. It draws from ecophilosophy and is informed by decades of art practice and exploration of the role of art in the world. Its taproot is love of place, especially of the wild, west coast of Canada, a community of myriad species in a green and fecund home-place. In it I reflect on questions of how art practices and aesthetic engagement in particular might hold the promise of change for the better in human–world, or nature–culture relations.

The conversation opens by proposing that the current eco-crisis is symptomatic of an ontological crisis, and further, that this eco-crisis presents an opportunity for a necessary shift in world view, or ontology, and hence to our understanding of the place of humans in the world. I argue that without such a shift we will perpetuate a dangerous alienation from a world in which we are, in every sense, environed—an alienation that has led to a continuum of unsustainable practices.

Canada is I think a good lens through which to view questions of ontology and sustainability, and I next discuss why I think this to be so. I then introduce the idea of aesthetic engagement, and of the arts as holding the promise of being key agents of change vis-à-vis a shift in world view. I bring this discussion to the Canadian west coast, where artists have opened conversational spaces that, among other things, accomplish what Marshall McLuhan long ago claimed as the role of art: "to create the means of perception by creating counterenvironments that open the door of perception to people otherwise numbed in a non-perceivable situation."[1] This opening, or widening, of our perception is, I argue, an opportunity for change.

In a Dark Wood

> *Midway upon the journey of our life*
> *I found myself within a forest dark,*
> *For the straightforward pathway had been lost.*
> —Dante, *The Divine Comedy*[2]

One might well ask whether we are indeed at the midway point of human life on this planet, or at the end. The clocks, both industrial and biological, are ticking. The story of humans on the planet might truly be "nasty, brutish and short," although for different reasons than Hobbes once argued, and on a scale that he could hardly have imagined. We are daily confronted with events and expressions of what is now widely accepted as a global ecocrisis—from species loss, deforestation, acidification of the seas and the poisoning of land, air, and water, to climate change. Yet while the crisis may be clear, responses to it are less so. Certainly, the question of how we as a species can more sensitively dwell within a world that we now experience as more finite and unstable than we had understood it to be is a critical one.

As philosopher Monika Langer pointed out as long ago as 1990, there have been many calls for a radical ontological shift if we are to make real and lasting change for the better in human–world relations.[3] I believe we have arrived at an ontological fork in the road, and in confronting the choice of direction before us it seems to me that we must reflect on which path might offer the best opportunity for such a necessary shift in ontology.

One fork leads along the road most often taken, travelling by way of a world view that perceives world and complex interrelationships as objects and systems and relies on developing further technologies to tinker with, repair, and retune the mechanisms of these. Commonly referred to as Cartesian, or dualist ontology, this way of being in the world is rooted in the belief of humans standing outside an external nature, or of the environment as object, with the human alone as subject. Having been with us for centuries this world view holds the comfort of the familiar; it fits our sense of continuity.

Taking a different path requires an active willingness to make a shift in ontology. Let us assume this alternate route to be a path of inclusivity which, rather than perpetuating the story of humans as standing outside a world, instead recognizes an intertwining of being—a vastly different perception of self and world from which very different responses and technologies might arise. Because such a shift requires opening to the acceptance of other agencies and powers as having a valuable voice in, and as co-genitors of, world futures—including human futures—the willingness to take this path also requires a surrender to uncertainty. This is an unfamiliar and more difficult-to-follow path. It may also prove to be a more challenging path to locate.

Since an ontology is so very basic to who we are and what we do in the world, calling for major changes to such a foundational structure is no small proposition. Hence, before considering what might facilitate our locating and proceeding along an alternate ontological path, it is helpful to first give some additional consideration as to why it is necessary to do so.

Lifeworld and Resource

Our current world view dangerously reduces all but the human to resource for the human: in Heidegger's terms, a "standing reserve." While there has been much speculation as to its genesis—whether it be Plato or the scientific and philosophical revolutions of the sixteenth and seventeenth centuries—we recognize that, to quote Monika Langer, "this reductionism effectively *uprooted* humans and rendered them homeless in a world perceived as radically *other*, essentially alien and meaningless—even hostile and threatening to what, by contrast, was defensively deemed of paramount importance—namely, the human 'mind' or 'soul.'"[4]

Anthropologist Tim Ingold considers this ontological entrapment to be the difference between perceiving our world in terms of either "globe," or "lifeworld." A globe is an object external to ourselves, which we can stand away from or view from afar, while a lifeworld is an encompassing place, wherein we are environed. When it comes to acting in the world in response to ecocrisis, Ingold observes "that the very notions of destruction and damage limitation, like those of construction and control, are grounded in the discourse of intervention. That is to say, they presume a world already constituted, through the action of natural forces, which then becomes the object of human interest and concern. But it is not a world of which humans are conceived to be a part. To them, it is rather presented as a spectacle. They may observe it, reconstruct it, protect it, tamper with it or destroy it, but they do not dwell in it."[5] This is similar to the comments of Lorne Neil Evernden in *The Natural Alien*, where he describes this perspective as "resourcism," the seeing of the world, whether with a view to save it or use it, as primarily resource for human interest. As such, the world outside the human is empty of meaning other than that projected or given by humankind, and so we need have no concern for its well-being in and for itself. Adopting this world view sweeps away potential ethical problems arising from human choices and behaviour, as well as any need for sensitive awareness or engagement.

In her essay, "Merleau-Ponty and Deep Ecology," Monika Langer considers Erazim Kohak's observation in *The Embers and the Stars* that we are now so embedded within this troubling ontology and its artifacts that everywhere we turn our attention we see it affirmed. From language to technologies, each aspect of our lives contains and enforces this ontology, and "even those wholeheartedly committed to eliminating the dualism of this ontology and reversing the degradation to natural systems frequently perpetuate the dominant paradigm themselves."[6] To paraphrase Marshall McLuhan's thinking, we shape our tools and thereafter our tools shape us. We are ourselves split; we embody dualism, a malady of separation. To once again quote Ingold, "Indeed, what is perhaps most striking about the contemporary discourse of global environmental change is the immensity of the gulf that divides the world as it is lived and experienced by the practitioners of the discourse, and the world of which they speak under the rubric of 'the globe.'"[7]

The Unsettled Land

> *Away with Canada's muddy creeks*
> *And Canada's fields of pine*
> *Your land of wheat is a goodly land*
> *But oh, it is not mine.*
> —"Settler's Lament" (traditional ballad)

To my mind, Canada is a nexus where important questions and issues of self and world, of place and belonging, of colonialism, resourcism, empire, industry—and, in particular, a clash of differing world views—are visible and foregrounded. Western Canada is of special interest in this conversation because it is the most recently settled in the push west by imperial interests, and also because it is the location of some of the last intact wild and functioning habitats in the world. Since the early Province of British Columbia did not recognize rights of Aboriginal title, it is also the case that little official title to the lands and waters west of the Rockies was ceded to colonial powers by Indigenous peoples.[8] It is a place of enquiry and experimentation, of seeking alternative ways of living within and belonging to a place. One has only to make a cursory historical survey to see that the coastal region has a long history of cultural, political, and environmental activism, of speaking out and taking stands against colonial-industrial powers and their practices.

There is within Canadian settler culture a deep, pervasive ambivalence in the relationship between self and place—an ambivalence and a tension that, although significant, is for the most part backgrounded. I understand this tension as a fault line lying deep under the Canadian psyche. It makes its presence felt from time to time, as does the earth when it adjusts its skin. Perhaps this tension and deep instability lie closer to the surface on the west coast, mirroring the real instability of living in a place of seismic activity. The ground is continually shifting beneath our feet. Settler culture is more unsettled than settled.

Our artists and writers, wakeful and attending to the subterranean murmuring and rumbling that disturb the cultural sleep, hear and seek to interpret and reveal the stories of the land and the self. Reflections on "place," of longing and belonging, have long been a focus of the artists of Canadian settler culture. While, for many, Canada offered a safe haven from persecution and cultural genocide, from the very early days of Scots settlers arriving, we find songs and poems of longing to be back in their home places.

Some years ago a long-time friend and colleague, Saskatchewan artist Sandra Semchuk, asked these two related questions in a public forum: "How do we, living here in Canada, having come from elsewhere, connect strongly with where we are? Because unless we can have that kind of connection, how can we be committed to where we are?"[9] I've hung onto these questions because I think them essential. They touch on the core of the issue, and I believe these are questions we must foreground. In essence, they enquire into the effects of a history of colonization on intimate place relationships within settler culture in Canada. They contain the premise that belonging—being deeply connected with where we are—is important, and they imply that the history of colonization, or even colonization itself, is what keeps us from being fully engaged with and committed to a place.

Canadians literary critic Northrop Frye and philosopher George Grant would agree. Grant commented: "That conquering relationship to place left its mark within us. When we go into the Rockies we may have the sense that gods are there. But if so, they cannot manifest themselves to us as ours. They are the gods of another race, and we cannot know them because of what we are, and what we did. There can be nothing immemorial for us except the landscape as object."[10] "*What we are*"—what do you suppose Grant means? Frye helps us out by elaborating on Grant's observations: "There are gods here, and we have offended them. They are not ghosts; we are the ghosts, Cartesian ghosts caught in the machine that we have assumed nature to be."[11]

Frye is linking colonization directly with Cartesian framing of the world as object and mechanism. Not only is the practice of colonization to blame for our disconnection, Frye takes this idea further, naming the children of colonizers in Canada as Cartesian ghosts haunting a place—disembodied, disconnected, not belonging. Ghosts in between worlds, here and not here at the same time.

And if we are *Cartesian* ghosts, then surely this not-belonging is more far-reaching than settler culture. What Frye seems to be saying is that it is not by virtue of colonization as we usually think of it that we are barred from belonging—not just because of what we did—but rather because of another kind of colonization: the colonization of our very beings by the idea or belief of not belonging in a deeper sense, of not being a part of an interconnected community of beings, dwelling within a place. Not just *because of what we did*, but because of *what we are*. In other words, we are stuck, as disembodied as ghosts, because of our beliefs. We imagined ourselves outside and because we believe it, our perceptual screening makes it difficult for us to perceive otherwise. Because we believe it, we are beings—to again borrow from Heidegger—doomed to perceive the non-human world and others as either tools, or as "standing reserve": as resource. What we are makes possible what we did—and what we do.

This suggests also that the deeper disconnect that the children of colonizers experience in Canada is not peculiar to colonizer, or settler cultures, but is rather the same disassociation endemic within Western European tradition. What is notable is that in predominantly colonial countries such as Canada, this disconnection is revealed, lit up by contrast with differing world views.

Without exploring a detailed account of First Nations ontology in Canada, which the length and scope of this chapter prohibits, I think it fair to say that these differ in important ways from the colonial, or settler world view insofar as relationship with world goes, and the place and role of humans in the world. Examples of such differences can be seen in approaches to, for example, governance, policies, and cultural practices in regard to what Western culture is wont to call "resource," but which the Haida, for example, understand as a complex and intertwined network of relationships—a lifeworld—within which the obligations of humans play a significant role. One often hears Coast Salish people speak of Cedar People, and Salmon People, and many coastal families are bound within familial interspecies clan relations, such as Raven and Bear clans. This expression of relationship, of being family, acknowledges this intertwining, formalizing, and reminding of clan and familial obligations vis-à-vis the more-than-human. Telàlsemkin/siyam (Squamish Nation Hereditary Chief Bill Williams) described *protocol* to me as "a path of balance that also consists of a code of proper conduct in relations with others, whether human or not. It is, for example, wrong to take anything without giving something back."[12]

The meeting of ontologies, like fault lines underneath our lives, produces a constant frisson of uncertainty. When one's world view is based on an ideation, and not well grounded in the material, perhaps it is less stable, more prone to a good shaking up. To my mind it is precisely this place where meanings meet and tussle that is both fecund and dangerous, holding great promise for a necessary disruption—and a creative reconstruction.

Walking the Fault Line

Considering the west coast of Canada through a cultural lens is akin to watching the sea where currents meet and clash, the water churning and rotating on itself. One can see that there is a great deal going on beneath the surface, in the depths, of which this roiling communion is but an indication.

Thrown into sharp relief against the world views of other, more long-term human inhabitants, here we witness the very accelerated transformation of the land—intact habitats becoming wastelands, species disappearing before our eyes. There has been no opportunity to grow into an acceptance of change through generations, and the distress burrows deep into us all, even as the cultural script calls for denial. Yet it seems very clear that when the world is injured—when the *Exxon Valdez* ran aground and flooded sea and shore with crude oil, or when the northern boreal forest is decimated for bitumen extraction and the clean waters spoiled—we, too, are injured. This injury is not just mechanical or biological—a case of increased cancer rates, for example. To paraphrase psychologist James Hillman, this is an injury to the soul, to the part of our souls that is also part of the *anima mundi*, or soul of the world; according to Hillman, this injury arises from anaesthesis, a disengagement that is the opposite of aesthesis, or aesthetic engagement.[13]

Dante's *Inferno* once again provides a helpful analogy: the souls in purgatory want to be there. Their difficulty, or torment, is due to their inability to imagine any other way of being. It is a failure of imagination and of perception. They are suffering not only from a dearth of vision—the inability to imagine—they are suffering from anaesthesis.

In order to think this through, it is helpful to visit the etymology of the word "aesthetic," and that is "to perceive by the senses ... to feel."[14] It follows that when we open to our feelings—or when our feelings open us—opening "the doors of perception" as McLuhan suggests, we are opened to the possibility of having a very different experience of world and self.

I explore these ideas at length elsewhere, and so will not do so here, but I will add Hillman's description of the moment of aesthetic engagement, or opening, as an "Aha!"—as a gasp, a sudden intake of breath, as literally an "inspiration."[15] One might also refer to this moment as "aesthetic arrest" or "epiphany," after James Joyce's character Stephen Daedalus in *Portrait of the Artist as a Young Man*.[16] But being perceptually opened by aesthetic engagement need not be so dramatic an experience. It can be a slow process of opening, so that one simply finds oneself opened, not quite knowing how one arrived at a place of altered perception, experience, and understanding.[17]

It is possible that we may open our perceptual field to find ourselves on the vertiginous edge of a place of unknown depth and possibility, and this could send us scurrying back to the familiar path. We may also open to a world of wonder, belonging, and possibility.

Getting to Where We Are

In her book of theory and criticism, *Art Objects*, writer Jeanette Winterson makes the following statement:

> But what is Nature? From the Latin *Natura*, it is my birth, my characteristics, my condition. It is my nativity, my astrology, my biology, my physiognomy, my geography, my cartography, my spirituality, my sexuality, my

mentality, my corporeal, intellectual, emotional, imaginative self. And it is not just myself, every self and the Self of the world. There is no mirror I know that can show me all of these singularities, unless it is the strange, distorting looking glass of art, where I will not find my reflection nor my representation but a nearer proof than I prefer. *Natura* is the whole that I am. The multiple reality of my existence.[18]

Within the Western cultural tradition and in its pre-Modernist sense, "nature is not only the uncultivated part of environment, but also the sum of living behaviours in general."[19] The Latin term *Natura*, common in the medieval, pre-Enlightenment, world is the more inclusive one, "the whole that I am. The multiple reality of my existence."[20] The distinction between the two may be thought of as follows: "nature" meaning the socially constructed ideation of an external, non-human, world, and "*Natura*" as that within which all is encompassed, or environed, where all life takes place. This latter understanding resonates with what anthropologist Tim Ingold calls "the dwelling perspective," which is "a perspective that treats the immersion of the organism-person in an environment or lifeworld as an inescapable condition of existence ... the world continually comes into being around the inhabitant, and its manifold constituents take on significance through their incorporation into a regular pattern of life activity."[21]

It was precisely this kind of thinking that informed the SongBird Project in Vancouver in the late 1990s. Co-founded by myself and writer-director Nelson Gray in 1997, SongBird was a creative sci-arts collaboration that resulted in a number of initiatives and events "designed to engage urban communities in understanding themselves to be part of a greater, non-human community and environing world."[22]

As mentioned above, the west coast of Canada is a place of upheaval and rapid change to both human communities and the community of the lands, waters, and non-humans that comprise this place. SongBird was developed in response to these changes and according to a set of beliefs that are central to this discussion. The first is my belief that artists, as cultural workers, have an essential role to play in cultural change of the kind we require if we are to "dwell" both more sensitively and more sustainably in the world and with those others with whom we share it. That belief is related to another: that it is by way of aesthetic engagement that we can be opened to change.

SongBird was about foregrounding and building the relationship between people and place and about building the knowledge and understanding of community as not restricted to the human. As Ingold put it, "Relationships among humans, which we are accustomed to call 'social,' are but a subset of ecological relations."[23] He also tells us, "Information is not knowledge, nor do we become any more knowledgeable through its accumulation. Our knowledgeability consists, rather, in the capacity to situate such information, and understand its meaning, within the context of a direct, perceptual engagement."[24] Over its five-year life, SongBird included four primary, or core, initiatives and events, most of which were annual. These were the Gardens of Babylon Habitat Challenge," the Living City Forum, the Dawn Chorus celebration, and the *SongBird Oratorio*.

The annual Gardens of Babylon Habitat Challenge was an initiative in partnership with the Institute of Urban Ecology at Douglas College, in New Westminster, BC, to encourage the creation within the city of balcony and rooftop habitats as important food and water sources for migratory pollinators. This not only engaged people in measurably valuable activities, it also allowed them to gain understanding and knowledge by way of doing. Habitat and awareness of being a part of habitat became incorporated into their daily lives.

The Living City Forum, held in 1998 and 1999, was a two-day forum with invited presenters from the arts, environmental, academic, governmental, and business sectors in conversation with the community at large. One of the presenters at the Living City Forum in 1998 was dancer/choreographer Karen Jamieson, who spoke about her community-engaged project, The River, as follows: "I became aware that I wanted the audience to experience the river in a way that would identify the watery flow of their own bodies, emotions and spirits with this now paved over, seemingly extinct, seemingly gone, but still there, river."[25] Time and again, what was clear in the communication of the artists at the Living City Forum was the depth and quality of engagement or communion they were seeking through their work with other-than-human parts of their immediate environment. This was amply revealed through their work and their expression of it, so that this opened the world for the members of the community who came to the forum and encouraged their greater engagement as well.

The annual Dawn Chorus Celebration was an event held on the first weekend of May. In neighbourhoods throughout the Vancouver area, people rose early to gather at a designated "listening site" staffed with a birder from the Vancouver Natural History Society. Everyone later assembled at a central location for food, shade-grown coffee, an unofficial community "bird count"—including a recounting of the morning listening—and performances. This event nurtured a powerful connecting of people with place and other species, engendering an expanded sense of community and belonging. There was always an aura of excited engagement about it.

The *SongBird Oratorio* was three years in development. We commissioned five composers to create five arias based on indigenous birdsong, which was but one part of the process. Perhaps the most challenging—and to my mind, important—aspect of the *Oratorio* development was the question of how to engage with the perceptual worlds of a species so different from our own, how to find a way for the voice of the other to be heard in the work in its own way. To some, this inclusion would seem as simple as incorporating recorded birdsong into the work, but this project and process was seeking something more inclusive and convivial. It is difficult to know how successful this was, and it is my thinking that in the end all our works can only be products of our own ways of perceiving and filtering world and other, however earnest our attempts to do otherwise. At the same time I found the most powerful aspect of the making of this work was this attempt to comprehend and allow space for the voice of the other. Certainly the work as it was performed seemed to embody this intention of conveying something of the voices and languages of birds as well as humans. In my paper "PRAXIS," I note the following: "As the idea of the Oratorio was hatched from the song of a robin, so in the final performance of the work, at twilight in a garden in the heart of the city, a robin arrived and burst into an evening song, performing a duet with the human singers."[26]

Artist and landscape architect Claire Bédat, working from the knowledge of birds, studied the architectural strategies of marsh wrens for the design and creation of a large, human-scale nest for the SongBird Project, which was constructed in situ outside the Roundhouse Community Centre in downtown Vancouver. Composed of cuttings from the local botanical gardens, it was strong enough to survive almost a year of weather in its location near the water. It was round, and when inside one was cocooned in a sweet, grassy space. Beloved of animals and small children, in the end it was removed by the Parks Board, which was concerned that someone was sleeping in it. Bédat made the following comments about the experience of building the nest:

> A nest is supposedly round; round like life, round like the body of a bird. The making was in its essence a very intuitive rendering as, dedicated as a bird, I used each part of my body to shape and build the nest. Unconsciously I was participating in the making of a shelter, a refuge, the home of my body. Growth is often assimilated to change—I changed during the making of this project and feel emotionally empowered and bounded to a greater cause: preserving biodiversity on Earth.[27]

This expression of not only being connected, but of being part of something larger than one's self or the strictly human community, was common among participants in the SongBird Project. A feeling of wonder was often expressed, as was the experience that "issues no longer seemed ... to be something happening 'out there' but were instead seen as arising within community relations, something people could ... engage with."[28]

Both the quality of the experience and the process of change that Bédat speaks about are not easy to pin down, and perhaps it is not necessary to do so. Not everything that comes to us is accessible to, or understood by, our cognitive framing—and this is a good thing, since through this bypassing of our preconceived notions and framing of experiences and world we can arrive at new understanding. It is worth noting that for Bédat the experience of the making itself seems to have been central to this change, and it follows that this kind of sensuous, embodied engagement may open us all to change.

The world is opened for us as an interrelational space, as it was in the material making of the Nest. Bédat's experience and that of others on the project resonate with the earlier mentioned comments of Ingold on the sort of engagement required in order to arrive at knowledge or knowing—that knowledge is situated and is arrived at perceptually,[29] or by way of aesthetic engagement.

In a public talk on Darwinian evolution and beauty, philosopher Denis Dutton noted that beauty lies in skill, in something done well.[30] This observation about doing also connects with Ingold's idea that doing provides a connection that leads to knowledge. Certainly in the case of the making of the Nest, the aesthetic, embodied engagement of doing, and doing well, led to appreciation and understanding. These gently subversive acts of engagement are also acts of subtle activism, more profoundly effective than any confrontational model.

Conclusion

The experiences we have as we move about our world have a significant impact on how we understand our relationships with—and in—that world. Earlier I mentioned Erazim Kohak's assertion that we are now so surrounded by and immersed within the artifacts of our ontology that this ontology is perpetually reaffirmed. These artifacts and the systems that support them seem massive and permanent. They are taken for granted in that we no longer really see them, as we move in, through, and around them. Yet regardless of this reinforcement, we stand on shaky ontological ground.

The internal splitting of the self by way of the disconnect between the lived experience of being in the world and the self-world relationship as framed by dualist ontology is culturally unacknowledged and is instead accepted as usual. Both this splitting and the denial of it are in Canada rendered more evident by coexisting and radically different ontologies in a state of constant interplay. The ontological narratives contained within the structures of settler culture can only be unstable, overlaid as they are on alternative and active world views with very deep roots indeed. These speak to an innate desire to belong to a place, and to the actual lived, embodied experience of being in and with a place. These voices also find a response in the deep cultural past of the dominant settler culture. If one searches back one finds, for instance, the tribal Celtic culture of the Irish, the Scots, the British, and indeed the Gallic roots of the French and other settlers. Those ancient Celtic peoples, whose deep cultural memories and blood runs through much of settler culture, would be as baffled as the coastal First Nations peoples of British Columbia by a notion that human and nature are two, rather than one. I find this particularly encouraging, since this suggests our current, problematic ontology may be destabilized rather readily and that this destabilization may, in fact, be inexorably occurring.

Regardless of the insistent voices of our structures and artifacts, the voice of experience, of lived being, is constant, and the voices and presence of alternate ontologies ubiquitous. It seems to me that we have been and are continually being changed. Like sea fog slipping under the back door, the knowledge and very being of this place finds its way into us.

All manner of things, such as the events and experiences offered by the SongBird Project, rattle the ontological cage we have created, foregrounding what is daily denied: our lived experience of the relationships we have with and in the world, and the necessity of these relationships to our very being.

It is important to consider that coming to know ourselves as deeply and inextricably intertwined within a world, to the extent that we at all times understand our behaviours and cultural artifacts as aspects of this relationship, does not guarantee that our choices or our artifacts will be friendly or sensitive to the interests of the non-human other or the whole. Foregrounding the relationship does insist on its recognition; recognition then asks us to consider our behaviour as members of a community—an ecological community, as Ingold has it. While recognition of relationship may not in itself change behaviour, I think it is the case that the recognition and acknowledgment of these relationships carries with it an ethical imperative to consider the well-being and interests of more than simply the human project. Certainly if one removes the myth of this project as existing outside of an environing ecosystem community, it becomes immediately apparent that the project of human flourishing must and can only take place within, and according to, the flourishing of the overall ecosystem community. Moreover, if it is indeed the case, as Hillman claims, that at the core of our dislocation and discontent, at the heart of this ecocrisis, which is also a crisis of self–world relationship, is a crisis of love—"that our love has left the world"[31]—then reawakening to this relationship can stir the heart. We nurture and protect what we love. With the understanding that loving self and loving world are not oppositional, but are rather absolutely together, we open to the other—to the choice of nurturing the whole, with ourselves nurtured by extension through being a part of that whole.

We can, it seems to me, be uncomfortably and dangerously stretched across a widening fault line until we collectively fall in, or we can simply let go of what may no longer be helpful to us, regardless of how familiar it appears. The alternative may prove more familiar than we think. In the words of ecophilosopher Isis Brook, "the reality of our situation is being environed, being engaged in an embrace, not as an optional extra—a lifestyle choice—but just how it is."³²

Notes

1. In Eric McLuhan and Frank Zingrone, eds., *Essential McLuhan* (Toronto: House of Anansi, 1995), 342.
2. Dante Alighieri, "Inferno," Canto I, *The Divine Comedy*, Digital Dante: http://dante.ilt.columbia.edu/comedy/.
3. Monika Langer, "Merleau-Ponty and Deep Ecology," in *Ontology and Alterity in Merleau-Ponty*, ed. Galen A. Johnson and Michael B. Smith (Evanston, IL: Northwestern University Press, 1990), 117.
4. Ibid., emphasis in original.
5. Tim Ingold, *The Perception of the Environment* (London: Routledge, 2000), 215.
6. Langer, "Merleau-Ponty and Deep Ecology," 117.
7. Ingold, *Perception of the Environment*, 215.
8. Early exceptions are Treaty Eight (1899), covering northeastern British Columbia, northern Alberta, northwestern Saskatchewan, and a part of the Northwest Territories (http://www.gov.bc.ca/arr/firstnation/treaty_8/default.html), and the extensive transfer of lands on Vancouver Island to the Crown, known as the "Douglas Treaties" and noted as being land sales through the agency of James Douglas of the Hudson's Bay Company (http://www.treaties.gov.bc.ca/background_history.html). There is also the recent Nisga'a Treaty (2000).
9. Beth Carruthers, "Hybrid Forms," video essay presented at Between Nature: Explorations in Ecology and Performance, Lancaster, UK, 27–30 July 2000.
10. Northrop Frye, "Haunted by Lack of Ghosts: Some Patterns in the Imagery of Canadian Poetry," in *The Canadian Imagination: Dimensions of a Literary Culture*, ed. David Staines (Cambridge, MA: Harvard University Press, 1977), 28.
11. Ibid.
12. Beth Carruthers, "PRAXIS: Acting as if Everything Matters," MA thesis, Lancaster University, 2006.
13. James Hillman, "The Practice of Beauty," in *Uncontrollable Beauty: Toward a New Aesthetics*, ed. Bill Beckly and David Shapiro (New York: Allworth, 1998), 261–74.
14. Douglas Harper, Online Etymological Dictionary, http://www.etymonline.com/index.php?allowed_in_frame=0&search=aesthetic&searchmode=none.
15. Hillman, "The Practice of Beauty." See also Beth Carruthers, "Through the Eye of the Heart: In Search of a Deep Aesthetics," a talk given at Thinking Through Nature, University of Oregon, 19–22 June 2008; and Beth Carruthers, "Call and Response: Deep Aesthetics and the Heart of the World," a talk given at the workshop Aesth/Ethics in Environmental Change, at the Biological Station of Hiddensee, University of Greifswald, Germany, 24–28 May 2010. Both talks are available at: http://ecuad.academia.edu/BethCarruthers. I expand and further explore the ideas presented in the latter talk in a chapter that appears in *Aesth/Ethics in Environmental Change: Hiking Through the Arts, Ecology, Religion and Ethics of the Environment*, ed. Sigurd Bergmann, Irmgard Blindow and Konrad Ott (Studies in Religion and the Environment/Studien zur Religion und Umwelt 7) (Berlin: LIT Verlag, 2013), 131–41.
16. James Joyce, *A Portrait of the Artist as a Young Man* (Harmondsworth, UK: Penguin, 1992), 222.
17. Carruthers, "Through the Eye of the Heart" and "Call and Response."
18. Jeanette Winterson, *Art Objects: Essays on Ecstasy and Effrontery* (Toronto: Alfred A. Knopf, 1995), 149–50.

19 Christian Rhor, "Man and Nature in the Middle Ages," paper given at Novosibirsk State University, 29 Oct.–1 Nov. 2002, http://sharepdf.net/view/14095/man-and-nature-in-the-middle-ages.
20 Winterson, *Art Objects*, 150.
21 Ingold, *Perception of the Environment*, 153.
22 Carruthers, "PRAXIS."
23 Ingold, *Perception of the Environment*, 5.
24 Ibid., 21.
25 Carruthers, "Hybrid Forms."
26 Carruthers, "PRAXIS."
27 SongBird Project, http://songbirdproject.ca/nest.html, and Carruthers, "Hybrid Forms."
28 Carruthers, "PRAXIS."
29 Ingold, *Perception of the Environment*, 24.
30 Denis Dutton, "A Darwinian Theory of Beauty," TED talk, given at TED 2010, http://www.ted.com/talks/denis_dutton_a_darwinian_theory_of_beauty.html.
31 Hillman, "Practice of Beauty," 266.
32 Isis Brook, "Can Merleau-Ponty's Notion of Flesh Inform or Even Transform Environmental Thinking?" *Environmental Values* 14, no. 3 (2005), 361.

CHAPTER 6

Sublime Animal

Maria Whiteman

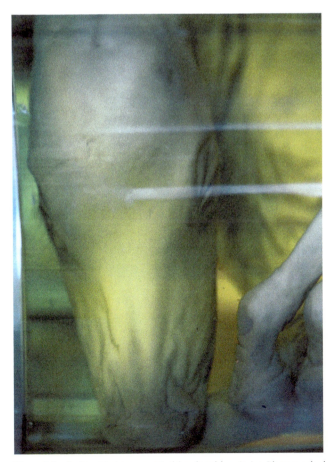

Figure 6.1 Embryonic horse, vertical head and legs in jar. Photographed at École Nationale Vétérinaire de Maisons-Alfort. Photo: Maria Whiteman.

As an artist, I am compelled by the ways in which animals are visualized in contemporary art and in the cultural sphere more generally, and by how distinct techniques of representation afford them differing degrees of cultural significance. This is the key dynamic that I have sought to explore in both my recent art practice and my writing on art, animals, and the discourse of animality. Underlying this investigation of the spaces in which animals are present and absent in the social imaginary is an affective presentiment concerning the fate of our co-habitation with animals on this planet; as a consequence, my work cannot help but tarry with empathy and mourning, that is, with the unfortunate evacuative of potentialities of being and belonging never given the appropriate conditions to become actual or practicable.

I evoke affect here at the outset with some hesitation. There's sometimes a tendency for art to too quickly embrace sentiment, feeling, and affect, and to do so in a manner that blunts the critical edge in which artistic investigations also most certainly engage. At the same time, there's a danger in not properly acknowledging how affect shapes artistic production *and* the encounter of viewers with artworks. Affect need not be the other of critical theory or philosophy, both of which inform and shape my work. I follow Eve Sedgwick in recognizing much of what passes for critical thought as a form of paranoid reading that trades in the hermeneutics of disclosure, that is, in the process of demystification or exposé through which the (supposedly) naive are exposed to truth. Sedgwick's point is not that paranoid readings are wrong, but that they constitute one mode or form of knowing, and that by assuming the position of the *only* way to know they have produced a "disarticulation, disavowal, and misrecognition of other ways of knowing."[1] I see my art and writing as crossing back and forth from philosophy to affect, without disavowing the importance of either in constituting knowledge and one's experience of the world—indeed, recognizing the necessity of both in the practice of art, and especially in art that tries to engage with animals: that essential Other against which human epistemologies and ontologies are produced.

Within this larger framing context of my art practice, the pieces I've included here have two broad aims. The first is to offer a record of a mode of visual-scientific practice in order to provide evidence of a form of knowledge production that is quickly becoming consigned to history. The photo series I am presenting records and examines the various modes of the visual in relation to animals placed on display to generate knowledge. I consider how this form of animal display participates in and informs ongoing discussions of animals and post-humanism. My second aim is to think about how contemporary art plays a role in post-humanist discourse. I am thus interested in the connections between animal displays in natural history museums and scientific practice on animal bodies *and* a philosophical inquiry into modes of knowing, which includes a consideration of the epistemic operations—its insights and well as its blindnesses—of the medium of photography.

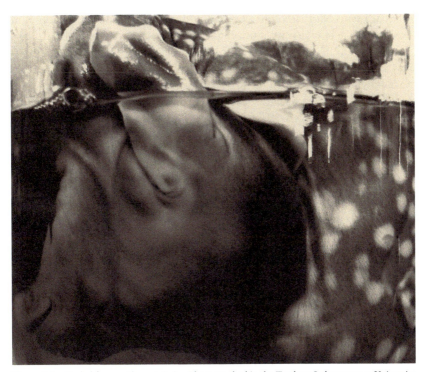

Figure 6.2 Curled fawn with spots in jar. Photographed in the Zoology Laboratory at University of Alberta. Photo: Maria Whiteman.

The series from which these photos are taken, *Taxonomia*, investigates the archive of animal bodies stuffed in jars, held in place by pins, wrapped up in string, and stuffed, mounted, and displayed in order to reduce the anarchy of the natural world to fit neatly into the strict categories of science. Knowledge of the animal world through biological taxonomy—domain, kingdom, phylum, class, order, family, genus, and finally species—is an ancient practice, with origins in the work of Aristotle (in the ancient world) and Linnaeus (the forefather of modern practices). It is also a practice fast coming to an end, as science shifts from learning about animals through visual display to the invisibility of the double helix of DNA. In place of genus and species, traced out through shifts in the colour of fur or markings of the skin, we get a sequence flashing up on a computer screen: A-C-G-T, for example.

Through the photographs, I am drawing attention to the central role played by animals put on display. These are not documentary records, or not only that. These photos are a form of aesthetico-critical practice. Their aestheticized surface is intended to draw attention to the frame of the visual in relation to animal bodies, something we consider too rarely. They put into play the function of vision in scientific knowing; they also highlight the difference between these forms of animal display and that which occurs in the other space in which we more commonly now confront animals: displays of taxidermy animals in natural history museums. Over the past quarter century, natural history museums have changed focus, becoming one of the few spaces in which members of the public come into contact with animal bodies and develop attitudes toward other living species. In museums and aquariums, animals are rendered as cute and cuddly as the toys that children long for in museum gift shops: the flesh and animality has been torn away in the effort to generate that ecological empathy that has become the function of the knowledge systems in natural history museums. What is lost when we forget the weight of flesh and the significance of its decay?

But I am also taken by the ontological demands made by these photos. Indeed, I feel as if my critical practice is driven by some need to capture what I have come to term the "sublime animal." The Kantian sublime names an experience of awe in the face of an otherness that it is intended only to domesticate in the end—an extreme difference whose extremity is only meant to highlight the power of human cognition to manage any experience that comes its way. The aesthetic plays a role in this game of superior knowledge in the dynamically sublime, which restrains imagination through the force of a cognition that can never be surprised or second-guessed. The origin of the zoology museum obeys this process. Bodies are captured whose very existence and otherness should be uncontainable in one's imagination—a mode of being that should be as alien to us as those creatures we imagine we might encounter among the stars, but whose physical preservation in quasi-embryonic fluid makes them sites at which knowledge is produced and organized in a matter that transforms Being into being. My photos are meant to challenge this fantasy of Enlightenment knowledge, cutting away at scientific taxonomy in order to reawaken the dissonance and wonder that comes from these suspended moments of Being; these photos are full of tenderness and dreamy with affect for a never-to-be-actualized life and possibility.

The concept of the sublime animal applies not only to what I see (or think I see) at the moment I am taking the photograph but also to a consideration of what the photographs themselves make possible in their representation of the animal. The photographs capture details: claws, paws, eyes, face, and even expressions. As I try to reveal or uncover a thing-in-itself, something beyond what is representable, what I understand myself to be doing is to assert Edmund Burke's idea of the sublime against its Kantian counterpart. For Kant, Burke's idea of the sublime as that which is "productive of the strongest emotion which the mind is capable of feeling"[2] constituted a danger to philosophical systems of knowledge. It is for this reason that Kant argues—or rather, asserts—there to be a second moment in the experience of the sublime, which he describes as "the feeling of a momentary inhibition of the vital forces

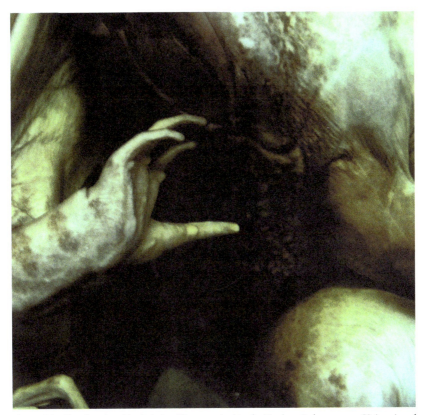

Figure 6.3 Frog hand and body in jar. Photographed in the Zoology Laboratory at University of Alberta. Photo: Maria Whiteman.

followed immediately by an outpouring of them that is all the stronger."[3] My photographs show the altered figurations in which the animal continues to function in the spatiotemporal world it shares with us: the lifeless life that goes on living even in its complete deadness. By capturing the dynamics of this space between stasis and growth—pickled animals that are preserved for science, but that are also placed in a quasi-embryonic fluid that positions the animal as a figure of fetal growth—my aim is to extend the "momentary inhibition of the vital forces" in a way that makes us experience and reflect on the practices and forces that have resulted in our encounter with animals to follow the pattern of Kant's sublime, rather than Burke's.

There is another way in which these photographs—and perhaps even more strongly, the videos that I've included in *Taxonomia*—participate in this undoing of existing ethical, aesthetic, and epistemological relations. The images capture the fixed gaze of animals lost in time and space. In "The Animal that Therefore I Am," Jacques Derrida explores the significance of the gaze of his cat, watching him as he undresses. Derrida is attentive to the depth of the unspoken drama of possession, ethics, and signification that is played out in this quotidian moment. I hope to have captured some of the energy of this disconcerting and displacing question of what it means to be figured in the animal's gaze. As the animal watches you and you look back at her gaze, my images generate a feeling of suspension of difference and the emergence of a (potential) sameness that is almost always undone by the presumptions with which we frame our relation to animals. This is repeated in the touch and caress of the hand on the bodies of wild animals—possible only because their death has made them available for me to engage them safely, less as others than as objects whose very existence confirms that these animals have been entered into a system in which they make possible the strengthening (à la Kant) of our "vital forces." I want viewers to take note of how we conceptualize animal Being in comparison with our own when we view them, as well as when we are viewed by them.

I am engaging with Burke and Kant not to separate out art and nature into the same categories that they do (the former, art, being always already beautiful, while nature is always already sublime). As I hope is clear, one of the aims of my pieces is to challenge these divisions, which continue to inform how we constitute knowledge—knowledge of animals, and in relation to the practice of art and our sense of what it means to be in the world at this time.

Figure 6.4 Embryonic sibling fawns wrapped together in jar. Photographed in the Zoology Laboratory at University of Alberta. Photo: Maria Whiteman.

Notes

1 Eve Kosofsky Sedgwick, "Paranoid Reading and Reparative Reading, or, You're So Paranoid, You Probably Think This Essay Is about You," in *Touching, Feeling: Affect, Pedagogy, Performativity* (Durham, NC: Duke University Press, 2003), 143.
2 Immanuel Kant, *Critique of Judgment*, translated by Werner S. Pluhar (Indianapolis: Hackett, 1987), 36.
3 Ibid., 98.

CHAPTER 7

The Becoming-Animal of *Being Caribou*: Art, Ethics, Politics

Dianne Chisholm

In 2003, Canadian wildlife biologist Karsten Heuer and filmmaker/environmentalist Leanne Allison spent five months and three seasons filming, writing, and tracking on foot a Porcupine Caribou Herd migration across the high Arctic of western Yukon and eastern Alaska. *Being Caribou* is the title they give their respective productions.[1] Deploying different artistic media, Heuer and Allison chronicle the vicissitudes of their passage from an oil-consumptive civilization to the caribou-charged tundra and back again. Their aim is to tell the story of the ordeal of migration that the caribou annually endure, and to add a whole other perspective to the controversial prospect of opening the Alaskan National Wildlife Refuge (ANWR) to oil and gas development. Though plans to drill on caribou calving grounds were part of this prospect, only human stakes in ANWR's future had made the news. As Heuer dissents in the Prologue to his book *Being Caribou: Five Months on Foot with an Arctic Herd* (*BC*):

> It was a classic development-versus-conservation dilemma, and it had attracted plenty of media attention. Cover stories had run in *Time*, *National Geographic*, *Vanity Fair*, and a host of other magazines, and numerous documentaries had aired on television. But as I read and watched all of these, I realized I wasn't hearing the voice of the caribou. It was always the experts doing the talking, citing numbers and statistics that can't really be compared: Six months' worth of oil versus 27,000 years of migration. The culture of about 4,000 caribou-eating Gwich'in versus the financial benefits to a handful of company executives and shareholders. Millions of

mammals and birds versus billions of barrels of oil. Nowhere was there a hint of what I'd felt out there on the tundra. Nowhere did I find the story of the caribou herd itself. (18)

For Heuer, the lay of the land lies beyond "numbers and statistics" that, in any case, cannot quantify and equate the different values at stake in urban economics and tundra ecology, or sustainable development and traditional hunting and gathering. He ventures, instead, to conjure the life and landscape that is immanent to being caribou, replacing his scientific point of view with an artistic frame of vision. In writing he figures a transference of being, thinking, and feeling that allows us to sense what it is to be caribou through the volatile seasons and phases of migration against a background of threatening, industrial encroachment.

Both book and film attempt to present "the story of the caribou itself" to make humans aware of the realities of migratory life that caribou so powerfully and precariously endure, as well as to provoke a sense of how this animal and territory will be affected by oil and gas development. But the book, I contend, goes further than the film in imaging the unfathomable metamorphoses that become the animal in migration and that become the two human caribou trackers in relation to the animal. Allison sharply foregrounds the physical extremes that caribou endure over their migratory trajectory, as well as the variations of body and emotion that Karsten and Leanne experience as they keep apace of the herd. Likewise, Heuer articulates passages of extreme travel and experience. But these passages tend to shift from a physical to a metaphysical direction, and the distance of difference between the human and the animal uncannily starts to blur. There is a sense that one gets from reading the book that one does not get from viewing the film that we are seeing, feeling, and thinking our way into a wholly different, non-human, yet no less real world where drilling pads and industrial grids have no viable place.

Heuer's version of *Being Caribou* is, I argue, a form of ecological activism that has little resemblance to conventional, representational, politics. At the end of his book, Heuer notes how miserably he and Allison fail to win bureaucratic attention when they visit Capitol Hill to lobby directly on the caribou's behalf. Yet, failure to translate their wilderness adventure into domestic politicking might be read as an attestation to their success in being caribou. On the tundra they are charged by an intensity of proximity to the animal in its territory, whereas in Washington, DC, their capacity to mobilize "members of the U.S. Senate and Congress who had never seen a caribou" is limited to the extent they are removed—"more than 4,000 miles from the range of the Porcupine Caribou Herd" (*BC*, 230). Conversely, Heuer is able to convey in writing that contagion of feeling that compels two humans to make a pact with caribou and form a trans-species alliance against George W. Bush's notorious "Drill, baby, drill!" campaign. Heuer's power to compose a landscape of sensation that sees how caribou affect and are affected by the things of their world, as well as to imagine a "bodies politic" of more-than-human forces, is a power worthy of critical investigation and the focus of this essay.

I begin this investigation, then, with the premise that the power of Heuer's writing derives from a capacity to articulate and mobilize caribou affects. Writing for caribou as a human ally rather than for the State as a wildlife officer, Heuer conjures strange, even otherworldly, ideas of "becom[ing] caribou":

> Guided by forces and knowledge we'd never known existed, we had stumbled into a dimension that neither university education, religious teachings, nor anything else in our Western upbringing had taught. It had taken a while, but for a few brief weeks we'd become caribou: content in our suffering, secure in our insecurity, fully exercising the wildness that had been buried within us all along. (*BC*, 221–22)

Though clearly no fantasy, the claim that "we'd become caribou" is somewhat preposterous, especially coming from a wildlife biologist. But suppose we take Heuer at his word: what visions and prospects does his testament evoke? With what ethical and political provocations? Should we take his word to be literal or metaphorical, and what difference does it make? To my mind, it makes all the difference. The change that Heuer's human characters undergo in proximity to the caribou is so intense that at times his first-person narrator has trouble telling them and the caribou apart. Such indiscernibility complicates the usual, anthropomorphic frame of view and cannot be attributed to empathic projection on the part of his narrating self, who is psychically undone by his caribou encounters. Nor is it by analogy or resemblance that he and Leanne become caribou, though often they appear as bone-weary and bug-ridden as the herd. They become caribou not because they *like*, or *are like*, caribou. Rather, they become caribou to the mysterious degree that the forces and stresses of migration pass through and between human and animal beings with powerful, intermingling effect. We see this best in scenes where Heuer's narrator closes in on the herd or the herd closes in on him, and he enters and is enveloped by the caribou's collective body. In these moments, we see Heuer see himself seeing caribou from a perspective that is no longer his own but that emerges from an affecting and disorienting contact with the animal. He and we are, then, afforded real possibilities of thinking, feeling, and acting beyond what it is to be human.

To understand the ontological import of Heuer's writing experiment, I turn to philosophy and, specifically, to the philosophy of Gilles Deleuze and Félix Guattari. Though Heuer nowhere indicates that he has read these philosophers, I contend that their idea of "becoming-animal" offers an auspicious framework for illuminating the metaphysics and minor politics of *Being Caribou*. Deleuze and Guattari understand Being to be a non-linear evolutionary process of "Becoming-other," and they are especially concerned to rethink philosophy's anthropocentrism with a radical focus on the non-human becoming of Man and Nature. They advocate an idea of "becoming-animal" to complicate the idea of human Being: "We believe in the existence of very special becomings-animal traversing human beings and sweeping them away, affecting the animal no less than the human."[2] Also, they elaborate this idea in a complex web, or "rhizome," of immanently connected ideas—ideas such as "nomadology," "territorialization," and "symbiogenesis"—to rethink the nature of Nature.[3] Deleuze further underlines the ontological connections between writing and becoming, and he analyzes the ethical and political potential that he discovers in the literary "fabulation" of various becomings-animal (e.g., Kafka's becoming-insect, D.H. Lawrence's becoming-tortoise, Melville's becoming-whale).[4] With Deleuze and Guattari in mind, I can make sense of Heuer's preposterous claim to have "become caribou."

To begin my investigation I explain a few of Deleuze and Guattari's ideas that I regard as key to reading Heuer's writing and wilderness adventure. First and foremost is their idea of "becoming-animal," or more precisely, "the becoming-animal of Man." This idea challenges the thinking of the majority in that it folds together, or inter-implicates, supposedly discrete categories of differentiated Being ("human" and "animal") in a singular process of "Becoming." For Deleuze and Guattari, Being is a complex matter of non-linear evolution, of individual beings becoming infinitely other than what they are. When a being (human or animal) undergoes "a becoming-animal," it partakes in a "creative involution" or "symbiosis" across species' lines. Ontology, that is, involves ethology, or the nature of animal behaviour. Ethology, in turn, involves ethics, an "immanent" ethics that takes account of how bodies behave in relation to other bodies in ways more or less beneficial to the "Life" of their mutual entanglement. The idea of becoming-animal is, moreover, political in that the being (human or animal) that undergoes a process of "becoming-animal" also undergoes a "becoming-minor" or "minoritization." In the event of "a becoming-animal of Man," the categorical hierarchy that divides the human from the animal is undermined and complicated by a "transversal" flow of (infra-, intra-, and inter-) relations across species at the "molecular" level. In becoming-animal, the human being is freed from capture by "molar" specifications that define and domesticate a vital, human potential to enter into creative, cross-species compositions ("heterogenesis" or "symbiogenesis"). Just as the human being who undergoes a becoming-animal is free to become other-than-Man in relation to animals and other beings, the animal that undergoes a becoming-animal is free to become other-than-the-generic animal in relation to other animals and beings.

The idea of "becoming-animal" is, at once, ontological, eth(olog)ical, and political, and, as such, it offers a critical framework for illuminating the contentious claims and radical articulations of Heuer's *Being Caribou*—claims and articulations that conservationist narratives do not typically deploy. First and foremost, I use the ontology of "becoming-animal" to explain the implicit metaphysics of Heuer's writing. I then turn my attention to the inter-implications of becoming and writing that Heuer insinuates into his composition of landscapes and events. Using Deleuze's concept of "fabulation," I illuminate the *art* of becoming-animal with which Heuer creates and mobilizes what we might call "caribou affects" and "tundra percepts." The remaining two sections of my essay investigate the activist component of Heuer's writing in light of Deleuze and Guattari's ethology/ethics and becoming-minor. Ultimately, I show how Heuer forms a conservationist thinking that deploys tactics other than representational politics. By giving vision to animal realities that elude categorization and measure, Heuer, I demonstrate, makes it possible to see the unfathomable degree to which drilling in the Alaskan National Wildlife Refuge will affect caribou well-being.

The Becoming-Animal of Being Caribou

Biological being is not the focus of Heuer's *Being Caribou*. It is not the species that captures the writer's imagination, even if Heuer is a scientist trained to look for variables of animal specification. His regard for the animal is stirred, instead, by a sense that, whenever he feels his body and the earth tremble with the sound of amassing caribou, or whenever he sees caribou flow across the tundra amid storms and floods, he and the caribou cross ontological boundaries. The caribou that populates his landscapes is not detached from other life forms. Nor does it remain the same, fixed entity through contingencies of climate, season, and terrain. It is not *Rangifer tarandus granti* that leaps across Heuer's page; rather, what animates his writing is a figural enfolding of caribou with other species and beings into something unspecifiable. What this "something" is defies taxonomy and calls for an exploration in ontology that Deleuze and Guattari help me conduct.

For these philosophers, there is more to Life than Being—namely, Becoming. Becoming-animal is a process whereby *a* singular, protean animal emerges from *the* animal type that biology classifies it to be. Unleashed from the stability of taxonomy, the animal becomes an unstable, mutable "haecceity."[5] Becoming-animal is not *the* animal fixed in its essence but *an* animal thrown into existence, where it is given to different material flows and intensities, and where it is drawn into heterogeneous couplings to unpredictable effect. It is a "virtual" process whereby actual, individual beings of different species and strata impact one another, transferring molecular bodies from one order to another in non-reproductive creation. Evolution, Deleuze and Guattari contend, occurs not only through filiation but also, if not more so, through "unnatural participations" across species lines (*A Thousand Plateaus* [hereafter *TP*], 242). Animals evolve in assemblage with other animals, as well as with manifold earth forces that animate their environment (239).

A ready example of such creative involution can be found in Rick Bass's *Caribou Rising*, where Bass imagines caribou hooves evolving in conjunction with snow and permafrost to become shovels for digging up snow-buried fodder and springs for launching off ice-bound earth.[6] Heuer imagines a similar hoof-tundra assemblage in a figure of "click[ing] [...] tendons like hundreds of clocks" that sound an animal passage of geological time (*BC*, 61). Bass's image helps us see how caribou anatomy evolves with tundra geography, while Heuer's image helps us see how tundra geology evolves with caribou tracking: "hours-old tracks give way to ruts underneath, carved deep into the rocky ground by millions of hooves" (105). Both Bass and Heuer compare the ecological imprint of caribou negatively to the industrial footprint of oil and gas development. But Heuer also sees caribou weave a rhizome of Life across the earth, where each rut "is a respun strand in an ancient blanket, an intricate stitch work covering four mountain ranges, two countries, and every valley, plain, and peak in a living, breathing, pulsing web" (105).

Heuer figures caribou less as an animal species than as an otherworldly realm, or a dimension of reality into which humans enter and are virtually transformed. He expresses astonishment at how he and Leanne are "guided by forces and knowledge we'd never known existed" and exposed to a life that "a Western upbringing" has not prepared them for (*BC*, 221–22). The precise mechanism that carries them into this "different dimension" eludes him, but he points to "the act of moving" and "the work of being caribou" as the material means of transport: "*June 25—Clarence River, Alaska—* ... it is the act of moving that has brought me here; the work of being caribou: the miles, the weather, the bears, and the uncertainty, hammering every extraneous thought, action, question, phone number, and song from my head. Cleansed, I am on the edge of something, some other realm of knowing, being pushed and pulled through the same physical world but in a different dimension of space and time" (164–65). By "working" to stay *with* the herd and to endure "the miles, the weather, the bears, and the uncertainty" that the herd, itself, endures, the two caribou trackers cross the border of the Other. Human intention and action are not the drivers. Rather, Heuer senses "being pushed and pulled" by some other forces—"forces and knowledge we'd never known had existed" (221). The world into where he is being pushed and pulled is imminently near: "the same physical world but in a different dimension of space and time." We might think of this same but different physical world as the metaphysical realm that Deleuze and Guattari call "the virtual." The virtual, they explain, is not another Reality; rather, it is a power of chaos that lends becoming a morphological force without shaping any actual thing. There are passages in *Being Caribou* where Heuer's narrating self and Leanne become virtually other when they match their circadian rhythms to those of caribou, though they do not turn into actual, molar animals with horns and hooves. Other passages see him become virtually entangled with caribou at extremely close range. One scene, for example, figures "heartbeats and footsteps mingling while we [he and three bulls] inhaled each other's breath"; the animal's presence is so intense that he virtually "experienc[es] caribou experiencing themselves" (196). In Deleuze and Guattari's words, we could say that he enters a "zone of proximity" where the human and the animal are no longer discernible (*TP*, 273). He and the caribou form not an actual hybrid but a virtual "body without organs" (156), or an impact and relay of bodily affects that has an intensive body of its own.[7]

In another passage, Heuer's narrator and Leanne adapt their regimens and routines to caribou life so intensively that caribou virtually become their existence:

> Pushed by the caribou, we fooled our bodies into doing more with less. We wolfed down dinner in the morning, scooped in handfuls of nuts or skipped lunch altogether, and paused for quick breakfasts in the middle of the night. Tired and confused, we moved beyond nagging hunger, beyond the blunt edge of exhaustion, beyond the limits of each day before. Awash in caribou, we came upon entire drainages—the Kolakut, the Kongakut, the Palokat, the Clarence—unanticipated and unnamed until we later looked at the map.... No longer did we know where we were or where we were going; caribou became our existence. (*BC*, 163)

Breathing, eating, and moving in time with caribou, the two caribou trackers walk off the map of human territory and jettison the grid of rationality. Heuer's narrator hears the beating of hooves "hammering every extraneous thought, action, question, phone number, and song from [his] head"; no longer do the earworms of quotidian communication occupy his brain (*BC*, 164–65). Losing themselves in a caribou world, they escape the anxieties of staying on track: "we'd become caribou: content in our suffering, secure in our insecurity, fully exercising the wildness that had been buried within us all along" (222).

Some passages of *Being Caribou* figure molecular caribou bodies virtually infiltrating the molar, human body.[8] In these passages the "old boundaries" that refrain consciousness are seen to erode: "Scat, hair, and the heavy scent of running, racing animals infiltrated everything—our clothes, our sleeping bag, our food, and the water we gulped down in great handfuls from the creeks. Dizzy and disoriented, we found that old boundaries began to blur, and the caribou that had dominated one realm of consciousness slipped into another, occupying our dreams" (*BC*, 163).

The Art of Becoming-Caribou

The caribou that slips from one realm of consciousness to another is the writer's cue to turn from science to art, and to try to articulate a becoming-caribou outside the biologist's disciplinary limits. This is not to say that Heuer indulges in fiction. Rather, he writes a form of creative non-fiction that Deleuze's concept of "fabulation" helps, I argue, to explain. For Deleuze, "there is no literature without fabulation."[9] Fabulation is not a representation of the actual but a *sensation of the possible*.[10] Using language, syntax, and technique, the writer invents figures, or "affects" and "percepts" with which we can feel, see, and think the otherwise imperceptible possibilities of becoming. Affects and percepts are, then, "possibles," or aesthetic figures that conjure forth abstract intensities—"sensations"—from actual affections and perceptions.[11]

Sensations are the stuff of artistic thinking, just as concepts and functions are the stuff of philosophical and scientific thinking. They do not spontaneously come to the writer; he must create them to think with them. Even on the tundra, Heuer composes fieldnotes to make sense/sensations of his caribou experiences that overwhelm his Western, scientific mind. He cites directly from these fieldnotes at threshold moments of narrative perception, thereby figuring "wildness" without taming it by imposing retrospective interpretation. *Being Caribou* is replete with non-narrative, a-signifying, and sensational passages that provoke readers to see and feel beyond what they know.

As the writer initiates a process of fabulation, fabulation initiates the writer in a process of becoming, or so Deleuze contends: "Writing is a question of becoming, always incomplete, always in the midst of being formed, and goes beyond the matter of any livable or lived experience. It is a process, that is, a passage of Life that traverses both the livable and the lived. Writing is inseparable from becoming: in writing, one becomes-woman, becomes-animal or vegetable, becomes-molecule to the point of becoming-imperceptible."[12] We can see the inter-implication of writing and becoming in *Being Caribou* whenever Heuer figures a crossing from one realm of consciousness to another or from one form of being (human) to another (caribou). Crossing is writing is becoming: a passage that is "always incomplete" and finds no original or final form.[13] In writing a becoming-caribou of Man, Heuer does not imitate the animal; rather, he creates sensations of intense in-betweenness that resembles nothing, for example, "heartbeats and footsteps mingling while we inhaled each other's breath" (*BC*, 196).[14]

Whatever experience Heuer occasions in the field, what lives on in his writing is something else: a "passage of Life that traverses both the livable and the lived." If lived sensations are ephemeral, artistic sensations are "monumental."[15] Yet, art's monumentality does not mortify life as taxonomy does. Fabulations of becoming-caribou elude capture by disciplined perception, and they eternally challenge our efforts to corral mystery into utilitarian definition.

Long after the two caribou trackers actually leave the scene, Karsten and Leanne are still out on the landscape of *Being Caribou* vibrating with sensations of becoming that have no perishable, sensory referents.

Crossing from a scientific to an artistic frame of view, Heuer radically alters his—and our—way of thinking. Caribou cue him to dream in place of scientific rationality, and to foreground the animal body over the human mind. His fieldnotes underscore this schizoid "split": "wildlife biologist turned dreamer and back again: a rational mind and a not-so-rational body discovering wildness for the first time. It's as though my spirit has split—cariboulike when I'm with the animals, humanlike when I'm not, crisscrossing the tundra in search of some middle ground" (*BC*, 180). Heuer's crisscrossing the tundra involves crisscrossing in thinking—from human/rational to animal/nonrational—that writing mediates. Writing, whether in the field or afterwards at his desk, is Heuer's conduit to "discovering wildness for the first time." A prosthesis of extrasensory perception, it allows him to view the events and terrain of migration *caribou-wise*. Deleuze and Guattari help to explain such onto-epistemological shifts (from mind to body, from human perception to caribou percepts) by contending that art and science think differently. Art thinks in "sensations" on a "plane of composition," whereas science thinks in "functions" on a "plane of reference."[16] Moreover, art can think "transversally" with scientific functions in artistic ways, just as science can think transversally with artistic sensations in scientific ways. I would argue that such transversal thinking is the possible "middle ground" that Heuer's figure of "crisscrossing the tundra" aspires to map.

Heuer's switch in thinking is acutely noticeable when his narrator stammers at the prospect of defining strange vibrations that caribou seem to emit but that the earth oddly amplifies. Unable to locate and identify their source, or to measure their audibility and interpret their meaning, he composes a "landscape of sensation."[17]

Instead of identifying functions and mapping coordinates, he fabulates sensations—affects and percepts—that make it possible to think what science cannot define. Some projection is involved. For just as the writer abstracts from the landscape sights and sounds that he actually sees and hears, he projects "visions" and "auditions" back onto the actual landscape.[18] Raising visionary from sensory landscapes, he is "a seer, a becomer."[19] Yet, as he projects himself into the landscape, the landscape absorbs him into the composition, so that it is more accurate to say that "the landscape sees," not the writer.[20] Heuer raises visionary from sensory landscapes, becoming a "dreamer" ("seer") instead of a wildlife biologist.

The most dramatic example of thinking artistically in *Being Caribou* occurs in landscapes vibrating with various animal, geological, and meteorological bodies. Whatever it is that the writer "sounds," it cannot be translated or interpreted as mere animal signs. We could say that a strange "audition" resonates at the limit of language and sense across Heuer's tundra in a running motif that he calls *"the thrumming."* His narrator hears the thrumming well before he sees any body to which he could attribute its origin. The thrumming depth-sounds a territory that earthquakes with caribou: "not hooves drumming—though there were those too—but something deeper, some infrasonic resonance on the edge of human hearing, humming an oscillating song. I closed my eyes and felt it spread through my body" (*BC*, 162). One night he feels "the land ... vibrating underneath me," awakening him but not Leanne, and "leaving [him] alone to puzzle whether the thrumming [he] heard was real" (106). As "a trained scientist" he is "vexed by its formlessness," but as an artist he "strained to tease out the baseline melody humming through the ground." Heuer sees himself become a seer who tunes into "some infrasonic resonance on the edge of human hearing" off the radar of Western science.[21] A "sorcerer," he senses "magic afoot" (161) on a caribou Earth that spirits him away with its "mysterious" confluence of tundra bodies:[22]

> The thrumming was still there when we set off again the next morning—not as loud but still buzzing in the background, a hum that spread across the next two valleys and spilled over the pass between. Leanne didn't hear it, but that didn't seem to matter. Schussing down the other side to the Kongakut River, we were in the flow of animals again, and that was enough. We were part of something larger, a communal push that was closing in on the mysterious place that had kept all of us moving for so long. (106)

A writerly fabulation, "the thrumming" is "impossible to capture on film" (*BC*, 188). Yet, it punctuates Heuer's narrative with revelatory palpability, especially during the chapter-seasons of "Late Spring Migration," "Post-Calving Aggregation," and "Summer Wandering." At its crescendo, more than a literary ear is involved. Proprioceptive, haptic, and seismic sensations compound auditions to compose unnerving, skin-tingling, bone-shaking landscapes. The thrumming figures a manifold, *an*organic Life, involving diverse animal and terrestrial bodies that vibrate with thousands of caribou bodies *ensemble*. "Animals descended into the fog in front of us," Heuer writes, "holes opened in the cloud, offering a glimpse of the great gathering of life" (188). His narrator sees naught but "a dark blot of brown sliding into greenness," focusing on nothing yet beholding life's greening overall. The landscape vibrates with the "rising" of a "communal" body, or what Deleuze and Guattari might call a "body without organs": "It was the thrumming. Even now, sitting on the rocky slope as the caribou amassed a thousand feet below, we could feel it—a potential that throbbed all around: in the animals pouring past our camp, in the throng of life below us, in the rocks, flowers, birds, even the tussocks, rising in goose bumps that crawled over our skin" (189).

The figure of thrumming develops over the course of Heuer's narrative. It is an evolving motif of interspecies cohabitation or creative involution. His narrator is surprised to reflect that not until the late spring migration does he "register" as the thrumming what at first was only a "subtle rumbling" (*BC*, 105): "I had felt hints of it during those first days when the caribou charged past us in the Richardsons, and again when large groups had filed past our camp in the Driftwood valley. But that had been early in the trip.... Now, however, camped in a corner of the Alaskan foothills with thousands of caribou and birds suddenly coursing all around us, it was more than a hint. The land was vibrating underneath me, as though the ground itself were alive" (105–6). By the end of "Summer Wandering," Leanne also feels the thrumming, and both she and the narrator tune into its cosmic volubility with animal prescience. The latter wonders, "Was the sound because of the convergence of all the animals, or was the convergence prompted by the sound? Did it come and go, or was it me who was changing, hearing something that had always existed, only noticing it for the first time?" (105–6). These are questions of becoming that might be answered thus: it is not that Heuer's narrator hears the earth thrum in the caribou's landscape but that the landscape hears him thrum in attunement to the caribou's earth. Heuer figures himself as a percept of sonority between caribou and earth, or what Deleuze and Guattari call the animal's "territorial refrain."[23] The thrumming's mounting intensities and spreading frequencies signal the cultivation of a becoming-caribou's musical ear. As his narrator attests, it is the thrumming's sonar, not "satellite collars, scientific reports, maps, or even tracks that had guided us" (189). Tuned into caribou and what caribou are tuned into, the two caribou trackers are guided to a "mysterious place" where "huge fronts of life" gather with "a potential that throbbed all around: in the animals ... in the rocks, flowers, birds, even the tussocks." Heuer's auditions sound a symphony of caribou ecology; they also augur a vitality that science and industry threaten to destroy in their search for oil and power.

The Ethology/Ethics of Becoming-Caribou

The primary task that Heuer sets himself in writing *Being Caribou* is to tell the story of the caribou and "bring it alive" (*BC*, 20). This poses several ethical challenges. One is to take into account the vital stakes that caribou might possibly have in the future of the Alaskan National Wildlife Refuge. Another is to tell the full story by framing into view the many and diverse bodies and powers that affect and are affected by the migration. Lastly, another is to convey the capacity of caribou to survive the migration and even to thrive in their severe, Arctic habitat. In short, I argue that Heuer challenges himself to write with a vitalist ethic. As his Prologue announces:

That's the story of the caribou, I thought to myself as arctic terns, king eiders, and a host of other birds splashed down between icebergs. That's the surge of life and death that all the magazine articles and television documentaries had failed to capture.... Four mountain ranges, hundreds of passes, dozens of rivers, countless grizzly bears, wolves, mosquitoes, and arctic storms—those were the measures, that was the story, and the time had come to put it all together and try to bring it alive. (BC, 19–20)

One of the ways that Heuer tries to bring the caribou's story alive is to emphasize the physical ordeal of migration. In doing so, he is careful to not superimpose a tale of human adventure.[24] He relates a few close calls that he and Leanne have with marauding grizzlies, flooding rivers, and bone-chilling blizzards but only as hazards that come with the territory and that caribou routinely endure. If he downplays his characters' ambition and athleticism, he emphasizes the finitude and susceptibility of a Western body and mind when transplanted onto caribou turf. Yet, in framing the limits of anthropomorphic reason and endurance, he draws out the sagacity and vitality of becoming-caribou. Repeatedly, he shows how caribou push him and Leanne beyond their human capacities. But as the humanness of his characters breaks down, the non-human powers of caribou emerge.

Like Deleuze's writer, Heuer concerns himself with health, not heroics. For Deleuze, the writer is not only a "becomer, a seer," but also a "physician," and he writes to divine/diagnose which "passages of Life" will be "interrupted, blocked, or plugged up."[25] As antidote to whatever might block these passages, he prescribes becoming-other but never a becoming-man, "insofar as man presents himself as a dominant form of expression that claims to impose itself on all matter."[26] For Deleuze's writer, "the world is a set of symptoms whose illness merges with man," and he takes it upon himself to ask, "What health would be sufficient *to liberate life* wherever it is imprisoned by and within man, and by and within organisms and genera?" He must have health, not a robust but a queer health, a hypersensitivity that allows him to be intensely affected by things, as well as to endure self-mortifying passages of becoming. "He possesses," Deleuze explains, "an irresistible and delicate health that stems from what he has seen and heard of things too big for him, too strong for him [...] whose passage exhausts him, while nonetheless giving him the becomings that a dominant and substantial health would render impossible. The writer returns from what he has seen with bloodshot eyes and pierced eardrums."[27] In tracking/writing the course of migration, Heuer composes a "passage of Life" that literally exhausts his character. By taking this passage beyond mortal limits, he is able to vision the caribou's powerful and precarious vitality. And, only then does he divine/diagnose the intense degree to which drilling in the Alaskan National Wildlife Refuge will block that ever-evolving "passage of Life" that is becoming-caribou.

Like Deleuze and Guattari, Heuer "avoid[s] defining a body by its organs and functions, [or by] its Species or Genus characteristics," opting instead "to count its affects" (*TP*, 257). Affects should not be confused with traits. They are fluctuating sensations of becoming-other, not fixable characteristics of being. For the philosophers, such an accounting of the animal body is "ethology [...] [in] the sense in which Spinoza wrote a true Ethics." It is not possible, they explain, to know a body, any body, without knowing "what it can do, in other words, what its affects are, how they can or cannot enter into composition with other affects, with the affects of another body, either to destroy that body or to be destroyed by it, either to exchange actions and passions with it or to join with it in composing a more powerful body" (257). Whether a body composes a more or less powerful body, or whether it impacts another body by "augmenting or diminishing its power to act," constitutes, for the philosophers, an immanent, ethological ethics. Likewise, for Heuer, the body is never more alive, never healthier nor better than when it joins other bodies to form a "powerful," composite body—and a body comprised of more than just caribou. His landscapes see an amalgamation of diverse tundra bodies from across the tundra. Consider, for example, this vision of the calving grounds:

> It was as if a switch had been thrown, releasing all the energy and potential that had gathered for weeks around us, sending it spilling onto the coastal plain. Killdeers, sandpipers, and other shorebirds flew in with brant geese on a wave of wings, while bees, butterflies, and swallows drifted and swooped in and out of view. It was a dance that everything was doing—the mergansers that glided in and plied the braided channels of the nearby river, the ground squirrels that darted in and out of their burrows, the pair of longspurs that, after days of subdued courting, copulated in front of our tent in a sudden flutter of wings. Even the new caribou mothers had fresh energy. (*BC*, 130)

At the calving grounds "everything seemed to be celebrating life." Life is "a dance that everything was doing" in a mutually invigorating embrace: "even the new caribou mothers had fresh energy." Later, at the scene of post-calving aggregation, Heuer has his narrator exclaim, "Is that real?" as thousands of bulls surge into confluence with as many cows and calves: "an entire hillside, was moving from the coastal plain toward us, shape-shifting in the last of the day's heat waves, tracking the mountains through ribbons of shadow and light" (*BC*, 168). What assembles before the narrator's eyes is a monstrous super-body, an unearthly becoming-caribou of *everything*: "Everything seemed to be lining up for a postcalving aggregation: the weather, the animals, the emerging bugs. We were about to reap the reward of ... becoming a part of a gathering of unfathomable force" (168). A biologist would not see a post-calving aggregation this way but, his thinking having shifted, Heuer sees differently. In place of statistical populations and taxonomic functions, variables that science and industry deem sustainable in a drilling environment, he visions caribou life as "a Life" of immeasurable and inextricable symbiogenesis.

Immersed in the climactic landscapes of the calving grounds and post-calving aggregation, the two caribou trackers sense themselves "becoming a part of a gathering of unfathomable force." That is, they do not discover what they are capable of becoming (beyond what it is to be human) until they join forces with the caribou. In "becoming a part" of the landscape of birthing and aggregating, their bodily regime is undone and, stripped of their human form, they enjoy an augmentation of *incorporeal* powers. Corporeal self-discipline is what makes them human, setting them apart from caribou bodies. But going with the flow of the herd's assembled movements and energies, they sense a release of their own bodily capacities. When, for example, they break their fitful and blinkered routine of "scheduled sleep" to amble with caribou at night, they become strangely becalmed and prescient: "there was a magic afoot at night that wasn't there in the daytime, a calmness and clarity that affected the animals and us" (*BC*, 161). At night they become less visible to themselves and to the creatures around them as imposing aliens, and they see, instead, "how the upland sandpipers circled over [their] stretched-out shadows.... And how the long-tailed jaegers that normally dive-bombed and forced [them] into tiresome detours simply watched as we stumbled past their camouflaged nests." Literally shadows of their restless, daytime selves, they bask in the calming affects of a nocturnal and communal animal.

Heuer's characters become caribou to the degree that their affective mutability and physical velocity changes in proximity to the herd. To use Deleuze and Guattari's Spinozist terms, we could say that Heuer "maps" the human body onto the "intensive" and "extensive" dimensions of the caribou body, or its axes of "latitude" and "longitude." The latitude of a body, the philosophers explain, consists "of the affects of which it is capable," while the longitude of body consists of "movement and rest, speed and slowness grouping together an infinity of parts" (*TP*, 256). Longitude involves extensive parts that fall into relation at a certain speed; latitude involves intensive capacities to affect or be affected: "To the [longitudinal] relations composing, decomposing, or modifying an individual in extension, there correspond [latitudinal] intensities that affect it, augmenting or diminishing its power to act." Deleuze clarifies that "a body can be anything; it can be an animal, a body, a sound, a mind or an idea; it can be a linguistic corpus, a social body, a collectivity."[28] Heuer places the bodies of the two caribou trackers in extension along the caribou's longitudinal axes. As their speed and movement is adapted to those of the herd, they "become a part of a gathering of unfathomable force." Their power to absorb and be mobilized by the affective powers of tundra life is enhanced to the degree that they are kinetically and kinaesthetically moved by caribou. Their vitality benefits from an affective coupling with caribou, just as the caribou's vitality benefits from an intensive intermingling with other bodies and forces.

To illustrate Heuer's mapping of intensified Life, let me return to the chapter on "Calving." The two caribou trackers have been racing to keep up with restive, pregnant cows, but they are stopped in their tracks when the cows suddenly pause to give birth. The latter encircle and entrap the former inside the nylon womb of their tent. Forced into still and intimate confinement with anxious mothers-to-be, the trackers themselves become pregnant with nervous anticipation and unbearable restlessness. But once birthing begins their agitation transforms into rhythmic accord: "We still tossed, turned, fidgeted, and readjusted, but it seemed smoother and more coordinated—as though our personal rhythms had fused into a subconscious dance" (*BC*, 135). A harmonizing of speeds and affects smooths relations between human and animal bodies, and it refrains insular, "claustrophobic" feelings into sensations of resonant ease. "The effect was magical: inside the tent all went quiet—and outside, as the cows and calves discovered their own system of body language all around us, the barks, grunts, bleats, and huffs that had dominated the last few days gave way to a soft, milling hush" (135).

The intensive affects that Heuer composes over the course of becoming caribou are not, however, always positive. At one point, his narrator observes that he and Leanne "had become part of something immense and immensely fragile" and that "behind every joyful moment lingered a kind of melancholy about the future, a dread that the two kinds of existence that we'd experienced—the one in Kaktovik and the one on the calving grounds—would soon collide" (*BC*, 131). "Kaktovik" is an Inupiat village and a remote outpost on the Alaskan coast to where Heuer and Allison fly midway during the migration to replenish their resources. During their brief stay in the village, they are exposed to and depressed by the degrading impact that industrialization and sedentarization have had on the Native population. The melancholy that overcomes them in Kaktovik still lingers when they return to the tundra, and it darkens the joyfulness of the calving grounds. According to Deleuze, "joy" and "sadness" figure, respectively, the vital, affective capacities of a body to be increased or diminished upon impact with another body.[29] Heuer's projection of the future of the Alaskan National Wildlife Refuge mixes the two affects—the joy of the calving grounds and the melancholy of Kaktovik—in a compound, diagnostic vision. An ethical disease, melancholy will spread across the Arctic, Heuer foresees, if the industrial West continues to "promote the uncoupling of people with their natural surroundings" and if it extends intensive, Kaktovik-like development to the calving grounds.

Heuer's "Epilogue" frames "Washington, DC" in a vision of visceral disaffection. In this scene, he and Leanne have left the tundra for Capitol Hill to lobby bureaucrats on behalf of "an Alaskan conservation group" and the Porcupine Caribou Herd (*BC*, 230). Just days since ending their journey in the Gwich'in village of Old Crow, and already 4,000 miles away from caribou, they are afflicted with a growing sense of "disconnection" (229). Distracted by "televisions flashing in lobbies," "billboards over the baggage carousels," "people pacing the halls with hands-free microphones, gesticulating madly as they talked and shouted to people no one else could see," he senses "parts of me that had taken months to open while moving with the caribou were already beginning to close down." And, he reasons, "They had to. Life in the modern technological world carries none of the subtleties of living with caribou. There's too much to absorb, too much for sharpened senses to do anything but go dormant if one wants to survive" (230). Heuer, the writer, diagnoses a blockage in his becoming-caribou due to "too much" communication, an ailment of contemporary society that Deleuze and Guattari also deplore.[30] Yet Heuer, the activist, fails to take the story of the caribou to Washington and "bring it alive"; his tale dies in the ears of congressional aides who are deaf to all but talk of "cheap gas" (231). What body politic, then, can he hope to stir with the visionary thrum of tundra life?

The Minor Politics of Being Caribou: Writing for "a New Earth and People"

Frustrated by Washington bureaucracy, Heuer defers to grassroots politicking: "'We need to work from the bottom up,' I said to Leanne. 'We need to mobilize the voters'" (*BC*, 231). But the populist environmentalism of his Epilogue strikes me as somewhat disingenuous. Heuer calls for a constituency of caribou stakeholders with no more commitment than he corrals the animal by its species and genus. To represent caribou politically he must appeal to a caribou-loving minority of American citizens, thus defeating his concern to mobilize people directly with caribou affects—affects that compose the animal's (as well as man's) becoming-caribou. To mobilize even a minority of voters, he must still play the majority game of representation. That is, as the caribou's representative he must represent caribou to humans as beings with representable traits worthy of their vote. Yet only weakly does his writing invoke the *molar* animal, and even more weakly does it rally man, the political animal. In writing, his passion is to fabulate a crossing of the great ontological divide that separates the human and the animal. In a world dominated by the territories of human settlement and stratification, the most radical struggle is not to form a political minority but, as Heuer strongly suggests, to perform a *becoming-minor* of Man.

Being Caribou is a "passage of Life" that traverses the sedimented and oppressive, ontological strata of Man's earth with a political aim to "bring it alive." To use the geophilosophical vocabulary of Deleuze and Guattari, I would argue that it "smoothes" the stratified hierarchies of power (between such established domains as Man and Animal, Self and Other, Nature and Culture) into two-way conduits of "molecular" flows.[31] Its landscapes thrum with a micropolitics of connection and alliance across species' lines and across other categorical borderlines. Auditions of the calving grounds and post-calving aggregation sound a coming together of tundra bodies in a groundbreaking, earth-shattering sublime-overcoming of man's neat onto-, bio-, and geo-logical taxonomies. Heuer's occasional rant against drilling in the Alaskan National Wildlife Refuge is drowned out by the "thrumming" of a vitality too otherworldly to prospect for human industry. Instead of harping on industrial nihilism and stressing the predictable—a degraded coastal plain populated by fragmentary remnants of the Porcupine Caribou Herd—Heuer presages the possible in visions that see caribou deterritorializing man's earth in lines of flight and weaves of new life, as well as caribou rising through the human stratum of organized, stratified, and codified flows in non-human, sweeping, and free-flowing assemblages.[32]

I read the "Calving" chapter as an exemplary passage of becoming-minor in *Being Caribou*. In this chapter, Heuer fabulates a radically delimited frame of view to figure the confinement that he and Allison experience when they are surrounded by pregnant caribou. Their usual mobile tracking skills now disabled, the biologist and filmmaker are seen struggling to locate a perspective outside their habitual vantage points. The two view-mastering humans are "forced to watch" from a minoritized position of utter stillness. At the same time, such delimitation intensifies their capacity to be affected by the caribou and the calving milieu. Immobilization of their usual mastery of vision foregrounds a non-human mobilization of life all around them. "We still couldn't move because of caribou, we were still relegated to sitting," his narrator reflects, "but it was no longer confining. Having been forced to watch, we couldn't help but appreciate what was happening around us. Because of the caribou, we'd stumbled onto the riches of being still" (*BC*, 130).

Heuer foregrounds this becoming-minor to a point of becoming-imperceptible. Human motions (emotions, commotions) are seen to harmonize with the heartbeats and bleats of caribou mothers and calves and to create a "zone of proximity" where difference between species becomes "indiscernible."[33] A harmony of speeds and affects becomes the basis of an alliance that involves and evolves with greater diversity than conspecies' affiliation. Moving in rhythm with the animal, breathing the same breath, embodied by the same web and womb of life, humans and caribou create momentarily what Deleuze calls "a singular life."[34] In "Calving," human bodies together with animal bodies (not just caribou bodies, but also wolf, bear, insect, and other bodies attracted to and connected to the herd) and bodies of the earth (geological, meteorological, climatological bodies across and through which the herd passes and changes) compose a body without organs, or what we might call a "bodies-politic," a physical, chemical, biological, neural, and social assemblage of mutually affecting and mobilizing things.[35] The politics of this coming together of bodies is, as Deleuze and Guattari would say, the force with which it captures and liberates, or "stratifies" and "destratifies" molecular flows of connection and creation.[36]

There are passages of becoming-minor in *Being Caribou* that open flows of connection and creation between anthropological, as well as biological strata. Heuer's narrative tracking of the Porcupine Caribou Herd ultimately brings Karsten and Leanne in touch with the Gwich'in, "the people of the caribou." A nomadic people, the Gwich'in have been adapting over decades, for better or worse, to the pressures of colonization—including sedentarizing pressures that force hunter-gatherers to relocate from the land to permanent villages built on an urban grid of government housing. To further sedentarize or stratify the Gwich'in, who have been moving with caribou for millennia, Western governments have tried to replace a language that "talks" to caribou and that divines their whereabouts and well-being with a heavily administered system of communication and orderly speech that strictly occurs between humans. Karsten and Leanne do not speak Gwich'in but by going nomad they dream of caribou, and they begin to cultivate a shamanlike, animistic prescience. Mapping the animal on its own plane of immanence, they can dispense with satellite radio. In realigning their motions and affections with the caribou, they also ally themselves with the people of the caribou—or with those among them who have not yet "forgotten" the "mysteries and beauties" that "Gwich'in medicine men, Inuvialuit shamans, and others had long ago discovered" (*BC*, 222). It is to this unspoken alliance that Heuer gives the last word. In the final scene of their return to Old Crow, a "changed" Karsten and Leanne are greeted by a Gwich'in Elder who, five months earlier, had seen them off (223). An exchange of looks lets the Elder know without being told that "we had talked to caribou, and caribou had talked to us" (226).

Becoming-minor aligns urban southerners with Native northerners through subliminal channels of communication that pass through the caribou. At some "infrasonic" level, these channels undermine geopolitical structures that divide the Arctic into a Western majority and an Indigenous minority. Heuer's nomadism deterritorializes colonial geographies. Moving with the caribou's alignment of flows, they traverse the boundaries of state-regulated resources and enjoy the confluence of non-harnessable energies. With nomadic eyes, Heuer tracks caribou across Yukon and Alaskan state borders beyond the purview of U.S. and Canadian patrols. If current political landscapes partition the migration into fragmentary and contradictory jurisdictions, like the Alaskan National Wildlife Refuge, a "protectorate" that the federal government threatens to open to oil and gas development, the literary landscapes of *Being Caribou* see the intensive mobilization of a mobile habitat across state and species' lines.[37]

Karsten Heuer mobilizes a minor politics through literary fabulation. Instead of representing that animal known to science and to the knowledgeable majority as the Porcupine Caribou (*Rangifer tarandus granti*), Heuer creates an animal landscape where we can see the power of caribou bodies to affect and be affected by other tundra bodies with which it enters into vital, ecological composition. Heuer's landscape of sensation strikes us with visions and auditions of a composite, at once powerful and precarious vitality that becomes a migratory "passage of Life." In place of representational numbers and predictable population statistics that State science deems sustainable, we sense the inviolable health of a collective body whose volatile constituency is ever changing, non-representable, and unpredictable. *Being Caribou* deploys art to articulate a different ontology from the ruling ontology that rationalizes Being into separate strata and that reifies the structure of human dominance. In place of Being, and human being as Being's prime representative, Heuer figures a Becoming-animal, an onto-*ethology*. And, in place of conventional, anthropomorphic representations of caribou he figures a becoming-animal of man. As seer, becomer, and meta-physician, he projects a becoming-caribou of man as an antidote to prospects of oil and gas drilling on the calving grounds, and other human encroachments.

From lived experience, Heuer abstracts caribou affects and tundra percepts, and he projects them onto a scene of vitality that is larger than life. He also brings into vision the diminished state of health of Arctic communities. Sadly, outside the villages of Kaktovik and Old Crow there are no people. The people of the caribou are missing from the land, having been coerced to leave by federal government and industrial powers. The earth is no less affected. The melancholy that overshadows the joy of the calving grounds presages an invasion of industrial clamour into the harmonious composition of tundra bodies. Ultimately, the most powerful vision of *Being Caribou* is, in the words of Deleuze and Guattari, that of "a people and an earth still to come."[38] The art of an anomalous scientist calls forth a bodies-politic that involves caribou and humans in mutually beneficial cohabitation.

Notes

This essay is reproduced with permission. It was previously published as Dianne Chisholm, "The Becoming-Animal of *Being Caribou*: Art, Ethics, Politics," in *Rhizomes: Cultural Studies in Emerging Knowledge* 24 (2012), http://www.rhizomes.net/issue24/chisholm.html.

1 Leanne Allison, *Being Caribou/Vivre comme les caribous*. DVD. National Film Board of Canada, 2004. Karsten Heuer, *Being Caribou: Five Months on Foot with an Arctic*

Herd (Seattle: Mountaineers, 2005). Further references to Heuer's book will be cited as *BC*.
2. Gilles Deleuze and Felix Guattari, *A Thousand Plateaus: Capitalism and Schizophrenia*, translated by Brian Massumi (Minneapolis: University of Minnesota Press, 1987), 237. Further references to this work will be cited as *TP*.
3. See esp. chapter 10, "Becoming-Intense, Becoming-Animal, Becoming-Imperceptible ..." (*TP*, 232–309).
4. For more on "fabulation," see Gilles Deleuze, *Essays Critical and Clinical*, translated by Daniel W. Smith and Michael A. Greco (Minneapolis: University of Minnesota Press, 1997). For more on "becomings-animal" in literature, see esp. chapter 10 of *TP*.
5. See Deleuze and Guattari, *TP*, 261. *Haecceity* is derived from the Latin *haec*, meaning "this." Instead of *the* generic animal in all its essence, Deleuze and Guattari want us to think of *this* singular animal in all its contingencies. For a discussion of "haecceity" in this context, see Brett Buchanan, *Onto-Ethologies: The Animal Environments of Uexküll, Heidegger, Merleau-Ponty, and Deleuze* (New York: SUNY Press, 2008), 151–86.
6. Rick Bass, *Caribou Rising* (Seattle: Mountaineers, 2004), 130.
7. Deleuze and Guattari coin the term "a body without organs" to refer to any heterogeneous assemblage of bodies that is not organized by the molar anatomies of Royal Science.
8. As Deleuze and Guattari explain, "becomings-animal plunge into becomings-molecular" (*TP*, 272).
9. Deleuze, *Essays*, 3.
10. "Aesthetic figures ... are sensations: percepts and affects, landscapes and faces, visions and becomings.... [Aesthetic] universes are neither virtual nor actual; they are possibles, the possible as aesthetic category." Gilles Deleuze and Félix Guattari, *What Is Philosophy?*, translated by Hugh Tomlinson and Graham Burchell (Minneapolis: University of Minnesota Press, 1994), 177.
11. "The aim of art is to wrest the percept from perceptions of objects and the states of the perceiving subject, to wrest the affect from affections as the transition from one state to another: to extract a bloc of sensations, a pure being of sensations" (ibid., 167).
12. Deleuze, *Essays*, 1.
13. Ibid.
14. "To become," Deleuze explains, "is not to attain a form (identification, imitation, Mimesis) but to find the zone of proximity, indiscernibility, or indifferentiation where one can no longer be distinguished from *a* woman, *an* animal, or *a* molecule" (ibid.); original emphasis.
15. Deleuze and Guattari, *What Is Philosophy?*, 164.
16. Ibid., 163–99. See also Dianne Chisholm, "The Art of Ecological Thinking: Literary Ecology," *ISLE* 18, no. 3 (2011), 569–93.
17. I am indebted to Ronald Bogue's illuminating explication of what he calls Deleuze's "landscape of sensation." See "The Landscape of Sensation," in *Gilles Deleuze: Image and Text*, ed. Eugene W. Holland, Daniel W. Smith, and Charles J. Stivale (London: Continuum, 2009), 9–26.
18. Deleuze, *Essays*, 5.
19. Deleuze and Guattari, *What Is Philosophy?*, 171.
20. Ibid., 169.
21. Heuer researches the thrumming but to no avail: "Since returning from our journey, I have read books on infrasonic communication with elephants, sifted through journal articles about whale song, and stumbled across the human accounts of similar phenomena, as in the poetry of Rilke. But I have found nothing about caribou." When "a thirty-year veteran of caribou biology" suggests that he pursue a doctoral project on the subject, Heuer decides that it is "best left in mystery" (*BC*, 233).

22 For Deleuze and Guattari, the writer/becomer is a "sorcerer": "If the writer is a sorcerer, it is because writing is a becoming, writing is traversed by strange becomings that are not becomings-writer, but becomings-rat, becomings-insect, becomings-wolf, etc." (*TP*, 240).
23 For Deleuze and Guattari, animals are artists that compose "territorial refrains" that resonate with expressive symbioses and creative couplings. See "On the Refrain" (*TP*, 310–50).
24 "Our goal," Heuer asserts, "was to be caribou, not human. The last thing either of us wanted was to turn what had started as a very special journey into just another cross-country hiking trip" (*BC*, 78).
25 Deleuze, *Essays*, 3.
26 "The shame of being a man," Deleuze queries, "is there any better reason to write?" (ibid.).
27 Ibid., emphasis added.
28 Gilles Deleuze, *Spinoza: Practical Philosophy*, translated by Robert Hurley (San Francisco: City Lights, 1988), 127.
29 As Deleuze explains, joy and sadness express the ethical affects (or existential modes) of impact that one body can have on another: "An existing mode is defined by a certain capacity for being affected. When it encounters another mode, it can happen that this other mode is 'good' for it, that is, enters into composition with it, or on the contrary decomposes it and is 'bad' for it. In the first case, the existing mode passes to a greater perfection; in the second case, to a lesser perfection. Accordingly, it will be said that its power of acting or force of existing increases or diminishes, since the power of the other mode is added to it, or on the contrary is withdrawn from it, immobilizing and restraining it. The passage to a greater perfection, or the increase of the power of acting, is called an affect, or feeling, of *joy*; the passage to a lesser perfection or the diminution of the power of acting is called *sadness*" (ibid., 49–50); original emphasis.
30 "We do not lack communication. On the contrary, we have too much of it. We lack creation. *We lack resistance to the present*" (Deleuze and Guattari, *What Is Philosophy?*, 108); original emphasis.
31 My thinking in this section draws from Deleuze and Guattari's chapters "The Geology of Morals" (*TP*, 39–74) and "The Smooth and the Striated" (*TP*, 474–500); and "Geophilosophy," in *What Is Philosophy?* (85–113).
32 Bass and Heuer create similar images of caribou rising and flowing freely in contrast to the trapping of oil and gas flows by the grid of techno-economic infrastructure on the human stratum. See Bass, *Caribou Rising*, 4–6.
33 Deleuze, *Essays*, 1.
34 Ibid.
35 I derive my idea of a "bodies politic" from Bonta and Protevi: "A key dimension of Deleuze and Guattari's work is the investigation of a 'bodies politic,' material systems or 'assemblages' whose constitution in widely differing registers—physical, chemical, biological, neural, and social—can be analyzed in political terms." Mark Bonta and John Protevi, *Deleuze and Geophilosophy* (Edinburgh: Edinburgh University Press, 2004), 10.
36 For more on the ethics and politics of de/stratification, see "The Geology of Morals" (*TP*, 39–74).
37 In Allison's film, Leanne and Karsten carry a George W. Bush doll, face-outwards on their backpack so that he can see the great gathering of life, and be moved to rescind his plan to drill in caribou territory.
38 "Europeanization does not constitute a becoming but merely the history of capitalism, which prevents the becoming of subjected peoples. Art and philosophy converge at this point: the constitution of an earth and a people that are lacking as the correlate of creation." Deleuze and Guattari, *What Is Philosophy?*, 108–9.

INTERLUDE

Creating Metaphors for Change

Lyndal Osborne

When I was a young girl in Australia I spent a lot of time outdoors walking the beaches and collecting. Over fifty years later I still do the same thing—walking around my home in the (once) rural area on the outskirts of Edmonton, and in my travels to other places. I see firsthand the dramatic changes that are taking place where I live. I read and think about the issues of global climate change and realize that it is going to take a huge shift in our lifestyle to alter the course of these events. Though discouraged in my belief that our species will be able to make this paradigm shift, my work for the past fifteen years has been driven by this central idea. I have tried to focus that experience by creating work through knowledge of my own locale, with materials collected or grown in my own garden and created and constructed in my own studio.

To me, discovering the energy that binds the global issues to local interpretation is found in making with my hands and collecting my own raw materials. Although I want to understand with my own eyes, to help me keep the idea clear, I am also looking for visual equivalents that keep me stimulated and that engage the viewers' imagination. I also leave lots of room for my own imagination by not trying to make literal interpretations of the ideas I am representing. I use my creative process to understand the bigger picture through looking at details and presenting them sensuously and provocatively. For example, in *Archipelago* (2008), the wire (DNA model rod-connectors) used in the river seemed important because it suggested a reference to laboratories producing fertilizers and pesticides that pollute waterways and affect the wild plant and animal life.

Many of my sculptures and installations are made from organic materials such as seed pods, plant roots, dried fruit skins, stalks, and shells. To complement the natural and foraged, other saved materials are utilized, such as discarded wire, used laboratory glassware and equipment, scrap plastic, and papier mâché. Combinations of materials can be altered through the application of colour, manipulating their original shape, and/or by placing the objects in unexpected contexts so that they develop new metaphorical meaning quite separate from the material's original history. When I make new works I sometimes reuse components from earlier work. Recycling and reinterpreting these materials interrupt the cycle of consumption and waste. This also keeps the work fresh.

Sometimes, as in *ab ovo* (2008) and *Endless Forms Most Beautiful* (2006–11), installations become very evocative and speak poetically and seductively of the forces of transformation within nature. I know that there are dark agendas by those who promote genetically modified organisms (GMOs) but, when focusing on the examination of issues such as GMOs, I make a conscious attempt to replicate the highly seductive promotion of agribusiness and consumer lifestyle by choosing to represent the ideas using a high degree of beauty.

In my research I read a lot, look at pictures, and visit various sites to experience them myself. Actual experiences and lots of reading help me to have a better grip on the big picture. Certain details resonate because they suggest opportunities to be so well expressed visually and sculpturally, and that helps me select the components for each work. For example, to research *ab ovo* I visited the Millennium Seed Bank at Wakehurst, in the United Kingdom, and observed how the newly acquired seeds were cleaned and fumigated in a glove box and then stored cryogenically in numbered plastic and glass containers. With this knowledge I constructed *ab ovo* using hand-blown glass cloches as containers for my sculpted seed forms, and I appropriated an ancient glove compartment as a visual element for the public to engage in a hands-on experience.

Because my practice is installation-based I use scale as a constant element to shrink down or to greatly enlarge components in my work. When something huge becomes very small I feel that a primitive power is released (miniature gardens in dried grapefruit skins are used as a metaphor for a river in *Archipelago*). In another work, *ab ovo* I saw electron microscope images of wild seeds stored in underground vaults. In the resulting work, the scale is enlarged by creating sculptural seed forms magnified hundreds of times their real-life size, moulded, and cast repeatedly. Playing with scale in this way makes the global local and makes the everywhere near.

I feel an affinity with notions of reusing and recycling materials such as twigs and industrial discards rather than always choosing highly manufactured products. This personal, local activity of recycling could have a far-flung impact on economic disparity, health, or climate in the societies where these products are produced. Because of the inundation of global news, we can sometimes feel overwhelmed and desensitized to the point where we are

Interlude ■ Creating Metaphors for Change 111

unwilling to change anything in our lives. Today, everywhere is near, and I believe that walking, collecting, classifying, and day-to-day living in an attentive way can have positive results.

In the creation of my environments I try to keep in mind the architectural space of the gallery and attempt to use it in a way that allows the viewer to interact with the piece as a "real place." I try to imagine how the viewers might see familiar things, but from unexpected angles. Often I find that the interaction of viewers and their response to the work mirrors my own sense of how the work came to be created and constructed.

Interlude Figure 1 *Archipelago* (2008), detail. Mixed media installation: sunflower stalks and grapefruit skins chine collé with lithograph drawings or painted, wire, glass beads, DNA model connectors, laboratory glassware, metal caps and Bunsen burners, sea balls, seed pods, Sculpey, silicone rubber, resin, papier mâché, paint, and dye. Dimensions: variable; approx. 2' x 30' x 25'. Photo: Mark Freeman.

Repetition of form is the most consistent element in all my work. Small hand-held forms accumulate to produce form(s) that grow into a metaphorical installation. In the process, much time is spent thinking about the work and how visual metaphors (like the laboratory setting for *Endless Forms Most Beautiful*) can be powerfully expressed. I enjoy working on the repeated construction of extremely detailed forms and the element of playfulness it can introduce into studio activities. The labour-intensiveness slows the process down, and so I engage fully with the work and am able to ponder the issues that I am concerned about. Sometimes the creative process takes months, but it can also span many years. The contemplative time spent on these activities helps me know when the work is finished.

Interlude Figure 2 *ab ovo* (2008), detail. Mixed media installation: glass, foam, wood, lights, clove-studded oranges, lime grass, gourds, lemons, sponge cakes, bananas, poinciana pods, DNA models, chestnuts, Port Jackson shark eggs, used grape skins, avocado skins, sunflowers, grapefruit skins, coconuts, fungus, ginseng, corks, corncobs, cycad seeds, eucalyptus seeds, seafoam, Tasmanian gumnuts, crab shells, bones, kelp, redwood pine cones, durian, beets, persimmon, moth flower, and laboratory glove compartment. Dimensions: 6' x 26' x 1'. Photo: Mark Freeman.

Interlude Figure 3 *ab ovo* (2008), detail. Mixed media installation: glass, foam, wood, lights, clove-studded oranges, lime grass, gourds, lemons, sponge cakes, bananas, poinciana pods, DNA models, chestnuts, Port Jackson shark eggs, used grape skins, avocado skins, sunflowers, grapefruit skins, coconuts, fungus, ginseng, corks, corncobs, cycad seeds, eucalyptus seeds, seafoam, Tasmanian gumnuts, crab shells, bones, kelp, redwood pine cones, durian, beets, persimmon, moth flower, and laboratory glove compartment. Dimensions: 6' x 26' x 1'. Photo: Mark Freeman.

Interlude Figure 4 *Endless Forms Most Beautiful* (2006–11). Mixed media installation: wire, papier mâché, spray foam, paint, glue, poppy heads, seaweed floats, iris stalks, Mexican vine, pipettes, sponges, horsetail, sand, paint, caps, ceramic disc capacitors, reflecting paint, lemons, gourds, chine collé, tea leaves, speaker wire, banksia leaves, cornflowers, seaweed, cycad seeds, carrots, chicken bones, shotgun shells, acacia seed pods, sedum seeds, ginseng, cornstalk, tea bags, steel stands, glass flasks, plastic tubing, coloured liquid. Dimensions: variable; 5' x 30' x 10'. Photo: Mark Freeman.

PART 2
CONSTRUCTING KNOWLEDGE

CHAPTER 8

Poetry, Science, and Knowledge of Place
A Dispatch from the Coast

Nicholas Bradley

> *The light beats on the stones,*
> *And wind over water shines*
> *Like long grass through the trees,*
> *As I set loose, like birds*
> *In a landscape, the old words.*
> —David Wagoner, "The Words"[1]

> *To compare across political barriers in this region is first to be aware of the great power—emotional and imaginative, as well as climatic—of the barriers of rock, the coastal ranges, which run north/south.*
> —Laurie Ricou, "Two Nations Own These Islands"[2]

The great power of rock and water is surely witnessed by those observers who, facing south from vantage points on southern Vancouver Island, look across the Juan de Fuca Strait at Cape Flattery, which appears to point into the Pacific vastness, and at the Olympic Mountains, which seem to pluck storms out of the sky. The power of wind and light must likewise be noticed by onlookers who, facing north from prospects on the Olympic Peninsula, gaze upon the Strait and the southwestern coast of Vancouver Island, a misty place where creek after creek pours oceanward from bluff and ridge. The Olympic Peninsula, in the state of Washington, and Vancouver Island, in the province of British Columbia, are separated by an international border, yet they share geological, climatic, and other environmental characteristics. They also have in common a vocabulary: the words that poets use to describe these places echo across the water that divides and links the locales.

This essay concerns two particular representations of landscapes of the Pacific Northwest, a region that extends across the border between Canada and the United States. My account, however, comes to the landscapes and their literary portrayals in a roundabout way. It compares two poems, David Wagoner's "On a Mountainside" and Don McKay's "Astonished –," in order to illuminate distinct yet related ways in which the Pacific Northwest has been figured in contemporary poetry. Wagoner, an American writer who has lived for decades in Washington, has memorably depicted locations on the Olympic Peninsula. McKay, a Canadian author who until relatively recently lived in Victoria, British Columbia, sketches in "Astonished –" and other poems in *Strike/Slip* (2006) the stretch of Vancouver Island's coastline that runs roughly from Jordan River to Port Renfrew.[3] The writers' works are regionally (or even microregionally) complementary, although they would normally be placed in discrete national categories. The broader bailiwick of my essay is the relation of regional poetry and criticism to environmental sustainability and activism. I thus begin by describing an ongoing concern with environmental relevance within the academic discipline of literary studies; I proceed to suggest that the function of poetry and literary criticism in an age of environmental crisis is neither self-evident nor easily established. And after having travelled some distance away from the Pacific Northwest, I come home to regionalism and the poems of McKay and Wagoner.

Ecocriticism, Interdisciplinarity, Activism

Interdisciplinarity and social relevance, understood as ideals or imperatives, have bedevilled the approach to the study of the interrelations of environment and literature (and culture more generally) known as ecocriticism. As an organized and institutionally recognized branch of literary studies, ecocriticism is more than twenty years old; its direct antecedents in the 1970s and 1980s give it a longer history still.[4] Diverse in subject and method, ecocriticism accommodates a range of objectives and interpretative techniques. It has always been, in Lawrence Buell's phrase, a "polyform ... movement,"[5] but it has been characterized, especially in its early stages, by a "commitment to environmentalist praxis"[6] and a nearly axiomatic belief in the importance of the critical engagement of science—and of ecology in particular.[7] The name of the principal ecocritical journal, *Interdisciplinary Studies in Literature and Environment* (*ISLE*), emblematizes the notion that ecocriticism is inherently interdisciplinary. The authority assigned by ecocriticism to scientific knowledge has sometimes been construed as a swerve away from the purported excesses of critical theories believed to have reduced nature to a mere sign.[8]

The materiality of nature, its capacity to be experienced phenomenally by the hiker in the back country or quantified by the scientist, and its existence outside language and ideology have been held as intellectually necessary principles. On such tenets rests the view that literary studies can address environmental crises in earnest.

Glen A. Love's *Practical Ecocriticism: Literature, Biology, and the Environment* (2003) provides a representative example of ecocritical concern with science as a method and body of knowledge through which literary studies can gain in "ecological relevance."[9] Love professes to be worried about "the sort of work we do in the real world as teachers, scholars, and citizens of a place and a planet"—"work" suggests one meaning of the titular term *Practical*—and he consequently aspires to a literary criticism that is deeply informed by science, chiefly biology.[10] The linked ambitions of interdisciplinarity and activism remain central subjects of ecocritical debate despite the field's increasingly theoretical bent, as was exemplified by a state-of-the-nation critics' forum in *ISLE* (2010),[11] in which appeared bracing statements: "ecocriticism is strongest when it is most *interdisciplinary*. We cannot afford to ignore the claims of science, as careful and limited as they may be."[12] The urgency of the phrasing reflects a continuing sense that criticism is most relevant (i.e., closest to praxis) when it engages the science that explains and measures the health of the physical world.

The importance to ecocriticism of interdisciplinarity and activism is perpetually open to scrutiny and contestation, in part because these matters lead directly to foundational metacritical questions. What is criticism for? What is its place in the world? What are the aims and methods of literary analysis? What is the relation of sophisticated, abstract, and sometimes arcane knowledge to complex social problems? The question of praxis obtains in all areas of literary study with overtly political and material aims, including the Marxist, feminist, and postcolonial approaches with which ecocriticism is sometimes aligned;[13] the question of interdisciplinarity obtains in the manifold approaches that seek to combine literary studies with other areas of inquiry.[14] Contemporary literary studies, moreover, are rarely purely formalist or aesthetic: they are typically multidisciplinary, nearly always inflected by non-literary bodies of knowledge. Yet because environmental issues involve science to a degree that distinguishes them from other problems of social inequity (even as environmental and social injustices may be intertwined), matters of interdisciplinarity and activism are acutely difficult in their ecocritical manifestations. Although ecocriticism habitually draws upon scientific knowledge, its capacity to contribute to scientific fields is unclear; when such contributions occur, they likely do so indirectly. A humane principle pertains: because literary works are of and about the world, knowledge of and about the world is pertinent to the critic. But literary criticism is not necessarily or obviously helpful to scientists as they conduct their investigations. Science, as a mode of inquiry, is less permeable than literary criticism, even if the individual ecologist or physicist is not immune to literary or otherwise aesthetic power. The waters of praxis have, in addition, been muddied by the institutionalization of ecocriticism. As it has become entrenched in the academic discipline of literary studies, ecocriticism has become highly self-referential, like any established critical mode, such that the world outside scholarly discourse is sometimes obscured by internal quarrels.

The case for a scientifically inflected ecocriticism has been relatively straightforward. Environmental science and especially ecology have provided notable metaphors for ecocriticism—among them are balance, harmony, interconnection, biomes, biodiversity, and edge effects. Here is evidence if not of deep interdisciplinarity then at least of critical awareness of non-literary fields and of interest in the physical world.[15] The case for ecocriticism's activist dimension has proven less tractable. The convictions of its practitioners notwithstanding, a specialized mode of writing with a typically circumscribed audience has a limited capacity to effect immediate political or social change.[16] Testaments to literature's power to move, inspire, frighten, and raise awareness may to a degree persuade, but they do not account for all forms of writing: not every novel is an *Uncle Tom's Cabin* or *The Jungle*, nor is critical commentary likely to exert socially or environmentally transformative power in the face of imminent crisis. In the conclusion to *Ecocriticism* (2004), a prominent overview of the field, Greg Garrard notes that "literature cannot provide specific solutions" and observes that therefore "ecocriticism must continue to adopt and adapt theories from feminist and Marxist traditions, enabling positive engagement in cultural politics."[17] Albeit in a different vein, Love likewise asserts the cultural and political relevance of English studies: "English teaching and research goes on within a biosphere, the part of the earth and its atmosphere in which life exists.... Teaching and studying literature without reference to the natural conditions of the world and the basic ecological principles that underlie all life seems increasingly shortsighted, incongruous."[18] But teaching and studying, however laudable these endeavours, are unpredictable means of creating change. The ambiguity of literary texts, the caprices of readers' responses, and the slow speed of careful reading and writing curtail the practical impact of ecocriticism and environmentally oriented literature.

The theoretical sophistication of ecocriticism in recent years has led to a reconsideration of central terms and concepts. The title of Timothy Morton's *Ecology without Nature* (2007) epitomizes such reconceptions; his study advocates abandoning a static, essentialist understanding of nature, and in so doing seeks to alter ecocriticism's parlance, favouring the language of the Frankfurt School and poststructuralism over the terms of the romantic and post-romantic literature that has been a long-standing subject of ecocritical inquiry. An early ecocritical emphasis on nature as a non-human, extra-discursive reality has been dislodged by a growing consensus that the conceptual tools of critical theory permit a useful re-examination of the lexicon and basic categories of the humanistic study of nature. In *The Future of Environmental Criticism* (2005), Buell distinguishes the priorities of a "first wave" of ecocriticism (including an emphasis on nature as conventionally perceived, an insistence on the authority of science, and a focus on literary realism) from those of a "second-wave" or "revisionist" approach that concentrates on urban and other non-wilderness spaces, on connections between environmental injustices and sociopolitical marginalization, and on science as discourse.[19] Garrard traces a similar shift by noting that Buell himself and Jonathan Bate, taken as Buell's British equivalent, replaced their early attention to the representative figures of Thoreau and Wordsworth, respectively, with "explicitly dialectical approach[es]."[20] Morton's critique hinges on related observations. He deems ecocriticism to be limited by its heavy thematic emphasis, burdensome "ideological baggage," overwhelming focus on the genre of the prose memoir ("nature writing"), anti-theoretical stance, and suspect politics.[21] *Épater l'écocritique* is Morton's unmistakable wish. Despite his subversiveness and theoretical ingenuity, however, he routinely appeals to relevance and science. His claim that "ecology is queer theory and queer theory is ecology," for example, is predicated on a desire for environmental-political relevance—"Let's do it," he writes, meaning to constitute the field of queer ecology, "because our era requires it"—and on faith in the fundamental significance of Darwin's writings.[22] According to Morton, "science is too important to be left to scientists"[23]—a witty line, to be sure, but also a position that ultimately places him not far from Love, who proposes that the life sciences are essential to an ecocriticism that would "contribute to the study of values in what we increasingly find to be a world where, to cite an ecological maxim, everything is connected to everything else."[24]

The theoretical directions pursued by Morton and others have not entirely displaced the field's constant concerns. The essays in *Coming into Contact* (2007), a volume published in the same year as *Ecology without Nature*, follow new lines of ecocritical inquiry but return to the topics of praxis (see Part 2, "The Solid Earth! The Actual World! Environmental Discourse and Practice") and interdisciplinarity (see Part 3, "Contact! Contact! Interdisciplinary Connections").[25] The volume's editors offer the admonishment that "looming environmental crises obligate ecocritics to remain alert to any opportunity where their unique blend of environmental imagination, criticism, and activism can be of help."[26] In some ecocritical quarters a stridently anti-theoretical position survives, one expressed succinctly by S.K. Robisch's pronouncement in a notorious jeremiad: "We write about literature under the influence of ecology. It's really not that complicated."[27] The claim subordinates critics ("We") to the scientific domain but simultaneously reduces ecology to a source of influence without explaining the shape or force thereof. Robisch's deferral to ecology is complemented by his insistence on the supreme importance of "trail cred," as if critical acumen were determined primarily by prowess in the wild: "We write too much. Shut up and go outside."[28] The screed prompted an extended response in *ISLE*—a "Special Forum on Ecocriticism and Theory."[29] The selection of short essays in that issue provides ample evidence that interdisciplinarity and activism endure as ecocriticism's bugbears. The claims made and the questions raised in the essays sound familiar refrains of praxis and relevance:

- "What are ecocriticism and theory *now*, to us, in an academy nested ... in an ecological and planetary crisis?"[30]
- "The point here is to sidestep the stereotype of theory as a realm of nitpicking intellectual games and indicate the functionality and interactivity that occur when we manage to ask useful questions."[31]
- "Which theoretical framework can justify ecocriticism as a discourse of cultural change and of social hope?"[32]
- "What can literature do to enhance ethical awareness and political inclusivity?"[33]
- "Particularly in light of the urgency and ubiquity of environmental crisis, ecocritics cannot afford to examine only those works which seem thematically most likely, but must critique particularly those works that are not explicitly environmental or nature-oriented."[34]
- "To find rational remedies to the ecological challenges we need both theory and praxis, both activism and philosophizing, both laws ... and environmental education."[35]
- "We have to remember that what we do matters, that the world we work in as scholars, teachers, writers, and citizens *is* the real world."[36]

Such statements and queries are doubtless sincere. They suggest that ecocritical scholars are commendably engaged in self-scrutiny, and demonstrate that anxiety about methods and relevance marks the field. Further examples abound—in the dedication of an ecocritical monograph to "environmental activists everywhere";[37] in the assertion, in a collection of ecocritical essays, of "environmental criticism's sophisticated engagement with science and technology studies through the discourses of evolutionary biology, biotechnology, cybernetics, medicine, and ecology";[38] and in the title of a comprehensive study of nature poetry, *Can Poetry Save the Earth?*[39]

In the preceding pages I have not attempted to supply a comprehensive account of ecocriticism's history or present state, although by referring to certain works and debates, I have provided, I trust, a glimpse of recurring tensions. While they have caused considerable consternation within the field, such concerns are probably of limited interest to those outside this particular corner of literary studies; the polemics and jostlings are typical of academic squabbles. But ecocriticism's perpetual self-consciousness also affords an instigation to think anew about what environmental literature and its analysis offer. What can or should literary critics say when they speak to experts in other fields or to the public, and when they attempt to explain the significance of their studies?

Poetry, Place, Knowledge

Although I hope not to advocate environmental quietism, I do not have ready answers to such questions. Perhaps ecocriticism, necessarily recondite, is only relevant in the same manner as other forms of humanistic study: stumblingly, primarily in the long term, and often personally rather than collectively. In what follows I will consider poetry as a type of knowledge, a means both of response and discovery. What do poets know about the environments of the Pacific Northwest? How can readerly interpretation of such knowledge lead to understandings of place? What do Don McKay and David Wagoner know about the region? Ecocriticism's chronic worrying about relevance can be intellectually invigorating. My modest contention, however, is that what may appear to be strictly aesthetic or apolitical accounts of place can be valuable on their own terms.

Some literary modes (satire, for example) are more obviously direct and publicly oriented than others, such as lyric. But poems that do not have palpable designs upon us, *pace* Keats, "proceed," as Eleanor Cook suggests, "by indirection, not being political or forensic or ecclesiastical or didactic oratory. As with all the arts, the relation with the outside world is both necessary and oblique."[40] Or, as Michael Riffaterre writes, "poetry expresses concepts and things by indirection. To put it simply, a poem says one thing and means another."[41] Indirection, ambiguity, and polysemy—poetic attributes that in terms of *techne* are often thought to be praiseworthy—pose problems for writers who would, *contra* Cook, view poetry as a form of political or didactic writing, or at least as having the capacity to speak directly. Environmentally oriented poets have, not surprisingly, made their own relevance an ongoing subject of examination.[42] McKay wrote, in a book published in 2001, that "admitting that you are a nature poet, nowadays, may make you seem something of a fool," suggesting, despite the comic tone, an uncertainty about the purpose of his poetry and the genre to which it belongs; "nowadays," in this instance, means "the last decade of the twentieth century, a decade in which I was determined to come to grips with the practice of nature poetry in a time of environmental crisis."[43] But even as impassioned an environmental advocate as Gary Snyder, an American poet with impeccable activist credentials, at times writes about the role of literature with less anguish than many ecocritical scholars: "Scholarship continually spades and turns the deep compost of language and memory; and creative writing does much the same but adding more imagination, direct experience, and the ineluctable 'present moment' as well."[44] Snyder also suggests that the primary function of "artists and writers" is to "'bear witness.'"[45] In both formulations, the environmental writer's vocation is taken to be retrospective and idiosyncratic, and tied not to praxis but to an archive of ideas. Snyder here echoes W.H. Auden's infamous (and commonly misinterpreted) claim that "poetry makes nothing happen."[46] If poetry is instead "A way of happening, a mouth," as Auden proposed in his elegy for William Butler Yeats (1939), it allows for or embodies forms of thinking that are not bound to an instrumentalist imperative.[47] In his essay "Freedom and Necessity in Poetry" (1969), Auden wrote, reiterating the sense of the passage of the elegy, that "poetry is gratuitous utterance."[48] Yet he asserted the fact that a poem is singular in origin and concern, while simultaneously claiming for poetry a public role. With a nod to T.S. Eliot's "Tradition and the Individual Talent" (1919), he suggested that "to say that a poem is a personal utterance does not mean that it is an act of self-expression. The experiences a poet endeavours to embody in a poem are experiences of a reality common to all men: they are only *his* in that this reality is perceived from a perspective which nobody but he can occupy. What by providence he has been the first to perceive, it is his duty to share with others."[49] In light of such characterizations of poetry in the abstract, it becomes possible (and appealing) to understand poetic depictions of place to contain specific, contingent, non-empirical, idiosyncratic—in short, personal—forms of environmental knowledge.

Auden's commonsensical view—versions of which are expressed elsewhere in his prose, notably in his essay on "Nature, History and Poetry" (1949/1950)[50]—perhaps assumes too easily both universal experience and distinct selfhood. Nonetheless it usefully avers a poet's sense of the importance of perspective: poems record the world in particular ways, paradoxically supplying by their very specificity glimpses of a wider world. To put Auden's idea in environmental-poetic terms, nature poems provide provisional accounts of experiences of place. Their relevance inheres not primarily in their advocacy of a given course of action, but instead in their connections to and departures from other such accounts, which together constitute, as a network or fabric, an understanding of the world that encompasses yet exceeds the human sphere: "To say that poetry is ultimately concerned with only human persons does not, of course, mean that it is always overtly about them. We are always intimately related to non-human natures and, unless we try to understand and relate to what we are not, we shall never understand what we are."[51] Auden's view is undeniably anthropocentric—self-knowledge is the stated aim—but an ecological sensibility is present in his recognition of the interconnection of human and non-human domains.[52] His later prose contains numerous passages that illustrate his growing interest in environmental matters, in environmentally oriented writers (including Loren Eiseley),[53] and in relations among different forms of knowledge. (He gave, writing in 1965, a clue to the autobiographical origin of this interest: "In my father's library, scientific books stood side by side with works of poetry and fiction, and it never occurred to me to think of one as being less or more 'humane' than the other.")[54]

Auden understood the human world to be simultaneously social and biological: "The existence of human beings is dual: as biological organisms made of matter, we are subject to the laws of physics and biology: as conscious persons who create our own history, we are free to decide what that history shall be."[55] His theory of the resolute personalism of poetry led him to distinguish the arts from the sciences. The distinction was examined by a host of philosophers and commentators in Auden's time, from Alfred North Whitehead to C.P. Snow to Mary Midgley; I.A. Richards concluded *Science and Poetry* (1926) with two cheers for poetry, which he deemed "capable of saving us; it is a perfectly possible means of overcoming chaos."[56] Auden suggested that "the job of poetry, of all the arts ... is to manifest the personal and the chosen: the manifestation of the impersonal and the necessity is the job of the sciences."[57] (The colloquial "job" undercuts the sententiousness risked in such pronouncements and implies that poetry and science are not rarefied pursuits but rather integral and allied aspects of human inquiry.) He further linked art and science by positing a common point of departure: "Like art, pure science is a gratuitous and personal activity, and I am convinced that the stimulus to scientific enquiry is the same as that of artistic fabrication, namely a sense of wonder."[58] Such statements propose that poetic knowledge does not necessarily sit at odds with, or far from, the powerful and often unsettling implications of modern scientific discoveries.

Yet to view poetry as a way of thinking or mode of inquiry comparable to science does not mean that poetry must be solemn. In his elegy, Auden wrote that Yeats was "silly like us."[59] Both McKay and Wagoner make the poet's ludic privilege a significant (although not omnipresent) part of their writing about places and environments; they play with wit and irony what Cook calls "poets' games, the serious games of all their indirections":[60]

> We can't get rid of figures of speech even if we want to. Dead or alive, they're part of the language. Poets are the great experts in this area and so can alert us to their power. To their pleasure too.
>
> This may come clearer if we recognize that poems are often answering the question "What is *x like*?" rather than "What is *x*?" "What is *x*?"

usually elicits the answer "*x* is—[a noun]." Poems answer "What is *x* like?" by means of figuration and fictive construct, and such answers can be just as valid and useful as answers to the question "What is *x*?"[61]

Works of all literary forms have environmental dimensions, and all writing employs figuration. Do poems represent a special case? Yes and no. "Yes," because of poetry's particular formal and linguistic emphases. As Cook notes, "figuration (figures of speech, tropes and schemes)" is "poetry's special domain."[62] But "no" because each literary genre (or artistic medium) has its own province: distinctiveness is a common attribute. Interdisciplinarity presumes the existence of disciplines; one disciplinary strength of literary studies is attention to figurative and other aesthetic uses of language and to the range of meanings that emerge therefrom. The task of the critic, on this view, is to read closely and deeply, to heed texts and the panoply of contexts to which they belong. The environmental capacity of criticism thus lies not primarily in action but in thoughtful observation.'

In his critique of ecocriticism, *The Truth of Ecology* (2003), Dana Phillips suggested that "ecocriticism is impatient with versions—impatient, that is, with texts not tied discretely to referents of fairly specific latitude and longitude."[63] He condemned what he perceived as naive and reductive understandings of mimesis and realism, focusing on Buell's *The Environmental Imagination* (1995). In the remainder of this essay, in which I am concerned with the coexistence of several literary versions of mappable places, and with the coincident but not identical versions of such places inside and outside poems, I hope to show that all representations are "versions," even those with specific geographical referents. To take Wagoner's "On a Mountainside" as a poem about a particular place in the Pacific Northwest is already an act of interpretation, an educated guess that, although there is no specific reference to a cartographic location in the poem, the work presents a version of the Northwest, a region that Wagoner's long career has charted in considerable detail. The poems of Wagoner and McKay are responses to locations—to two closely related places, it seems. What then do they say these places are like? What is the validity, to use Cook's term, of the poems' visions?

Poets, Puns, Perspective

Two poems of the Pacific Northwest, each of which illustrates a landscape, form a curious pair. One of the poems is by Don McKay, who is, according to my rough assessment, one of the Canadian poets most often discussed under the rubric of ecocriticism. He has for good reason referred to himself, with a measure of self-deprecation, as "Mr. Nature Poet."[64] The other poem is by David Wagoner, an American author, professionally accomplished and highly regarded, whose affinity for the landscapes of Washington is well known.[65]

My pairing depends upon the geographical relation of the two poets' works: their landscapes appear to face each other across an international border and a body of water, as if places and poems were mirror images. McKay suggested the illusory proximity of the Olympic Peninsula to Vancouver Island in the title essay of *Deactivated West 100* (2005): "The fog lay below him, opulent and plump as blown-in insulation, filling the whole strait between the clearcut where he stood and the Olympic Mountains, which stuck up, it seemed, about a hundred yards away to the south."[66] But my pairing rests as well on an odd coincidence, namely the presence in two poems of the same verbal game. The lexical connection yokes the geological poems together.

McKay's "Astonished –," a short poem in *Strike/Slip*, a book generally concerned with the landscapes and geological history of the Juan de Fuca region, begins with a list of consonant terms, each of which is a synonym for the word that provides the poem's title: "astounded, astonied, astunned, stopped short."[67] The repeated *st-* sound and especially the presence of *ston-* in "astonished" and "astonied" suggest that the listed words are semantically linked to *stone*, a word that appears in the poem's second line:

> astounded, astonied, astunned, stopped short
> and turned toward stone, the moment
> filling with its slow
> stratified time. Standing there, your face
> cratered by its gawk,
> you might be the symbol signifying eon.

The observer, the "you" to whom the poem is addressed, is surprised and made speechless by his recognition of the stone's imponderable age. His mouth, open in disbelief, resembles a zero, the cipher required (many times over) to express geological time numerically. (Perhaps the mouth also resembles the letter *Q*, the geological symbol for the Quaternary [thus *Q*] Period, the most recent of the periods of the Cenozoic Era; it includes our own time.) The apparent etymological association of *stone* and *astonished* implies that the observer has been turned to stone, rendered immobile, made a part of the landscape that so transfixes him. Motionlessness stands in contrast to the immeasurably slow geological changes that he cannot see but nonetheless knows are always taking place:

> Somewhere
> sediments accumulate on seabeds, seabeds
> rear up into mountains, ammonites
> fossilize into gems.

The observer's befuddlement (or, to maintain the poem's insistent, accusatory use of the second person, *your* befuddlement), his sense of the relative fleetingness of time measured on a human scale and of the transitoriness of human achievements ("Cities / as sand dunes, epics / as e-mail"), leads to a perception of the divided self. One part is stunned by the geological sublime, while the other is impelled into movement toward the ocean, which is itself a reminder that the permanence of rock is chimerical: "Someone / inside you steps from the forest and across the beach / toward the nameless all-dissolving ocean."

McKay's portrayal of the coastal landscape emphasizes the geological indicators of the tremendous age and inexorable flux of the earth. The strike-slip fault to which the book's title alludes is a sign of regional geological instability. The Pacific Northwest in *Strike/Slip* can terrify the observant bystander. The seriousness of the theme is leavened, however, with the comic tone of many of the collection's poems. One of the subtle jokes in "Astonished –" is that the etymological link between *stone* and *astonished* is in fact false. No true connection can be found between the *ston-* of *stone*, a Germanic word that derives from the Old English *stán*, and the *-ston-* of *astonished*. *Astonied* comes to English from the Old French, with the Latin *tonāre* ("to thunder") at its heart. (Compare the modern French *étonner*.) The poem's glossolalic opening lines suggest, if the reader listens closely, the tricks that language plays on its users. In "Apostrophe," thirty-seven pages later in *Strike/Slip*, McKay similarly conveys the failure of geological terminology to catch the impassivity of the rock formations that the speaker confronts:

> Protero, palaeo, meso, ceno:
> I had, I thought,
> a thing to say as I approached
> the columns of angular basalt. But all those
> rough chateaux were shut, their epochs
> slammed, their hours immured.
> The spells of textbooks
> echoed, so much gabbling Greek.[68]

"So much gabbling Greek": it's all Greek to me. The cliché refers not only to a foreign language but to a state of incomprehensibility. The speaker babbles in order to express his baffled response to place, his linguistic precision (a poet's job requirement) stymied by the scale of the landscape. The search for the right words leads McKay to depict a humorous near-speechlessness. His version of the coastal wilderness combines a jovial sense of the limitations of mimetic language with a genuine desire for comprehension. As Malcolm Woodland notes, Loss Creek, a waterway that lends its name to another poem in *Strike/Slip*,[69] "murmurs 'husserl husserl' to its phenomenologically questing listener."[70] The speaker hears the philosopher's name (Edmund Husserl) in the sound of the water; the joke expresses both a longing to comprehend environmental phenomena and the inability to escape language. McKay's personae cannot find the right words, yet they have only words.

McKay's wry use of the *stone/astonished* resemblance has a counterpart in Wagoner's "On a Mountainside," first published in the journal *Poetry* in 1993[71] and then collected in *Walt Whitman Bathing* (1996) and *Traveling Light* (1999).[72] Wagoner's poem is also in part about being turned to stone— about becoming petrified, rendered immobile by fear, fatigue, and cold, three forces that reliably impede a mountain climber's progress. As John Taylor contends, one of Wagoner's "poetic aims" is "an examination of man's apartness from nature, a separation staged dramatically in 'On a Mountainside.' In this poem, an amateur alpinist stranded on a cliff ... seeks to grasp the lessons he might learn from his, perhaps fatal, predicament."[73] Being turned to stone is a familiar literary trope, and it is tempting to see the climber's snaking ropes as the hair of Medusa. Wagoner's version of the alpine Northwest is decidedly more grim than McKay's comic coastline. The poem does not reveal whether the cold, scared climber descends safely; Wagoner leaves the reader hanging, so to speak. But the evocation of a geological sublime is eerily similar to McKay's:

> Here on the north face, on a slab of granite
> With a fractured overhang like a lean-to,
> You take your time
> To look at the horn peaks and truncated spurs
> Of another ridge whose three cirques tilt their glaciers
> As if to pour them
> Over their brims and down to a hanging valley.
> You stare that way and try hard to remember
> How you once thought
> But instead sit down astonished, stunned, astounded—
> Your words all formed from *stone* and turning to stone
> Again like your lips and tongue
> While you catch your breath as slowly and painlessly
> As you can.

There is no obvious reason to think that McKay meant to allude directly to Wagoner. Literary associations of astonishment and *stone* date from the Middle Ages; both contemporary poets are at once using a conventional trope (as in Elizabeth Barrett Browning's "Bereavement"—"And I astonied fell and could not pray") and making an old joke anew.[74] McKay has not, to my knowledge, referred to Wagoner's "On a Mountainside" in his published works; nor have I found evidence in McKay's unpublished papers that he has read the poem. Both poets, however, are included in an anthology of poetry of the Pacific Northwest: David Biespiel's *Long Journey* (2006; it contains McKay's "Astonished –"). It is plausible that McKay had already seized upon Wagoner as a regionally significant poet whose works had some bearing on his own. In any case, my aim is not to posit direct influence. Instead I am concerned with the repetition of an extended verbal figure in two poems that portray geographically proximate locations, one on either side of the international border.

Wagoner employs numerous words to refer to shapes or surfaces of the mountain ("face," "slab," "overhang," "horn peaks," "spurs," "ridge"), kinds of rock ("granite"), and glaciological features ("cirques," "glaciers," "ice"). They serve as a realist backdrop for the poem's punning vocabulary ("perdurable," "stony") and darkly funny dismissals of the importance of Keatsian truth and beauty to the imperilled climber. As does McKay's "Astonished –," "On a Mountainside" combines accurate description and technical knowledge with a quasi-visionary sense that Northwestern landscapes are simultaneously beautiful and terribly overwhelming. And Wagoner senses, again like McKay, that language fails to capture the thing itself and that perception depends on perspective: "the closer you come / To any mountain, the harder it is to see." In "Mapmaking" (the first poem in *Sequence: Landscapes*, the suite to which "On a Mountainside" belongs), Wagoner draws a portrait of the poet as cartographer, intimating that map-making, like poetry, is an inevitably imprecise art:

> remember no one
>
> Really depends on you
> To do away with uncertainty forever.
> Your piece of paper may seem in years to come
> An amusing footnote
> For wandering minds[.][75]

But if "mapp[ing] out / Some share of the unknown" is a desire never to be realized,[76] Wagoner nonetheless persists in the attempt, as he writes of the cartographer, "To confirm your bearings, / To reconcile what you saw with what you see, / Comparing foresight and hindsight."[77]

The old words that Wagoner and McKay set loose in the landscape link the poems to each other. "Astonished –" and "On a Mountainside" equally link out-of-the-way places in the Pacific Northwest to long-standing traditions of representing place. Behind Wagoner and McKay stands, for instance, Robinson Jeffers, another poet of the Pacific, who in "Oh Lovely Rock" (1937) provided a description of the coastal geology of the Ventana Creek area of California that anticipates, in its theme of astonishment, the descriptions of the Northwestern poems:

>   it was the rock wall
> That fascinated my eyes and mind. Nothing strange: light-gray
>   diorite with two or three slanting seams in it,
> Smooth-polished by the endless attrition of slides and floods; no
>   fern nor lichen, pure naked rock … as if I were
> Seeing rock for the first time. As if I were seeing through the
>   flame-lit surface into the real and bodily
> And living rock.[78]

The poems of McKay and Wagoner remain slippery, less securely tied to place than they may seem. Did Wagoner know that the Olympic Peninsula has no granite bedrock? The "Mountainside" could well be in the North Cascades, although Wagoner's body of work invited the present reader to locate it in the Olympics. Or it could be in no range at all. Does the geological fact matter, aside from its disruption of my neat pairing? Once again, yes and no. "Yes," because scientifically informed poets, such as McKay and Wagoner (and Jeffers), value accuracy. And "no" because a representation of place is not the place itself, and the Northwest, for both poets, is imagined as well as described. Wagoner's granite slab exists in close textual and thematic relation to his various Olympic poems, even if the mountain is to be found elsewhere.

When I wrote to Wagoner to ask whether the mountain in the poem corresponded to a particular location, he replied that the poem's landscape had an important psychological and spiritual dimension, and was not based on a single, actual peak. In other words, no: the mountainside does not match a real mountain. In a sense, the poem is regional by association, gathering significance from its connections to Wagoner's other works, including many poems with clear geographical correspondences, and to ways of depicting place that emerge over literary-historical time. What is perhaps most germane is the method by which both poets represent their attraction to and knowledge of related places, real or fictive. Northwestern environments, on the evidence of the two poems, are elusive and terrifying. They test the limits of imagination and lead poets to humanize their hard-earned knowledge of geological scale with wit and puns.

I have touched upon the intellectual, emotional, and even spiritual value of poetry that, the authors' verbal trickery and sense of humour aside, is essentially contemplative. But what of poems written in direct response to environmental crisis that veer away from lyric tranquility? Rita Wong's "resilience, impure, forms," from the collection *forage* (2007), concerns an urban neighbourhood—unnamed but suggestive of part of Vancouver—in which local commerce and culture meet global capitalism.[79] Wong is a much younger poet than either Wagoner or McKay, and a different kind of writer, expressly political and aligned with a tradition of formally innovative writing. McKay's "Astonished –" does contain the contemporary word "e-mail" and Wagoner's "On a Mountainside" includes several words—"hardware," "blisters," "Calluses," "guts"—that, while not newly coined, are colloquial or worldly enough to suggest that the poem is less cloistered than my "contemplative" implies. But Wong's poem eagerly deploys the language of public relations ("corporate spin"), computers ("microchips," "cookies," "video glare"), and consumerism ("'purchase decisions,'" "gadgets"). The poem is centred on a culturally diverse city, not a remote landscape; it concludes by naming birds but does not venture far into their world. Wong's poem, however, has deeply traditional elements despite its contemporary idiom. It echoes the praise poems of Gerard Manley Hopkins—"vessels maintain & trim, all husk & hue, hollow & watertight" evokes, by rhythm and vocabulary, "God's Grandeur" and "Pied Beauty"—and alludes to the lyric's generic functions of prayer and bearing witness. It ends with an expression of hope for a hale, peaceful, and biodiverse world: "may branches hold and restore marbled murrelets, ducks, geese, / shelter ibises, grateful swallows, egrets, peace."

The poems in *forage* depart from the lyric-descriptive mode of "On a Mountainside" and "Astonished –," but they are occasions for the expression of perspective in Auden's sense. The radical politics of the depictions of local places in the context of global capital lie in attention to *radix* and *polis*—to roots, to the polity, to community. As Wong's poem shows, the world is always present in local places.[80] The Pacific Coast has often been figured by writers as a place apart—as, in Jack Hodgins's memorable phrase, "the Ragged Green Edge of the World."[81] The metaphor of marginality, undoubtedly vital to the literature of the Pacific Northwest, does not mean that the coast is truly separate from the rest of the world. The trope of the edge suggests connection as well as extremity.

Coda: Perspective

I used to rent an apartment that gave a view of one edge of downtown Victoria and of Mount Baker, the massive, snow-covered volcano that lies about one hundred and twenty kilometres away, as the crow flies, in Washington. From my eleventh-floor perspective—high up in low-rise Victoria—the mountain seemed much closer than it really is, as though it were just over the line of low hills in the near distance. The shifting patterns of light on ice, snow, and rock drew my gaze throughout the day. I often thought of a passage in Snyder's *Danger on Peaks* (2004) about volcanoes of the Pacific Northwest: "In a gentle landscape like the western slope, snowpeaks hold much power, with their late afternoon or early morning glow, light play all day, and always snow."[82] (Gentle, as Snyder knows, only until the pastoral is violently interrupted, until "white-hot crumbling boulders lift and fly in a / burning sky-river wind of / searing lava droplet hail.")[83] Because optical illusion placed the mountain just beyond the city, the world as I saw it from my home strangely, simultaneously included alpine snows, coastline, cityscape, and a considerable amount of non-idyllic traffic. Wagoner warns in "On a Mountainside" of the error of "Confusing elevation with mastery." I attempted to be attuned to the distortions and provisionality of my perspective, even as I was rather more securely ensconced than the poem's desperate climber. Directly across from my window was the gravel-covered roof of an office building, the substrate for a small ecosystem of puddles and mosses, an unplanned green world. By contrast, the planned green roof of the building kitty-corner was an ordered garden of shrubs and long grasses. From my vantage I could attest that the regional poets named in this chapter—McKay, Wagoner, Wong, Snyder—coloured my sense of local place and my perpetual rereading of the scene, of its fusion of constructed and wild environments. My making sense of concrete and cherry blossoms depended in part on the sense-making of regional writers.

Perhaps poetry's strongest claims to environmental significance consist in its showing how places have been imagined and understood, and in its shaping of readers' imagination and knowledge of places—in its forging of community among poets, readers, and places. Poets, like alpinists and *flâneurs* alike, are explorers. They report on what they have found. Pay attention, poems tell readers, pay attention. Ecocriticism and regional literary studies have not failed by failing to stop unsustainable logging in the Northwest or to prevent new pipelines from being built. On the contrary, these critical approaches succeed if they direct attention to the centrality of environment in all cultural activities; bearing in mind that standard, I have attempted, in addition to revealing certain technical and contextual aspects of the poems in question, to engage the poetry sufficiently to demonstrate that the writers' imaginings of place are compelling, powerful, and complex. The praxis of poetry is making sound and sense. The environmental praxis of poetry is making sound and sense of place.

And yet. The area that McKay depicts in *Strike/Slip* was under acute threat, during the time in which I wrote this essay, from commercial development. The Juan de Fuca Marine Trail extends along forty-seven kilometres of beach and coastal forest contained within a provincial park. The path intersects Loss Creek and Parkinson Creek, both named in *Strike/Slip*, and encompasses landscapes that McKay evokes but leaves unnamed.[84] The construction of a large resort in the area surrounding the park would have changed the character of the region, environmentalists feared (and I with them), by abetting the transition from wilderness to exurban sprawl. In teaching *Strike/Slip* while the future of its setting was a subject of local debate, I suggested to my students that the connection between our study and the so-called real world was strong. But equally powerful was the suspicion that our keen interest in McKay's poems, even if they were something more than elegies, was too little and too late.

As it happened, a series of public hearings seemed to have swayed the minds of the local officials responsible for determining land use in the Juan de Fuca region, although earlier hearings and protests had had little evident effect. Zoning bylaws would not be changed, according to the reports, and the commercial development would not proceed as planned. As a result, Juan de Fuca Provincial Park would be undisturbed and "The coastal trail" would still be, as McKay writes in "First Philosophies," "a line of thought for those / obsessed with origin, fugitives from history's / inland labyrinth."[85] The news was good, I believed, but it confirmed the differences between poetry, which in this case made nothing happen, and other forms of discourse, which depend less on indirection. And now years after the fracas over the park, news reports warn that the Juan de Fuca Strait will in the near future grow busy with tankers carrying oil across the storied seas. Poetry often resides at a distance from public matters; its imaginative power stems from its being private and unpredictable. But its force is not therefore diminished. Like Wagoner's climber or McKay's margin walker, readers find their expectations eluded, their perspectives transformed by the wildernesses into which they have ventured.

Notes

I am grateful to Dianne Chisholm for her remarks on an earlier version of my essay and to Iain M. Higgins for his comments on the historical frequency of the association of astonishment and stone (and of the words "astonishment" and "stone").

1 David Wagoner, *Traveling Light: Collected and New Poems* (Urbana: University of Illinois Press, 1999), 3.
2 Laurie Ricou, "Two Nations Own These Islands: Border and Region in Pacific-Northwest Writing," in *Context North America: Canadian/U.S. Literary Relations*, ed. Camille R. La Bossière (Ottawa: University of Ottawa Press, 1994), 53.
3 Wagoner (b. 1926 in Massillon, Ohio) moved from the American Midwest and Northeast to Washington in 1954. McKay (b. 1942 in Owen Sound, Ontario), who has lived in various places in Canada and written about many of them, is not a strict regionalist. The case for McKay as a Northwestern poet has been made by anthologists such as David Biespiel, in *Long Journey: Contemporary Northwest Poets* (Corvallis: Oregon State University Press, 2006), and critics including Robert Bringhurst, in *Everywhere Being Is Dancing: Twenty Pieces of Thinking* (Kentville, NS: Gaspereau, 2007), 262.
4 The Association for the Study of Literature and Environment (ASLE), the major professional organization for ecocriticism, was founded in 1992. *The Ecocriticism Reader: Landmarks in Literary Ecology*, edited by Cheryll Glotfelty and Harold Fromm, was published in 1996; the anthology is often cited as the indicator of ecocriticism's arrival in the critical mainstream and as the "standard point of entry" to the field. Lawrence Buell, *The Future of Environmental Criticism: Environmental Crisis and Literary Imagination* (Malden, MA: Blackwell, 2005), 112. The early and mid-1990s also saw the publication of signal ecocritical books by Jonathan Bate, Karl Kroeber, Lawrence Buell, and others. The term "ecocriticism" originates with William Rueckert,

in his "Literature and Ecology: An Experiment in Ecocriticism," *Iowa Review* 9, no. 1 (1978), 71–86. Of course, the examination of nature in literature long predates ecocriticism. As Glen A. Love writes, "the study of literature's relationship to the physical world has been with us, in the domain of the pastoral tradition, since ancient times." *Practical Ecocriticism: Literature, Biology, and the Environment* (Charlottesville: University of Virginia Press, 2003), 1. Scholarly analyses of the character, purpose, and effects of environmentally oriented literary and cultural criticism appear with some frequency. See, e.g., Gillen D'Arcy Wood, "What Is Sustainability Studies?," *American Literary History* 24, no. 1 (2012), 1–15; and Daniel J. Philippon, "Sustainability and the Humanities: An Extensive Pleasure," *American Literary History* 24, no. 1 (2012), 163–79.

5 Lawrence Buell, *Writing for an Endangered World: Literature, Culture, and Environment in the U.S. and Beyond* (Cambridge, MA: Belknap Press of Harvard University Press, 2001), 3.

6 Lawrence Buell, *The Environmental Imagination: Thoreau, Nature Writing, and the Formation of American Culture* (Cambridge, MA: Belknap Press of Harvard University Press, 1995), 430. For a critique of Buell's appeal to praxis, see Dana Phillips, *The Truth of Ecology: Nature, Culture, and Literature in America* (New York: Oxford University Press, 2003), 161–63. On anxiety about irrelevance, see also Phillips, ibid., 4–5, 161.

7 On the view that ecocriticism has always been interdisciplinary, see Scott Slovic, "Part II: Elements of This New Alliance," *ISLE* 17, no. 4 (2010), 757–58.

8 On ecocriticism's "challenge to ... postmodern critical discourse," see Love, *Practical*, 1. Anthony Lioi suggests in contrast that much early ecocriticism "engaged Continental philosophy." "Part I: An Alliance of the Elements," *ISLE* 17, no. 4 (2010), 756.

9 Love, *Practical*, 8.

10 Ibid., 7, 9–10.

11 "Special Forum on Ecocriticism and Theory," *ISLE* 17, no. 4 (2010), 754–99.

12 Jim Warren, "Placing Ecocriticism," *ISLE* 17, no. 4 (2010), 771; original emphasis. Cf. Glen A. Love, "Ecocriticism, Theory, and Darwin," *ISLE* 17, no. 4 (2010), 773–75.

13 Timothy Morton, however, distinguishes ecocriticism from other political approaches: "ecocriticism ... consciously blocks its ears to all intellectual developments of the last thirty years, notably ... feminism, anti-racism, anti-homophobia, deconstruction." *Ecology without Nature: Rethinking Environmental Aesthetics* (Cambridge, MA: Harvard University Press, 2007), 20. Greg Garrard makes a similar claim, critiquing ecocriticism's "crypto-theological" dimension and advocating a rapprochement with the "rich tradition of Marxist literary theory." "Literary Theory 101," *ISLE* 17, no. 4 (2010), 780.

14 My discussion draws on an extensive debate about interdisciplinarity in literary studies published in *PMLA*. "Forum," *PMLA* 111, no. 2 (1996), 271–311.

15 On metaphors, see Greg Garrard, *Ecocriticism*, in the New Critical Idiom series (Abingdon, UK: Routledge, 2004), 174–75.

16 See Phillips, *Truth*, 161.

17 Garrard, *Ecocriticism*, 176.

18 Love, *Practical*, 16.

19 Buell, *Future*, 21, 17, 40, 138, 30, 112, 19.

20 Garrard, *Ecocriticism*, 177. On the same point, see Lawrence Buell, "Foreword," in *Environmental Criticism for the Twenty-First Century*, ed. Stephanie LeMenager, Teresa Shewry, and Ken Hiltner (New York: Routledge, 2011), xiii–xiv.

21 Morton, *Ecology*, 2, 5, 11–12, 20, 114.

22 Timothy Morton, "Guest Column: Queer Ecology," *PMLA* 125, no. 2 (2010), 281, 273, 278.

23 Ibid., 275.

24 Love, *Practical*, 6, 7.
25 Annie Merrill Ingram et al., eds., *Coming into Contact: Explorations in Ecocritical Theory and Practice* (Athens: University of Georgia Press, 2007).
26 Annie Merrill Ingram et al., "Introduction: Thinking of Our Life in Nature," in ibid., 8.
27 S.K. Robisch, "The Woodshed: A Response to 'Ecocriticism and Ecophobia,'" *ISLE* 16, no. 4 (2009), 701.
28 Ibid., 705, 702. "Deferral": cf. Garrard: "Ecocritics … are in the unusual position as cultural critics of having to defer … to a scientific understanding of the world" (*Ecocriticism*, 10).
29 Robisch's essay was also contested by Serpil Oppermann in "Ecocriticism's Theoretical Discontents," *Mosaic* 44, no. 2 (2011), 161–63. But not all assessments of the essay have been hostile; see Love, "Ecocriticism," 775.
30 Lioi, "Part I: An Alliance," 754; original emphasis.
31 Slovic, "Part II: Elements," 757.
32 Serenella Iovino, "Ecocriticism, Ecology of Mind, and Narrative Ethics: A Theoretical Ground for Ecocriticism as Educational Practice," *ISLE* 17, no. 4 (2010), 760.
33 Ibid.
34 Astrid Bracke, "Redrawing the Boundaries of Ecocritical Practice," *ISLE* 17, no. 4 (2010), 765.
35 Serpil Oppermann, "Ecocriticism's Phobic Relations with Theory," *ISLE* 17, no. 4 (2010), 769.
36 Warren, "Placing," 772; original emphasis.
37 Stacy Alaimo, *Bodily Natures: Science, Environment, and the Material Self* (Bloomington: Indiana University Press, 2010).
38 "Introduction," in *Environmental Criticism for the Twenty-first Century*, ed. Stephanie LeMenager, Teresa Shewry, and Ken Hiltner (New York: Routledge, 2011), 1.
39 John Felstiner, *Can Poetry Save the Earth? A Field Guide to Nature Poems* (New Haven, CT: Yale University Press, 2009).
40 Eleanor Cook, *Against Coercion: Games Poets Play* (Stanford, CA: Stanford University Press, 1998), xii.
41 Michael Riffaterre, *Semiotics of Poetry* (Bloomington: Indiana University Press, 1978), 1.
42 Salient examples include two essays by Gary Snyder, "Ecology, Literature, and the New World Disorder: Gathered on Okinawa" and "Writers and the War against Nature," in Gary Snyder, *Back on the Fire: Essays* (Berkeley, CA: Counterpoint, 2007), 21–35, 61–71.
43 Don McKay, *Vis à Vis: Field Notes on Poetry and Wilderness* (Wolfville, NS: Gaspereau, 2001), 25, 9.
44 Snyder, *Back*, 32.
45 Ibid., 63.
46 W.H. Auden, "In Memory of W.B. Yeats," in *Collected Poems*, ed. Edward Mendelson (New York: Vintage, 1991), 248.
47 Ibid.
48 W.H. Auden, "Freedom and Necessity in Poetry," in *The Place of Value in a World of Facts: Proceedings of the Fourteenth Nobel Symposium, Stockholm, September 15–20, 1969*, ed. Arne Tiselius and Sam Nilsson (New York: Wiley, 1970), 139.
49 Ibid., original emphasis.
50 W.H. Auden, *Prose*, vol. 3, *1949–1955*, ed. Edward Mendelson (Princeton, NJ: Princeton University Press, 2008), 229–30. The essay was written in 1949 and published in 1950. Auden's ideas tend to recur in various essays; he often recycled passages, with the result that similar wordings appear in essays published in some cases in close succession and in others years apart. Although I refer in this chapter to only a few of

Auden's works, I have drawn generally on his interrelated statements, published in various venues, on imagination and knowing, including "Making, Knowing and Judging," "The Dyer's Hand," *The Enchafèd Flood* (especially the sections on "Romantic Aesthetic Theory," "The Stone: The Romantics and Mathematics," and "The Artist as Don Quixote"), and "Freedom and Necessity in Poetry."

51 Auden, "Freedom and Necessity," 139.
52 On Auden's anthropocentrism, see Rainer Emig, "Auden and Ecology," in *The Cambridge Companion to W.H. Auden*, ed. Stan Smith (Cambridge: Cambridge University Press, 2004), 212–25. See esp. 215.
53 W.H. Auden, *Forewords and Afterwords*, selected by Edward Mendelson (New York: Random House, 1973), 464–73. The essay in question is "Concerning the Unpredictable."
54 Auden, *Forewords*, 497.
55 Auden, "Freedom and Necessity," 141. Cf. Auden, *Prose*, 228.
56 I.A. Richards, *Science and Poetry* (London: Kegan Paul, Trench, Trubner, 1926), 82–83.
57 Auden, "Freedom and Necessity," 139.
58 Ibid.
59 Auden, *Collected Poems*, 248.
60 Cook, *Against*, xi.
61 Ibid., xii; original emphasis.
62 Ibid.
63 Phillips, *Truth*, 17.
64 McKay, *Vis à Vis*, 28.
65 My interest in Auden has a distant biographical connection to Wagoner: Auden was the best man at the wedding, in 1953, of Theodore Roethke, Wagoner's teacher and mentor, to Beatrice O'Connell.
66 Don McKay, *Deactivated West 100* (Kentville, NS: Gaspereau, 2005), 113.
67 Don McKay, *Strike/Slip* (Toronto: McClelland & Stewart, 2006), 3. All subsequent quotations of the poem correspond to this reference.
68 Ibid., 40.
69 Ibid., 7.
70 Malcolm Woodland, "Poetry," *University of Toronto Quarterly* 77, no. 1 (2008), 53.
71 David Wagoner, "On a Mountainside," *Poetry* 163, no. 1 (1993), 1–3.
72 David Wagoner, *Walt Whitman Bathing: Poems* (Urbana: University of Illinois Press, 1996), 71–73. All subsequent quotations of the poem correspond to Wagoner, *Traveling*, 258–59.
73 John Taylor, review of *Walt Whitman Bathing: Poems*, by David Wagoner, *Poetry* 171, no. 3 (1998), 230.
74 Elizabeth Barrett Browning, *The Works of Elizabeth Barrett Browning*, ed. Sandra Donaldson, vol. 2, *Poems, 4th Edn (1856), Continued*, ed. Marjorie Stone and Beverly Taylor (London: Pickering and Chatto, 2010), 66.
75 Wagoner, *Traveling*, 256–57.
76 Ibid., 257.
77 Ibid., 256.
78 Robinson Jeffers, *The Wild God of the World: An Anthology of Robinson Jeffers*, ed. Albert Gelpi (Stanford, CA: Stanford University Press, 2003), 163.
79 Rita Wong, *forage* (Gibsons Landing [Gibsons], BC: Nightwood, 2007), 64–65. Subsequent quotations of the poem correspond to this reference.
80 Buell observes that "Questions of place-attachment and place-(re)construction have been central to ecocriticism since the movement's inception, but the high valuation it initially set on local or bioregional allegiances has been seriously roiled by its recent engagement with postcolonial and 'ecocosmopolitan' models of thinking." "Foreword," xvi.

81 Jack Hodgins, *The Resurrection of Joseph Bourne, or A Word or Two on Those Port Annie Miracles* (Toronto: McClelland & Stewart, 1998 [1979]), 9.
82 Gary Snyder, *Danger on Peaks: Poems* (Washington, DC: Shoemaker & Hoard, 2004), 5.
83 Ibid., 11.
84 McKay, *Strike/Slip*, 7, 60, 72.
85 Ibid., 26.

CHAPTER 9

Deception in High Places: The Making and Unmaking of Mounts Brown and Hooker

Zac Robinson & Stephen Slemon

This book asks the question How might interdisciplinary critical knowledge enable ethical action on behalf of western Canadian environments? With a view to understanding something of the cultural politics that underwrite the way in which a particular kind of environment—mountain landscapes, in this instance—comes to be conceptually located, and why it is that foundational understandings of landscape, once historically embedded, prove so hard to change, we turn to a moment in mountaineering history that takes place before mountaineering itself begins as a formal and consolidated practice. The larger narrative we seek to unpack pertains to the curious persistence of "knowledge" in the social archive, and to how dominant assumptions about landscape and its meanings come to overwrite the socially layered *petits récits* of cross-cultural agency and the relations of class. The story begins in a moment that would go on to prove itself as the most notorious example of mismeasure in mountaineering history—and one that would put Canadian mountaineering, literally, on the map.[1]

In the spring of 1827, David Douglas (1799–1834), a botanical collector in the employ of the Horticultural Society of London,[2] was on his way back from what he hoped would prove to have been a career-making expedition, gathering plant specimens in western Canada and the United States. He had already prepared and shipped seeds and specimens of many kinds: a flowering currant, a yellow lupine, a purple-and-yellow peony, and, most crucially for his imagined future, some cones from a giant "sugar pine" that he had

Figure 9.1 David Douglas (1798–1834). From *Curtis's Botanical magazine; or flower garden display*, vol. 63 (London: Samuel Curtis, 1836). Lithograph by R. Martin & Co; sheet 156 x 253 mm.

come across in Oregon. Douglas's highest hope was that at least some of his botanical finds would turn out to be "originals"—plants as yet unknown in Europe—but of this he couldn't really be sure. For although Douglas had sent many specimens back to the Horticultural Society, ones that seemed new to him, he had not actually analyzed and classified those specimens. That kind of intellectual work belonged to professional botanists, men of the educated upper crust, and Douglas was a self-taught mason's son. Seven years earlier, while working as a gardener at Glasgow University, Douglas had been taken in hand by William Jackson Hooker (1785–1865), professor of botany. Hooker had discovered an aptitude in the young Scot, trained him in the art of flower pressing and drying, and sent him down to the Horticultural Society of London with a view to carrying out exploratory fieldwork. Soon after, Douglas was shipped out to Philadelphia and began his new career by collecting furiously. His botanical specimens, however, had so far met with minimal success back in London. A chance at redemption came in 1824, when the Hudson's Bay Company agreed to sponsor a botanical collecting expedition along the Columbia River, and Douglas—again, with help from Hooker—secured the position. And so on 1 May 1827, Douglas found himself at Athabasca Pass, travelling east along the fur-trade route over the Great Divide, and harbouring hopes for a very different type of upward mobility than the kind for which he was about to become so disturbingly famous.

"I set out," Douglas wrote later, in his 1828 manuscript titled *A Sketch of a Journey to the North-Western Parts of the Continent of America during the Years 1824, 1825, 1826, and 1827*, "with the view of ascending what appeared to be the highest peak" guarding the height of land.[3] Why he did so remains unclear. Professionally, Douglas's interest in mountains pretty much ended at the treeline. Though in Douglas's day people did hike up mountains for exercise or leisure, mountain climbing itself, as technique and sport, was hardly a consolidated activity. The birth of alpine club culture was still decades away in England. But Romanticism, and the Grand Tour in Europe, had made mountain viewing fashionable, in part for the capacity of mountains to evoke a sense of awe in the face of the sublime.[4] Whatever the case, Douglas's moment of Romantic wanderlust on that 1 May would produce what many have called the first mountaineering ascent in Canada.[5]

"The height from its apparent base exceeds 6,000 feet, 17,000 feet above the level of the sea," Douglas continued, "1,200 feet of eternal ice. The view from the summit is of that cast too awful to afford pleasure—nothing as far as the eye can reach in every direction but mountains towering above each other, rugged beyond all description." And then the Romanticism in Douglas's writing surrenders to the prose of social climbing: "This peak, the highest yet known in the Northern Continent of America, I felt a sincere pleasure in naming **MOUNT BROWN**, in honour of R. Brown, Esq., the illustrious botanist, no less distinguished by the amiable qualities of his refined mind. A little to the south is one nearly of the same height, rising more into a sharp point, which I named **MOUNT HOOKER**, in honour of my early patron the enlightened and learned Professor of Botany in the University of Glasgow."[6] As anyone familiar with the Rockies knows, nothing in the range rises to anywhere near 17,000 feet above sea level.[7] The peak now named Mount Hooker—and there's ample evidence to suggest that Douglas's Mount Hooker was, in fact, the nearby (and significantly lower) McGillivray's Rock[8]—rises to a reasonably respectable 10,781 feet, eighty-fifth highest in the range. At

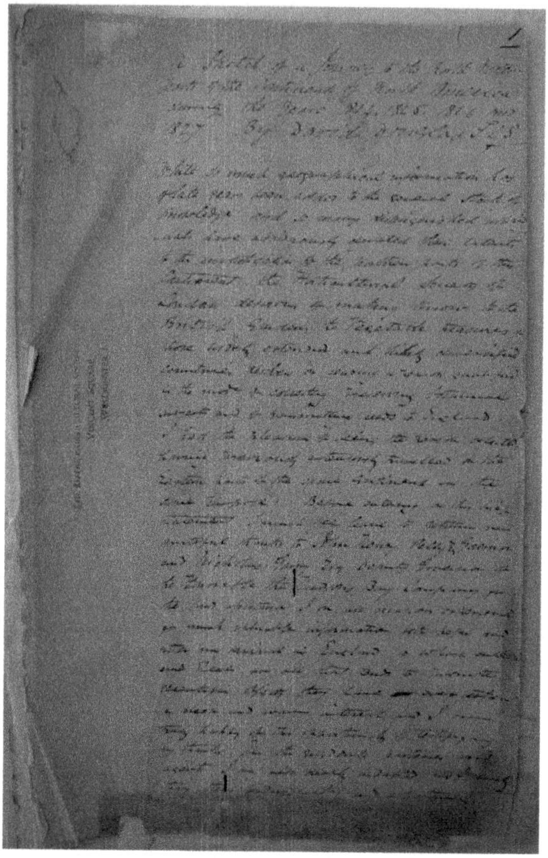

Figure 9.2 A manuscript page from David Douglas's 1828 narrative, *A Sketch of a Journey to the North-Western Parts of the Continent of America During the Years, 1824, 1825, 1826, and 1827*, prepared for, but never submitted to, publisher John Murray. Housed in the archive of the Lindley Library of the Royal Horticultural Society, London. Photo: Zac Robinson.

9,184 feet, Mount Brown looms to only about 600 feet higher than Mount Lady Macdonald, a pleasant and popular day hike just north of Canmore in the eastern front ranges.

But it's not always the facts that make history. Hope and pity play their own compositional part in this tale. David Douglas returned to London to discover that many of his samples *had* proven to be "originals." Within months, he was elected to membership in the Linnean Society, the Zoological Society, and the Geological Society with the usual membership fees waived. John Murray (1808–1892), the famous publisher of Albemarle Street, awarded him a book contract—it was to be *the* book of the year—and Murray wanted a ripping yarn. It was an extraordinary honour. Murray specialized in books of travel, exploration, and adventure (like, for instance, Captain John Franklin's *Narrative of a Journey to the Shore of the Polar Sea, in the Years 1819, 1820, 1821, 1822* [1823] and Charles Darwin's *The Origin of Species* [1859]), but he had never before considered a work by an ordinary botanical collector. Douglas, however, wanted to add botanical classification to his exploration memoir, and so he threw himself into scientific self-training in the Linnaean system. And here his social ascent ended. He was invited to read a paper to the Linnean Society, and would have done so himself, without the usual professional elocutionist, had he not succumbed to a paralyzing nervousness on the day. Overwhelmed by feelings of misgiving and inferiority, he delayed on the Murray book contract, as the self-education continued. The manuscript—*A Sketch of a Journey ...* —stalled out at a mere fraction of the length of his fieldnotes. He never submitted it for publication.[9] Broken, the would-be scientist accepted a contract from Hooker to help prepare the map for the professor's forthcoming magnum opus on the plant life of North America, *Flora Boreali-Americana* (1829).[10] He departed soon after on another specimen-collecting

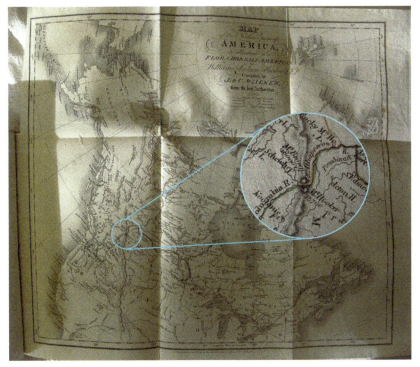

Figure 9.3 The first map showing Douglas's mountain giants in Sir William Jackson Hooker's *Flora Boreali-Americana; or, The Botany of the Northern Parts of British America*, vol. 1 (London: H.G. Bohn, 1829).

expedition to the west coast of North America, and never returned to England. That Douglas died in 1834 under bizarre circumstances—his body was found, lifeless and trampled, at the bottom of an open pit dug to trap wild bulls near Mauna Kea in Hawaii—fuelled speculation of foul play, even suicide.[11] He was thirty-five years of age.

Hooker privately wondered about Douglas's mountaintop measurements, suspecting they were "egregiously overrated."[12] In the end, the heights were both reduced by about 1,000 feet. But he felt sufficiently grateful to his specimen collector to include Douglas's Rocky Mountain giants on the *Flora Boreali-Americana* book map: mementoes, it would seem, of a life that had not reached its professional summit. That map, printed in October 1829, first presents the two high mountains in the Canadian Rockies, each with Douglas's hopeful name, each with only a little taken off the top: Mount Brown at 16,000 feet, Mount Hooker at 15,700. *A Sketch of a Journey* (Douglas's failed travelogue) was published, two years after his death—not by John Murray but by Hooker himself—as a mere essay in the *Companion to the Botanical Magazine* (1836).[13]

Atlas makers steal information from each other—to the extent that most commercial map publishers today include a fictional "trap street" or two on their urban maps in order to catch their thieving competitors out.[14] In the mid-nineteenth century, physical information about western North America was scarce, and publishers had to make a living. And so it was that Douglas's mountains, first documented in an 1829, highly technical, botanical opus became the dominant trap streets of nineteenth-century cartography. They remained the highest points on any map of British North America for almost three-quarters of a century, and so became the siren call for Canadian mountaineering exploration. These giants in the Rockies *had* to exist, for by the turn of the twentieth century every atlas and geography book showed them as existing—somewhere.

The problem was that nobody else had actually seen them.[15] "A high mountain," wrote Arthur P. Coleman (1852–1939) in 1911, "is always seductive. A mountain with a mystery is doubly so.... When I studied the atlas and saw Mount Brown and Mount Hooker, the highest points in the Rockies, standing on each side of the Athabasca Pass, I longed to [find] them.... My eyes turned to them irresistibly whenever I looked at the map, and my mind was soon made up to visit and, if possible, climb them."[16] Born in Lachute, Quebec, and raised in rural Ontario, Coleman had trained in chemistry, mineralogy, botany, and zoology, and obtained his PhD in geology from Breslau University, Germany, in 1881. While a student abroad, he visited the North Cape of Norway. The trip sparked in him an interest in glaciers and a lifelong passion for mountains. He returned to Canada in 1882 and was appointed

Figure 9.4 Atlas map showing Mount Brown and Mount Hooker. *British Columbia and the North West Territory in the Dominion of Canada*, London Atlas Series (London: Edward Stanford, 1901). Since first appearing on a map in 1829, Mount Brown and Mount Hooker remained the highest points on any map and atlas showing North America until the early years of the twentieth century.

professor of geology and natural history at Victoria College (his alma mater) in Cobourg, Ontario. Later, in 1891, Victoria College became a federated college within the University of Toronto, and Coleman became a professor in the School of Practical Science. In time, Coleman would become a leading protagonist in the Alpine Club of Canada, founded in Winnipeg in 1906. As its inaugural vice-president, he would lead the first expeditions to the unclimbed Mount Robson—the *actual* highest peak in the Canadian Rockies.[17]

Coleman's first of eight expeditions to the Canadian Rockies took place in 1884, a year before the transcontinental Canadian Pacific Railway (CPR) laid its steel over the Continental Divide at Kickinghorse Pass, near Laggan (Lake Louise). Conveniently, Lucius Coleman (b. 1854), Arthur's younger brother, had become a rancher near Morley, the Stoney Nakoda Reserve, just west of Calgary at the mouth of the Bow Valley. And the Coleman brothers heavily relied on local knowledge.

The "Mountain Stoney" had lived for more than ninety years in the eastern front ranges. They played an active role not only in laying the original rail line through the Rockies in the 1880s, but also in the early construction of walking trails for the tourism boom that followed the completion of the CPR and the establishment of the Banff Springs Reserve (later Banff National Park) in 1885.[18] This tourist boom brought with it the rise of a commercial outfitting industry, which, within a very few years, displaced First Nations work in the newly formed parklands. Stoney Nakoda peoples, once central to the area's nascent tourist trade, suddenly found themselves redefined as "poachers" in the eyes of park officials, tourists, and tourism developers, and excluded in the name of "game" (not wildlife) conservation and sport hunting.[19] They would be allowed again into the Banff area economic infrastructure, but only as colourful backdrops to the tourist industry, in full ceremonial regalia, during the weeklong festivities that constituted what the CPR would call "Banff Indian Days" (1902–45).

But in the 1880s, as earlier, so important were Stoney Nakoda outfitters to travel in the area, and Aboriginal knowledge in general to the collective cognitive "map" of the Rocky Mountains, that Arthur Coleman himself undertook to learn Cree.[20] He was keenly interested in local lore, and transcribed stories he heard on his travels in his journals. He sought Stoney advice on routes across the mountains. He even named several passes and lakes after his Indigenous guides.[21]

It should be hardly surprising, then, that in 1893, because of a map drawn for him by Chief Jonas Goodstoney of the Stoney Nakoda Nation, Coleman did at last find mounts Brown and Hooker: found them, that is, to be "frauds."[22] By then, Coleman had spent three summers searching for Douglas's mounts Brown and Hooker. His first expedition, in 1888, embarked from the CPR line at Beavermouth (a small lumber town northwest of Golden, BC), followed the Columbia River northwards on the west side of the Rockies, and then foundered in the thick bush and "diabolical mosquitoes" they encountered west of the Continental Divide. Sore in body and dejected in mind, the group gave up after two hard months. Coleman's second expedition, in 1892, followed a route through the eastern front ranges, from Morley north to Fortress Lake, near present-day Jasper. Coleman initially mistook Fortress Lake for the "Committee's Punchbowl"—the tarns at the summit of Athabasca Pass, below the mythic Mount Brown, where fur traders famously dipped their cups and toasted the governing committee of the Hudson's Bay Company. The tarns had taken on mythic proportions themselves. Coleman's third expedition, in 1893, proved the successful one. Of the Committee Punchbowl tarns, he would write: "some of the maps make the Punchbowl a lake ten miles long, but here in real life it was only a small pool less than two hundred yards long." The canvas boat they had freighted for over 400 kilometres, with a view to rowing it around the "lake," "remained in its pack cover of green canvas." And as for mounts Brown and Hooker: "That two commonplace mountains," Coleman would write, "… should masquerade for generations as the highest points in North America, seems absurd.... How could any one, even a botanist like Douglas, make so monumental a blunder ...?"[23]

"Blunder." "Fraud." A "masquerade." Coleman, thanks only to Jonas Goodstoney, may have cartographically got things right, but by 1893 Douglas's mountain "giants" had towered above the imaginative skyline for two-thirds of a century, and other travellers remained unconvinced. "Knowledge" about geographical as well as social otherness, once embedded in the archive, necessarily proves adamant. A new kind of "explorer" was now arriving in the Rockies, brought close by the new convenience of the railway, members of what the CPR's general manager termed "the class that travels."[24] These new travellers practised a highly codified form of leisure, distinct from the ways that earlier scientific expeditions or Romantic travellers had engaged mountain landscapes. They were part of a consolidated, metropolitan, professional, and mostly male community—"mountaineers," who after the Alpine Club came to be founded in London, in 1857,[25] found themselves part of a self-globalizing middle class that, as the renowned British climber Geoffrey Winthrop Young (1876–1958) would put it, aspired to their "own territory and ... [their] own prophetic book of adventure." "And of them all," Young continued, "perhaps, Norman Collie was the man of the greatest natural endowment and the man most exclusively devoted to mountains."[26]

A fellow of the Royal Society and the Royal Geographical Society, as well as an equally accomplished chemist, J. Norman Collie (1859–1942) was likely Britain's finest amateur climber of the day. In time, he would become president of The Alpine Club (1920–22). But in 1897, four years after Coleman's dispiriting trip, Collie, along with Swiss guide Peter Sarbach (1844–1930), found himself called upon to join a "memorial climb" on Mount Lefroy, near Lake Louise on the Continental Divide, which a year earlier had been the scene of a fatal accident involving the president of the Appalachian Mountain Club. Collie agreed to go, on the provision that Sarbach's services be made available to him for what remained of his time in the Rockies. Collie and Sarbach made easy work of Mount Lefroy, and then Mount Victoria, as well: both were first recorded ascents. But for Collie, the view to the north that these ascents provided him—of a seemingly boundless ocean of high mountains straddling the Divide—caused him to question Coleman's disappointing finding regarding Douglas's mountain giants, published two years earlier.[27] Collie's original plan had been to travel south to the unclimbed Mount Assiniboine. He instead resolved to head north. Through what remained of the summer of 1897, he and Sarbach pursued their dogged search for mounts Brown and Hooker, travelling by pack train farther and farther north from the rail line, always hoping that each new discovery of a grand peak might yet vindicate the myth of the mountain giants.

High on Mount Freshfield, Collie caught a glimpse of what would compel his second visit to the range a year later: "From the highest point reached, 10,000 feet, a very lofty mountain—probably 14,000 to 15,000 feet—was seen lying 30 miles away in a northwesterly direction. Only the peaks north of [Mount] Lyell are marked on the map, and these are Mount Brown and Mount Hooker, which are supposed to be 16,000 and 15,000 feet high respectively; consequently we at once took it for granted that we had seen one of them."[28] But the map Collie was using, unlike Coleman's, was not informed by Stoney Nakoda knowledge. The commercialization and formalization of Banff-area travel had by now taken hold, and mountaineering expeditions in the region were now outfitted by "professionals," such as the ex-railway surveyor Tom Wilson (1859–1933) of Banff, who employed mostly newcomers. Collie's understanding of Rockies geography was, therefore, profoundly archival and textual—indeed, he relied for the most part on John Palliser's map and "Journals," which detailed the 1857–60 British North American Exploring Expedition and the search for railway routes across the Cordillera. Palliser's map showed the lands south of Mount Freshfield, Howse Pass, and the North Saskatchewan River, but it left the area to the north geographically incomplete—a "blank space on the map," to use a then oft-quoted phrase by Clements Markham (1830–1916), president of the Royal Geographical Society.[29] What filled that "blank space"—for Collie, as for most others—was therefore what was there, already, in the cognitive colonial archive: mounts Brown and Hooker. "They next year (1898)," he would later quote—from Kipling's "The Explorer"—"were the 'something lost behind the ranges' that we sought."[30]

And, a year later, on the summit of Mount Athabasca, wrote Collie, "We halted. The view that lay before us in the evening light was one that does not often fall to the lot of modern mountaineers. A new world spread at our feet; to the westward stretched a vast ice-field probably never before seen by human eye, and surrounded by entirely unknown, unnamed, and unclimbed peaks."[31] The Columbia Icefield, as he called it, easily became "the apex" of the Rockies "for the melting of its snows descend into three great river-systems, flowing into three different oceans—to the Columbia and thence to the Pacific; to Hudson's [sic] Bay via the Saskatchewan; and by the Athabasca to the Arctic Ocean."[32] In two summers, Collie and his outfit mapped much of the Wapta, Waputik, Freshfield, and Columbia icefields, climbing and naming numerous peaks along the way. On the Columbia Icefield, to a mountain just north of "The Dome," Collie gave the name "Peak Douglas." Perhaps it seemed fitting that Douglas's name be inscribed on the great height of land.[33]

Collie's 1898 expedition did more than any other single expedition had yet done—or has done since—to consolidate mountaineering activity in the Canadian Rockies. But it did little to clarify the whereabouts of Brown and Hooker, and Collie returned to England mystified: "There was no pass between the two highest peaks we had seen," he wrote. "And where was the Committee's Punchbowl that should lie between them?"³⁴

Unsurprisingly, given our argument, Collie's answer to the puzzle of mounts Brown and Hooker came to him ultimately not from direct observation in the field, but at "home," where he had started, in the archive. Pouring over everything he could find on the Canadian Rockies in The Alpine Club Library, and in the British Library, Collie came across a reference in Bancroft's *History of British Columbia* (1887) to Hooker's *Companion to the Botanical Magazine* (1836)—the journal that contained Douglas's *A Sketch of a Journey*. With Douglas's failed travelogue at last in hand, Collie was able to conclude, and concede: "[That] Douglas climbed a peak 17,000 feet high in an afternoon is, of course, impossible ... to Prof. Coleman belongs the credit of ... settled accuracy."³⁵

Charitable writers have ever since sought ways of understanding Douglas's spectacular mountain deception as being, somehow, innocent. Jerry Auld's introspective *Hooker and Brown* (2009), which attempts to understand the story at the level of character, and through the narrative possibilities of historical fiction, is the latest in a long line of Canadian mountain-history speculation.³⁶ Most commentators now agree that Douglas's miscalculation likely derived from an altitude estimation made in 1826 by surveyor Lieutenant Aemilius Simpson (1772–1831), who had been hired by the formidable Sir George Simpson (1787–1860), governor of Rupert's Land (and a distant cousin by marriage), as a hydrographer and surveyor for the Hudson's Bay Company.³⁷ The least charitable moment came in 1927, on the centenary of Douglas's alleged ascent of Mount Brown, and it came from the most distinguished alpine historian, writer of the region's first mountaineering guidebook, and later president of the American Alpine Club, James Monroe Thorington (1894–1989).

After making a trip to England with the express purpose of comparing Douglas's original fieldnotes—a hefty ledger of 131 pages, with entries covering the entire 1824–27 expedition—with *A Sketch of a Journey*, the shorter prepared, but never submitted, manuscript, Thorington questioned whether Douglas actually got to the top of Mount Brown at all. A trip up the mountain with guide Conrad Kain (1883–1934) in the summer of 1924 confirmed Thorington's suspicion.³⁸ The altitudes, while grossly exaggerated, were not where the deception lay. Fur-trade records indicated already "a tradition of height in the region." Everyone, according to Thorington, believed the mountains in the area were somewhere between 16,000 and 18,000 feet high.³⁹ The deception was Douglas's claim of an ascent.

In his fieldnotes, Douglas didn't name or attribute elevations to Brown or Hooker. These inventions, Thorington discovered, were created later in England in the shorter manuscript prepared for John Murray.[40] Furthermore, in his fieldnotes, Douglas described the view by saying "nothing, as far as the eye could perceive, but mountains such as I was on, and *many higher*"[41] (emphasis added). The latter part of this sentence is dropped in the Murray document and replaced with "the view from the summit" and "the highest yet known in the Northern Continent of America."[42] In fact, the only suggestion in the fieldnotes that perhaps puts Douglas on the actual summit of Mount Brown is a sentence that reads "the ascent took me five hours; descending only one and a quarter"[43]—and this is assuming, of course, that Douglas's use of the word "ascent" implies actually getting to the top. It's a big assumption for 1827. Again, in Douglas's day, mountaineering as sport does not exist. And so it is difficult to say with certainty where exactly Douglas was standing when, in his fieldnotes, he wrote as follows: "I *remained* 20 minutes, my Thermometer standing at 18°; and night closing fast in on me and no means of fire, I was reluctantly forced to descend"[44] (emphasis added).

High on Mount Brown, Douglas's fieldnotes in hand, Thorington could make little sense of the actual terrain in relation to the notes. The steeper cliffs near the top, for instance—terrain that would challenge anyone wearing snowshoes, as Douglas was—are not mentioned at all. Moreover, Douglas's time of five hours hardly jived with the realities of spring conditions and snow. Travel at that time of year is just not that fast during the afternoon. Lower on the mountain, Douglas complained about "sinking on many occasions to the middle."[45]

It was these details and others that led Thorington to suggest that, if we are to take the fieldnotes at face value, Douglas likely "reached the snow plateau on the southern shoulder" and "it should not be forgotten that this was a time in mountaineering history when many a man 'climbed' a mountain without attaining the very summit. It was only necessary that one should reach a considerable height." Thorington's conclusions were published in his *The Glittering Mountains of Canada* (1925) and, again, in the 1926–27 *Canadian Alpine Journal* (*CAJ*). But to fully understand the story of mounts Brown and Hooker, fragmentary and uncertain as it remained, Thorington challenged the *CAJ*'s readership to simply "remember the man," David Douglas, "who created it a hundred years ago."[46]

Arthur O. Wheeler (1860–1945), the obstinate and fiery long-time director of the Alpine Club of Canada, wanted none of it, and a heated debated ensued between the two titans in the journal.[47] Looking back on the exchange, writer/climber Bruce Fairley, in his wonderful *Canadian Mountaineering Anthology* (1994), surmised that Wheeler "simply could not conceive that so famous an explorer and scientist [Douglas] could simply have fabricated the details of his

Figure 9.5 The summit of Mount Brown (*centre left*, in the distance) from high on the mountain's meandering southwest ridge. Photo: Zac Robinson.

historic climb out of whole cloth."[48] If Fairley's right, Wheeler missed Thorington's point—because it shows that Wheeler knew little about David Douglas himself. Ironically, this is no small part of the Brown-Hooker problem.

Contemporary mountaineering writers have largely disregarded or misread Thorington's thesis. And they've all imagined Douglas in contexts befitting only what now seems to be the standard stock-in-trade creation myth of North American mountaineering. For instance, in both Andy Selters's *Ways to the Sky* (2004) and Chic Scott's *Pushing the Limits* (2000), attention is given to Douglas's exaggerated heights, but his summit achievement is taken for granted. Douglas is refashioned as both a great man of science—a "botanist-explorer," writes Selters—and an actual climber.[49] "His elation and joy upon reaching the summit," says Scott, "can still be understood by mountaineers today."[50]

In *Climbing in North America* (1976), Chris Jones goes further to claim that Douglas "does not give us science, botany, or geography, but he has stated what makes a mountaineer: a person who, without qualification, *desires* to climb peaks. We see in him the archetypal mountaineer." Jones continued, "If we understand what it was about those wintery peaks at Athabasca Pass that drew him to them, we have a grasp of mountaineering." Here, Douglas has been wholly remade as not only a climber—"he was our first mountaineer"—but as one of early mountaineering's exemplary figures, a fantastical sort of George Mallory à la coureur de bois.[51]

To Douglas now goes the hefty honour of establishing mountaineering culture itself in Canada, or so any keen scrambler might interpret from the summit register atop Mount Brown. A note written by Robert Sandford—the author of *The Canadian Alps* (1990)—which was taken to the top by a group of Jasper park officials in 2002, reads: "On this, the 175th anniversary of David Douglas' ascent, our expedition aims to commemorate the importance of ... the role David Douglas played in the creation of this country's mountaineering culture."

To take Thorington's challenge seriously is to consider Douglas in the context of his place and time. And to do so perhaps tells us more about the exclusive class-based world of Victorian science than it does about an emergent mountain culture in North America. Douglas was not ahead of his time, but rather a sad product of it. And if 1 May 1827 was a foundational moment for Canadian mountaineering, a serious appraisal that puts geography, literature, and history in direct conversation with one another is necessary. It's almost certain that Douglas did not climb to the summit of Mount Brown. It is probable, however, that he ascended to a highpoint somewhere on the mountain's long, meandering southeast ridge—just above that point, perhaps, where the Interprovincial Boundary Survey would build their camera station ninety-three years later. An old bolt and a cairn still mark the spot where the surveyors measured and Douglas maybe mused.[52] But "fraud" is too strong a word for that complex process of botanical, geographical, and literary intermingling that put Douglas's spectacular mismeasurement into the imaginative colonial archive.

In a sport where the false claim has occasioned a special fascination among writers and readers—consider the whole Robson saga,[53] for example, or Fredrick Cook's mendacious account of a first ascent on Mount McKinley[54]—the Brown-Hooker problem fails to rise to the level of fraudulent deception. Here's why. While Douglas's claim puts him squarely on the summit of the highest point on the continent, it has little to do with mountaineering achievement, and even less to do with sensationalism. Thorington was mistaken to conclude that "the creation of Mt Brown and Mt Hooker and their altitudes ... were introduced for purposes of personal publicity."[55] An examination of the *entirety* of Douglas's two handwritten texts—the 131-page fieldnotes and the 56-page manuscript—tells a different story. Murray awarded Douglas the contract because he presumed the collector would confine himself to the narrative portions of his fieldnotes: colourful day-to-day accounts of expedition travel interlaced with descriptions of scenery, and amusing or adventurous anecdotes, dangerous encounters with wild animals, equally dangerous encounters with stereotypically wild "Indians." And, in fact, Douglas's fieldnotes are *stuffed full* of that kind of narrative material—stories of the kind that a travel publisher like Murray and his reading public yearned for. But what remains of Douglas's unhappy, and incomplete, *A Sketch of a Journey* proves that Douglas had no intention of writing that popular, sensational travel memoir that Murray thought he had commissioned. In fact, those anecdotes that could have formed the basis for the book Murray wanted—a bear-shooting incident, an encounter with scary "Indians," and the like—are actually *removed* from Douglas's book attempt. In their place remain the sullen outlines of stories Douglas did not want to have to tell, some dry attempts at professional botanical classification, and an echoing homage to his scientific betters, Brown and Hooker: lions of a community into which he could never fully ascend.

Beyond the legacy of two chimeric mountain giants, David Douglas is best known for another taxonomic persistency: the "sugar pine" tree he found along the Columbia. Perhaps unsurprisingly, Douglas's "sugar pine" also resonates through history as a story of failed definition and mismeasurement: it's still known, again wrongly, as the "Douglas *Fir*."

Notes

We owe thanks to many who have assisted us in thinking through this essay, especially Pamela Banting, Frank Geddes, Sean Isaac, Conrad Janzen, Ian MacLaren, Peter Murphy, Jack Nesbitt, and Liza Piper. We are especially grateful for support from the Eleanor Luxton Historical Foundation of Banff.

1. An earlier version of this essay appeared in the *Canadian Alpine Journal* (hereafter *CAJ*) 94 (2011), 12–17.
2. Formed in 1804, the Horticultural Society (later, in 1861, the Royal Horticultural Society) gave institutional presence to the English landed gentry's enthusiasm for the scientific cultivation of their gardens. Founded, in part, by Sir Joseph Banks (1743–1820)—horticultural adviser to kings and long-time president of the Royal Society, the most influential scientific body in England—the Horticultural Society comprised a select group of men. As biographer William Morwood noted, "among the ninety-three new members admitted, ten were either peers or sons of peers. By 1809, when the membership totaled 576, including 'many of the most distinguished names in the kingdom,' the heady aura of nobility was such that the rank-in-file designation of 'member' was upgraded to 'Fellow,' an appellation still employed." William Morwood, *Traveler in a Vanished Landscape: The Life and Times of David Douglas, Botanical Explorer* (New York: Clarkson N. Potter, 1973), 12.
3. *Journal kept by David Douglas during his travels in North America, 1823–1827* (New York: Antiquarian Press, 1959 [1914]), 71.
4. David Robbins wrote that "mountaineering was invented by the British in the middle decades of the nineteenth century. Prior to this it is not possible to draw a distinction between mountaineering and other activities, science and tourism, of which it was an aspect. The earliest ascents of the highest peaks of the European Alps had been undertaken for the purposes of scientific research and cartography and it subsequently became fashionable for adventurous tourists to include an ascent of Mount Blanc or some other notable viewpoint in the itinerary of the European or Alpine tour. In the 1850s, however, the practice of visiting the Alps specifically to climb the peaks and

cross the passes was for the first time recognized by participants as a distinctive form of activity." David Robbins, "Sport, Hegemony and the Middle Class: The Victorian Mountaineers," *Theory, Culture, and Society* 4, no. 3 (1987), 583–84.

5   See Chris Jones, *Climbing in North America* (Seattle: Mountaineers, 1997 [1976]), 20; Andy Selters, *Ways to the Sky: A Historical Guide to North American Mountaineering* (Golden, CO: American Alpine Club, 2004), 7; Chic Scott, *Pushing the Limits: The Story of Canadian Mountaineering* (Calgary: Rocky Mountain Books, 2000), 33; Athelstan George Harvey, *Douglas of the Fir: A Biography of David Douglas, Botanist* (Cambridge, MA: Harvard University Press, 1948), 131, 246; Ben Gadd, *Handbook of the Canadian Rockies* (Jasper: Corax, 1992 [1986]), 751.

6   *Journal kept by David Douglas* (1959 [1914]), 72.

7   There are some fifty-plus peaks in the Canadian Rockies that exceed 11,000 feet (3,352 m) and only four that surpass the 12,000 foot mark (3,657 m). Of the latter, Mount Robson is the highest at 12,972 feet (3,954 m).

8   See James Monroe Thorington, *The Glittering Mountains of Canada* (Philadelphia, PA: John W. Lea, 1925), 300–302.

9   Writing in 1836, William Hooker summarized Douglas's 1827 reception in London as follows: "Qualified, as Mr Douglas undoubtedly was, for a traveler, and happy as he unquestionably found himself in surveying the wonders of nature in its grandest scale, in conciliating the friendship (a faculty he eminently possessed) of the untutored Indians, and in collecting the productions of the new countries he explored; it was quite otherwise with him during his stay in his native land. It was, no doubt, gratifying to be welcomed by his former acquaintances, after so perilous yet so successful a journey, and to be flattered and caressed by new ones; and this was perhaps the amount of his pleasures, which were succeeded by many, and, to his sensitive mind, grievous disappointments. Mr Booth remarks, in his letter to me on this subject, 'I may here observe, that his appearance one morning in the autumn of 1827, at the Horticultural Society's Garden, Turnham Green was hailed by no one with more delight than myself, who chanced to be among the first to welcome him on his arrival, as I was among the last to bid him adieu on his departure. His company was now courted, and unfortunately for his peace of mind he could not withstand the temptations (so natural to the human heart) of appearing as one of the Lions among the learned and scientific men in London.... Flattered by their attention, and by the notoriety of his botanical discoveries, which were exhibited at the meetings of the Horticultural Society, or published in the leading periodicals of the day, he seemed for a time as if he had reached the summit of his ambition. But alas; when the novelty of his situation had subsided, he began to perceive that he had been chasing a shadow instead of reality.' As some further compensation for his meritous services, the Council of the Horticultural Society agreed to grant him the profits which might accrue from the publication of the Journal of his Travels, in the preparation of which for the press, he was offered the assistance of Mr Sabine and Dr Lindley: and Mr Murray of Albemarle-street was consulted on the subject. But this proffered kindness was rejected by Mr Douglas, and he had thoughts of preparing the Journal entirely himself. He was, however, but little suited for the undertaking, and accordingly, although he laboured at it during the time he remained in England, we regret to say, he never completed it. His temper became more sensitive than ever, and himself restless and dissatisfied; so that his best friends could not but wish, as he himself did, that he were again occupied in the honourable task of exploring North-west America." William Hooker, ed., *Companion to the Botanical Magazine* 2 (London: Samuel Curtis, 1836), 142.

10 Sir William Jackson Hooker, *Flora Boreali-Americana; or, The Botany of the Northern Parts of British America*, vol. 1 (London: H.G. Bohn, 1829). *Flora Boreali-Americana* incorporated the botanical discoveries of Douglas; Thomas Drummond (1790–1835), who was recommended by Hooker as assistant naturalist to John Richardson

(1787–1865) on Captain John Franklin's second expedition to the Arctic; Richardson himself; and Capts. Frederick Beechey (1796–1856) and William Parry (1790–1855), who both explored the northern reaches of Hudson and Davies straits. On the map, Franklin, Richardson, and Drummond's route was marked in red (with a side trip of Drummond's in yellow), Parry and Beechey's in blue, and Douglas's in green.

11 See Harvey, *Douglas of the Fir*, 237, 251–53.
12 Writing to Richardson, who also collaborated on the map, Hooker commented as follows: "I only wish Douglas had left out his Mount Brown and Mount Hooker, which he has surely most egregiously overrated as to height." Letter, 22 Jan.1829, WJH/2/7 folio 76, William Hooker collection, Library and Archives, Royal Botanic Gardens, Kew.
13 Hooker's misgivings over Douglas's heights are further evidenced in the *Companion* version of *A Sketch of a Journey*. Douglas's line, "the height from its apparent base exceeds 6,000 feet, 17,000 feet above the level of the sea," was changed to "its height does not appear to be less than 16,000 or 17,000 feet above the level of the sea." See Hooker, *Companion*, 136.
14 In his book *The Island of Lost Maps*, Miles Harvey writes of how major corporations in the mapping industry attempt to manage the cartographic theft industry by including a fictional roadway on their residential street maps. The name they use for these fictional roadways is "trap streets": "We place the trap streets in areas that would be relatively harmless and would not mislead someone," a Rand McNally representative tells Harvey. "Just a cul-de-sac at the end of some development.... This allows us to do a quick spot check of our competitors' maps to see if they have stepped on our toes." Miles Harvey, *The Island of Lost Maps: A True Story of Cartographic Crime* (New York: Broadway Books, 2000), 140–41.
15 Daniel Kyba attributes the persistence of this spectacular mismeasure through much of the nineteenth century to the relative inaccessibility of Athabasca Pass. See Daniel Kyba, "Chasing the Giants," *Alberta History* 59, no. 1 (2011), 23. However, Thomas Richards, in *The Imperial Archive: Knowledge and the Fantasy of Empire* (London: Verso, 1993), attributes the persistence of data of this kind to the social work of the imperial archive. Knowledge assembled from far-flung outposts of empire, Richards argues, came to be organized within the imperial archive as a form of anxious symbolic management of actual imperial relations.
16 A.P. Coleman, *The Canadian Rockies, New and Old Trails* (Toronto: Henry Frowde, 1911), 79–80.
17 Arthur O. Wheeler, "Expedition to Mount Robson," *CAJ* 1, no. 2 (1908), 100–102.
18 Tolly Bradford, "A Useful Institution: William Twin, 'Indianness,' and Banff National Park, c. 1860–1940," *Native Studies Review* 16, no. 2 (2005), 78.
19 See Theodore (Ted) Binnema and Melanie Niemi, "'Let the Line Be Drawn Now': Wilderness, Conservation, and the Exclusion of Aboriginal People from Banff National Park in Canada," *Environmental History* 11, no. 4 (2006), 724–50.
20 A.P. Coleman. "Cree Words" (Morley, June 25, 1892) in *Notebook* 12 (1892), 23–25. http://library2.vicu.utoronto.ca/apcoleman/rockies/first_nations.htm. Accessed 29 June 2012.
21 James White, *Place-Names in the Rocky Mountains between the 49th Parallel and the Athabaska River: Transactions of the Royal Society of Canada, Section II* (Ottawa: Royal Society of Canada, 1916), 501–35.
22 Coleman, *Canadian Rockies*, 173–74, 203. With Brown and Hooker deposed, Coleman's consolation—at least for the rhetorical purposes of his book, *The Canadian Rockies, New and Old Trails* (1911)—became the so-called "discovery" of Fortress Lake. Nine years passed before his next visit to the Rockies.
23 Ibid., 203, 208.

24 The company's London agent, Alexander Begg (1825–1905), referred to them more precisely perhaps as "the better class of people such as tourists and others of that character." A. Begg to C.C. Chipman, 7 June 1886, file 1129/1886–1143/1886, no. 1233, vol. 42, Records of the Canadian Government Exhibition Commission, Library and Archives Canada.

25 See Stephen Slemon, "The Brotherhood of the Rope: Commodification and Contradiction in the 'Mountaineering Community,'" in *Renegotiating Community: Interdisciplinary Perspectives, Global Contexts*, ed. Diana Brydon and William D. Coleman (Vancouver: UBC Press, 2008), 236–37.

26 G. Winthrop Young, "John Norman Collie," *Alpine Journal* 54 (May 1943), 62.

27 See A.P. Coleman, "Mount Brown and the Sources of the Athabasca," *Geographical Journal* 5, no. 1 (1895), 53–61.

28 Norman Collie, "Exploration in the Canadian Rockies: A Search for Mount Hooker and Mount Brown," *Geographical Journal* 13, no. 4 (1899), 343.

29 Clements R. Markham, "The Present Standpoint of Geography," *Geographical Journal* 2 (1893), 481.

30 J.N. Collie, "The Canadian Rocky Mountains a Quarter Century Ago," *CAJ* 14 (1924), 83.

31 Hugh E.M. Stutfield and J. Norman Collie, *Climbs and Explorations in the Canadian Rockies* (London: Longmans, Green, 1903), 107.

32 Ibid., 121.

33 Peak Douglas was renamed "Mount Kitchener" in 1916. "Collie may not have been aware," writes the author of the website Peakfinder.com, "that Mount Douglas in the upper Red Deer Valley had been named after David Douglas." See http://www.peakfinder.com/peakfinder.ASP?PeakName=mount+kitchener. Accessed 29 June 2012. George Mercer Dawson (1849–1901) performed that act of naming in 1884, the year before he was named director of the Geographical Survey of Canada.

34 Collie, "Exploration in the Canadian Rockies," 351.

35 Ibid., 354.

36 Jerry Auld, *Hooker and Brown: A Novel* (Victoria: Brindle and Glass, 2009). Also see Kyba, "Chasing Giants"; Don Beers, *Jasper-Robson: A Taste of Heaven, Scenes, Tales, Trails* (Calgary: Highline, 1996), 196; and Peter J. Murphy with Robert W. Udell, Robert E. Stevenson, and Thomas W. Peterson, *A Hard Road to Travel: Land, Forests and People in the Upper Athabasca Region* (Durham, NC: Forest History Society, 2007), 151.

37 With a broken and therefore useless mercury tube, Aemilius Simpson reported the altitudes near Jasper "based upon his general feeling of how far he had climbed from York Factory" and thus likely began the hyperbole in 1826. He passed this and other geographical information along to Drummond, who had joined Simpson's party in the Upper Athabasca Valley. Drummond included Simpson's geographical data in his report: "The height of one of the mountains, taken from the commencement of the Portage, Lieut. Simpson reckons at 5,900 feet above its apparent base, and he thinks the altitude of the Rocky Mountains may be stated at about 16,000 feet above the level of the sea." Thomas Drummond, "Sketch of a Journey in the Rocky Mountains and to the Columbia River in North America," in William Hooker, ed., *Botanical Miscellany* (London: John Murray, 1830), 190. Douglas met Aemilius Simpson later that fall in Fort Vancouver. See *Journal kept by David Douglas* (1959 [1914]), 239. If the height speculation wasn't passed to Douglas then, it was surely received second-hand from Drummond himself, who sailed with Douglas from Hudson Bay to England in 1827. See Kyba, *Chasing Giants*, 18–19. And it was likely Hooker's association with Drummond—remember, Drummond, like Douglas, contributed to the *Flora Boreali-Americana* (1829) book map, as well—that further motivated Hooker to lower Douglas's initial elevation for Mount Brown from 17,000 feet to 16,000 feet on both

the 1829 book map and in his subsequent 1836 *Companion*. Confusing the matter of elevation was David Thompson's published narrative, first printed in 1916. A mapmaker and surveyor for the Northwest Company, Thompson (1770–1857) became the first non-Native to cross Athabasca Pass in 1811. Later, writing in 1840, when he was seventy years of age, Thompson made two claims concerning the height of Athabasca Pass. The first comes from the entry for 10 March 1809, when he states that "at the greatest elevation of the passage across the mountains by the Athabasca River, the point of boiling water gave 11,000 feet, and the peaks of the Mountains are full 7000 feet above this passage, and the general height may be fairly taken at 18,000 feet above the Pacific Ocean." *David Thompson's Narrative*, ed. J.B. Tyrell (Toronto: Champlain Society, 1916), 403. The second comes from Thompson's entry for 10 January 1811, the famous day he crossed the pass itself: "The altitude of this place above the level of the Ocean, by the point of boiling water is computed to be eleven thousand feet (Sir George Simpson)" (448). The editor, Joseph B. Tyrell (1858–1957) carefully compared Thompson's *Narrative* with his field diary in 1915 and found that the mapmaker's original notes did not contain any reference to the altitude of Athabasca Pass, nor did they contain any mention of an attempt to determine the height, for that matter. Thompson, writing in 1840, like Douglas in 1828, thus added the grand heights when preparing his fieldnotes for publication. But the bracketed note—"Sir George Simpson"—suggests that Thompson obtained the estimated altitudes from the governor. Did Thompson, in his old age, confuse Aemilius for George? It is difficult to say with any certainty. The connection between the exaggerated heights and Sir George Simpson has never been located in any register beyond David Thompson's bracketed note from 1840. Although the governor *did* cross Athabasca Pass on his return journey across the North American continent in the spring of 1825, no mention of altitude is found in his published journals. See Frederick Merk, ed., *Fur Trade and Empire: George Simpson's Journals* (Cambridge, MA: Harvard University Press, 1931), 144–45. Interestingly, though, Douglas did meet Sir George Simpson along the fur-trade route in June 1827, at Norway House, in present-day Manitoba. See *Journal kept by David Douglas* (1959 [1914]), 273.
38 See Thorington, *Glittering Mountains*, 165–82.
39 Thorington cites the estimations of Ross Cox and Thomas Drummond, but mistakenly attributes the hyperbole to an "incorrect boiling point determination" made by David Thompson (see n37). J. Monroe Thorington, "The Centenary of David Douglas' Ascent of Mount Brown," *CAJ* 16 (1928), 186.
40 E.W.D. Holway (1853–1923) first suggested the theory in "New Light on Mounts Brown and Hooker," *CAJ* 9 (1918), 47.
41 *Journal kept by David Douglas* (1959 [1914]), 259.
42 Ibid., 72.
43 Ibid., 259.
44 Ibid.
45 Ibid.
46 Thorington, "Centenary of David Douglas' Ascent," 188.
47 See Arthur O. Wheeler, "Mounts Brown and Hooker," *CAJ* 17 (1929), 66–68; J. Monroe Thorington, "Mounts Brown and Hooker: A Reply," *CAJ* 17 (1929), 69–70.
48 Bruce Fairley, ed., *The Canadian Mountaineering Anthology: 100 Years of Stories from the Edge* (Edmonton: Lone Pine, 1994), 173.
49 Selters, *Ways to the Sky*, 7.
50 Scott, *Pushing the Limits*, 34.
51 Jones, *Climbing in North America*, 19–20; original emphasis.
52 The bolt and cairn is located 89°25'46" at a height of 2,545m (8,347 ft). Its position was fixed by a survey team led by surveyor Richard W. Cautley (1873–1953), one of three commissioners appointed to map the boundary between British Columbia and

Alberta, in 1920. See R.W. Cautley and A.O. Wheeler, *Report of the Commission Appointed to Delimit the Boundary between the Provinces of British Columbia and Alberta*, Part II, *1917 to 1921, From Kickinghorse Pass to Yellowhead Pass* (Ottawa: Office of the Surveyor General, 1924), 151.

53 See Zac Robinson, "Storming the Heights: Canadian Frontier Nationalism and the Making of Manhood in the Conquest for Mount Robson, 1906–1913," *International Journal for the History for Sport* 22, no. 3 (2005), 415–33.

54 See Bradford Washington, *The Dishonorable Doctor Cook: Debunking the Notorious McKinley Hoax* (Seattle: Mountaineers, 2001).

55 Thorington, "The Centenary of David Douglas' Ascent," 192.

CHAPTER 10

Escarpments, Agriculture, and the Historical Experience of Certainty in Manitoba and Ontario

Shannon Stunden Bower & Sean Gouglas

In nineteenth-century Ontario, farmer E.D. Smith looked down toward Lake Ontario from his property above the Niagara Escarpment in Saltfleet Township. Smith was dedicated to agriculture, believing that careful crop management and attention to markets would surely lead to farming success. Over time, Smith came to understand that moving down below the escarpment, to the fertile lands that benefited from the moderating effects of the lake, would allow him to grow more soft fruit with less risk of frost.

In twentieth-century Manitoba, farmer L.C. Wilkin gazed up at the Manitoba Escarpment, toward what he perceived to be the origin of his area's water problems. The Manitoba lowlands where he farmed had long suffered from surface water flooding. Those living in the lowlands believed land clearing and agricultural intensification above the escarpment exacerbated the situation, as these processes changed the rate and timing of runoff.

Though separated by significant differences of time and place, the stories of Smith and Wilkin shared an important common element: the definitive presence of an escarpment. The term "escarpment" refers to a steep slope of considerable length that dominates a section of landscape.[1] The Niagara Escarpment extends northwestward from south of Rochester, New York, in the United States. It crosses into Ontario at Queenston on the Niagara Peninsula and runs to Tobermory at the tip of the Bruce Peninsula. It is a major physiographical feature of southern Ontario, dividing south-central Ontario and peninsular Ontario to the southwest. At Hamilton, in the area where E.D. Smith farmed, the average elevation is about 150 metres above sea level. The Manitoba Escarpment (known in the United States as the Pembina

Escarpment) runs northwestward from South Dakota, cuts diagonally across southwestern Manitoba, and crosses into Saskatchewan near the town of Kamsack. Dividing the Manitoba lowlands from the southwest uplands, the escarpment varies in height above the lowlands from about 300 metres to about 600 metres. The high points are known as the Porcupine, Duck, and Riding mountains. Similarly, the portion of the Niagara Escarpment that extends through Hamilton is known locally as "the Mountain." Such language, whether colloquial phrasing or accepted terminology, might seem to exaggerate the height of these topographical elements, but it does reflect the landforms' local significance in what are otherwise relatively flat regions.

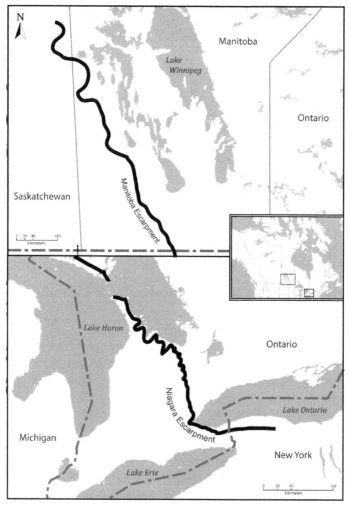

Figure 10.1 Map of Manitoba and Ontario scarp landscapes. Created by the authors.

The focus of this chapter is on the question of environmental knowledge within the scarp landscapes of Manitoba and Ontario. Within environmental history, Richard White was among the first to focus on knowledge itself. In *The Organic Machine* and also in "'Are You an Environmentalist or Do You Work for a Living?': Work and Nature," White explores the social divisions that ensued at least in part from differences in the predominant ways individuals and groups encountered nature.[2] The key divide in White's work is between knowing nature through work or through leisure. Many scholars have picked up on White's analysis, further exploring this divide as well as offering alternative interpretations.[3] A smaller number of scholars focus not on how nature is known, nor on emergent social divides, but instead on the quality of the knowledge itself. In the United States, scholars such as Conevery Bolton Valenčius and Alan Taylor offer extended explorations of how specific people in the past thought about their environments, with attention to how these people understood the knowledge they had.[4] Was their knowledge complete? Reliable? Flawed in some way? As Valenčius puts it, the goal for these scholars was to "describe and work within" the understandings of their subjects.[5] Canadian historian Joy Parr's work is inspired by scholarship from outside the field of history, but in its focus on historical questions of knowledge in specific landscapes, it shares some elements with the contributions of Valenčius and Taylor.[6] Also relevant is the work of Liza Piper, which probes the ways individuals and groups have understood their environments and activities while involved in the industrialization of subarctic Canada.[7]

In their works on environmental knowledge, both Taylor and Valenčius focus on themes of fear and dislocation. They are concerned with environmental knowledge in the context of changed, changing, or unfamiliar environments, in which anxiety and uncertainty come to the fore. Recently, scholars who address environmental knowledge possessed by people at risk have picked up these themes. Indeed, in a 2004 special issue of the annual journal *Osiris* focused on exposure to environmental hazards, editors Gregg Mitman, Christopher Sellers, and Michelle Murphy identify uncertainty as a key theme running throughout the included articles.[8] The editors of the special issue describe uncertainty as "a historical artifact," something "produced by particular ways of apprehending the world or by clashes between different versions of the world."[9]

Important recent scholarship focuses on uncertainty as a historical artifact, a function of human experiences; in contrast, this chapter focuses on certainty as a historical product—a result of both human and environmental factors. There were many differences between agriculture in the scarp region of Manitoba and the scarp region of Ontario. As will become clear throughout the chapter, these are places that were settled at different times, that had distinct climates supporting distinct crop regimes, and that faced disparate environmental challenges. Shared between the agricultural populations of the two regions in the periods considered, however, was a profound confidence in local environmental knowledge. What were the factors that contributed to this common experience of certainty?

Importantly, this chapter examines the production of certainty as experienced by historical actors, not the production of accurate knowledge. Indeed, some of the environmental knowledge considered here diverges from current understandings of environmental processes. But whether or not certainty amounted to accurate knowledge, the historical experience of certainty, like the historical experience of uncertainty, affected how people engaged with the world and the sorts of choices they made. Questions of certainty and uncertainty are also relevant in the present day. Unwillingness to accept the existence of anthropogenic climate change, which persists in some quarters despite widespread scientific consensus, is an example of a contemporary issue that can be better understood through examination of the various factors that contribute to the construction of certainty or uncertainty.

Farmers in both the Manitoba and Ontario scarp regions were obliged to come to terms with the agricultural significance of the escarpments. These large-scale landscape features bore on factors such as climatic conditions and water-flow patterns throughout the regions they influenced, with important consequences for agricultural production. For environmental historians, scarp areas offer an opportunity to examine fairly large areas affected by one key environmental factor. Because of this, they are particularly good landscapes in which to assess the significance of environmental conditions in the production of environmental knowledge. Scarp regions provide excellent opportunities to explore the possibility that certainty was not so much an artifact (a purely human creation) as a product of both human and environmental factors.

The comparisons in this chapter span time as well as space. The Ontario story takes place primarily in the nineteenth century, while the Manitoba tale plays out mainly in the early twentieth century. This different periodization does, however, address early settler agriculture in both provinces, as Manitoba and Ontario were settled by agriculturally oriented newcomers at different points in time. There are two parts to this chapter. In the first, we examine how residents perceived and responded to differences between areas above and areas below the two escarpments. Significantly, the environments in question appeared relatively stable, changing in limited and predictable ways in the periods studied here. This allowed settlers to gain confidence in their understandings of the landscapes they inhabited. In the second part, we examine the convictions of those who lived near these escarpments that they possessed particularly accurate and important information about the local landscape. Ultimately, this chapter suggests that certainty as experienced by historical actors can result from long-term shared experiences of environments that change in familiar and predictable ways.

Relatively Stable Landscapes

The Niagara and Manitoba scarp regions were subject to various environmental pressures. In both landscapes, both human and non-human factors operated to modify aspects of the landscapes. Historically, as well as in the present day, processes under way along one or both escarpments included erosion, gravel mining, soil exhaustion, and changes to tree cover. Still, during the periods studied here, the changes in both Manitoba and Ontario scarp landscapes were relatively small in scale. These changes nested within a larger context of apparent stability, one that allowed local residents to gain confidence in their understanding of the local environment.

Both the Niagara and the Manitoba escarpments were dramatic breaks in otherwise comparatively flat terrain. These changes of elevation complicated travel to all who had to cross them and posed a hazard to those who lived near them. For instance, in 1873, Saltfleet Township resident Abram Lee reported that a horse had been lost over what he called "the mountain," a heavy loss for a mid-nineteenth-century farmer.[10] Yet it was not so much the risks or difficulties associated with crossing the escarpments as the different agricultural conditions above and below these landforms that bore on the experiences of those who lived nearby. In both Manitoba and Ontario, the particular challenges to successful agriculture related to land position above or below the escarpment.

Surveyed and opened for settlement in 1792, Saltfleet Township (part of Wentworth County) straddled the Niagara Escarpment immediately southwest of the western edge of Lake Ontario. In this region, the lands below the Niagara Escarpment received more sun and more rain than the lands above. Also, the escarpment concentrated the warming effect of Lake Ontario in the area between the escarpment and the lake. As a consequence, those who lived below the escarpment enjoyed a moderated climate. According to thirty-year climate normals published by Environment Canada for two locations adjacent to Saltfleet Township, one above and one below the escarpment, the areas below experienced 2,345.9 growing degree-days as opposed to 2,092.1 for lands above. Also significant were winter temperature extremes, with the average minimum below the escarpment at $-27.2°$ C and the average minimum above at $-29.4°$ C. In the Saltfleet area, people who lived geographically proximate occupied climatically distinct areas, depending on where they were situated in relation to the Niagara Escarpment.[11]

The relatively small average temperature differences above and below the Niagara Escarpment gained significance in relation to the particular growing conditions required by fruit trees. While hardy apples were grown throughout the areas both above and below the escarpment, more delicate (and for farmers, more profitable) fruit such as grapes, melons, and pears were grown primarily below.[12] For fruit farmers, how fruit grows raised the stakes. Soft fruit trees required three years to mature before they began to produce significant quantities of fruit. A winter-killed fruit tree meant a reduced harvest not for one year, as with an annual crop such as wheat, but for a number of years. Even the temperatures below the escarpment had the potential to kill fruit trees, but the odds for fruit farmers were better in the warmer area. Farming was always a gamble. But with respect to fruit, a crop for which the stakes were particularly high due to growth patterns, the odds differed significantly between the regions above and below the escarpment.

The early settlers of Saltfleet Township recognized that better agricultural opportunities were available on the lands below the escarpment. As a result, the lower portions of the township filled up more quickly. Initially, farms below the escarpment were considerably larger than farms above. But as specialization in expensive fruit meant fewer acres were necessary for a profitable farm, and in response to growing demand for land due to provincewide population increase as well as in response to declining soil fertility due to overfarming and extensive clearing, the situation began to change. By 1861, the average farm below the escarpment was only twenty-six acres larger than the average farm above. By 1871, the size difference was minimal. By 1890, the average farm below the escarpment was considerably smaller than the average farm above.[13]

This was a complete reversal from the area's early days. Farms below the escarpment started bigger and dropped in size more rapidly. Smaller farms below the escarpment could still support decent livelihoods as long as lucrative crops such as soft fruit were grown. Farms above the escarpment started smaller and declined more gradually, as larger acreages were necessary to support profitable farms in this area. The distinct trajectories of farm size above and below what the locals called "the Mountain" illustrate how, even within a single township, the escarpment played a major role in defining farmers' choices. Because these areas retained their distinctive environmental conditions over time, those living above and those living below the escarpment had significantly different opportunities and constraints. Agricultural trajectories in Saltfleet Township diverged substantially between the areas above and below the Niagara Escarpment; in the context of a relatively stable environment, these divergences became more pronounced as years passed.

The Manitoba Escarpment marks the boundary between the flood-prone Manitoba lowlands and the higher, drier southwest uplands. Those living above the escarpment and those living below experienced a different set of surface water conditions. Farmers above the escarpment enjoyed relatively well-drained soils and generally adequate precipitation, together adding up to an environment that, at least once the land was cleared and broken, proved conducive to agricultural success. In the Red River Valley, impermeable soils and flat land meant that surface water was slow to drain away. Following the spring melt or any heavy precipitation, it sat on farmers' fields and interfered with agriculture. While this surface water enriched the local soil, creating a region in which bumper crops were possible, it could also drown out a year's harvest. Flooding posed a significant barrier to agricultural success throughout much of the lowlands of the Red River Valley.

The escarpment had an effect on climate, with its highest reaches having somewhat distinct climate patterns.[14] But Manitoba's continental climate, with temperatures very low in winter and very high in summer, was not moderated in a way that had notable consequences for agricultural production. There was no particularly significant overlap between the climatic influence of the escarpment and the growing conditions of a particular crop, as there was with the Niagara Escarpment. Notably, the climatic differences associated with the Manitoba Escarpment were expressed partly through greater rates of precipitation in higher areas, which contributed to the runoff that exacerbated flooding in areas down below.[15]

In a bid to improve the surface water situation, an extensive network of surface drains was constructed through the Manitoba lowlands, as well as through the province's lakeshore areas that were also subject to flooding. A system of drainage districts was created, under which the province would pay construction costs up front and recoup the expense through levies paid over a period of decades by drainage district residents. The most intensive period of drainage was between 1896 and 1914, with some twenty-one drainage districts created in this period. Three more districts were created by the late 1920s. Districts ranged greatly in size, with the largest encompassing nearly 450,000 acres and the smallest including about 4,800.[16]

Size differences aside, residents' persistent dissatisfaction united drainage districts. Faulty construction, inadequate maintenance, and unreasonable expectations combined to ensure that even as lowland surface water patterns were changed, drainage problems were not solved. Indeed, many lowland residents felt flooding was becoming more frequent and more severe. Some blamed this on land change above the escarpment, arguing that clearing and cultivating in higher areas caused more water to run off more quickly. In their view, such altered flow patterns were overwhelming the lowland drainage systems that had been constructed to manage the water problems that existed prior to land use changes. Lowland residents began arguing that highland residents should bear some financial responsibility for the costs of expanding drainage systems to cope with the new situation. They went so far as to band together in a lobby organization to pressure the provincial government to take action to protect their lands. At a relatively small scale, the flows of surface water meant the situation in lowland Manitoba was constantly in flux. But at a higher scale, at least one thing was stable: the perception among those who lived in the lowlands that flooding was due to actions on the highlands.

If in Ontario those above and those below the escarpment occupied distinct agricultural trajectories, in Manitoba those below perceived their agricultural ambitions to be on a collision course with the ambitions of those who lived above.

In both scarp regions considered here, the escarpments amounted to key environmental factors for those who lived nearby. These large-scale landscape features were not, strictly speaking, unchanging. They were subject to a variety of environmental processes, some anthropogenic and some non-anthropogenic. Still, considering the effect on local agricultural conditions, particularly with respect to temperature in Ontario and particularly with respect to water in Manitoba, the escarpments affected significant areas in similar ways over long periods of time. From this perspective, these escarpments were relatively stable landscapes. Such apparent stability was a key ingredient in the production of what agricultural residents would experience as certain knowledge.

Long-Term and Shared Experiences

Both the Niagara Escarpment and the Manitoba Escarpment were large-scale landscape features, representing dramatic changes in elevation and extending over significant distances. The influence of these landforms also seemed to stretch through time, affecting their regional environments in similar ways over lengthy periods. This section will illustrate how, in the context of apparent stability, the knowledge of local residents was consolidated over time spent in the area and through experiences shared with others, eventually becoming what locals perceived as certainty. While both factors were in play in both regions, the Ontario example highlights the importance of the passage of time, while the Manitoba example underlines the role of like-minded neighbours.

In Ontario, close examination of two generations of the Smith family provides a window on experiences in the scarp region. Damaris and Conrad Smith settled 170 acres atop the Niagara Escarpment in 1853. Though the land they purchased had some improvements already, they set about investing labour and capital in their farm by clearing trees, uprooting stumps, and draining swampy lands.[17] Together, they established a traditional mixed-wheat farm complemented with fruit and vegetable crops as time and space allowed.

During this period, both Conrad and Damaris were learning of the particular possibilities and constraints of the region they farmed. Damaris Smith perceived her husband's careful observation of his farm's qualities and recorded her thoughts in a written remembrance. She saw in observations Conrad made such as "this is a great field for barley" or "what a crop of wheat I can get off that field" evidence that "he understood his business and to a great extent the nature of the soil."[18] As his wife's comments suggest, Smith was studying the land, making careful judgments, and thereby acquiring a store of environmental knowledge. As Conrad fine-tuned his grain production, Damaris continued to experiment with fruits and vegetables. Initially, planting currants in the garden may not have made "much of an appearance," as the only evidence early on was "a little stick about an inch or so above the ground," but eventually the berries proved so prolific that they crowded out the vegetable garden. In the fourth year, Damaris sold some nine dollars' worth of currants. Encouraged, she expanded her efforts and enjoyed further successes like a crop of hubbard squash worth thirty dollars.[19] Both Conrad and Damaris invested intellectual as well as physical energy in their farm, thinking critically about how to improve their situation through more effective farming practices and acquiring knowledge about the environmental conditions in the region they inhabited.

Over time, as soil exhaustion and agricultural pests cut into profits, families such as the Smiths relied even more on their acquired local knowledge. These difficult conditions forced the Smiths and their fellow farmers to think hard about how best to maximize yields, and many entertained ideas of crop diversification or specialization according to their farm's particular environmental characteristics.[20] Farmers who personally or through family connections had a history of farming at or near a particular location had a distinct advantage through this process. In the Saltfleet region, greater time spent on the land was associated with greater agricultural success.[21] Presumably, this was because long-time farmers had acquired a store of environmental information about their land, and they understood its particular strengths and limitations far more fully than did most newcomers.

Ernest D'Israeli (E.D.) Smith, son of Conrad and Damaris, took over the family farm in 1874. He began immediately, as he put it in his diary, to look for some "more remunerative way of farming."[22] If he had acquired his father's penchant for critical farming, he had also inherited his mother's appreciation for the agricultural potential of fruit. The younger Smith's plans included an apple orchard and a vineyard, and he "commenced at once by setting out 100 grape vines."[23] By the late 1870s, he had expanded further into fruit, believing that a diversified crop would provide greater stability of income.[24] Beyond achieving greater stability, Smith was also eager to take advantage of what he perceived as an emerging economic opportunity. By the early 1880s, he was poised to capitalize on what he saw at one point as "an almost unlimited market"[25] for fruit. Smith laid out a three-year expansion plan that included strawberries, grapes, quinces, gooseberries, raspberries, plums, and peaches.[26]

Climatic conditions above the escarpment proved a barrier to Smith's efforts to focus on fruit. In 1881, weeks of extreme cold beyond the −23°C tolerated by peach trees led Smith to suspect his entire orchard had been damaged.[27] The early frosts of 1883 devastated Smith's grapes, leaving him with nothing to do but to pick the ruined crop and send the grapes to market "to sell for what they will."[28] Smith recognized that to succeed as a fruit grower he would have to move to a more suitable location. In the context of the relatively stable scarp landscape, Smith knew where to look for what he wanted. In 1889, he purchased an eighty-acre farm below the Niagara Escarpment. He described these lands, after sufficient under-draining removed the cold springs and wet bottoms, as "the very best soil for many purposes."[29] It was at this location that Smith established his highly successful fruit growing and nursery endeavours, eventually expanding into canning. He enjoyed local and then commercial success: in 2010, E.D. Smith Foods Ltd celebrated 125 years of jam making.

If, through generations of a remarkable family, E.D. Smith reflects the construction of reliable knowledge of possibilities and constraints within the scarp environment, the development of a local agricultural society in which Smith participated reflects how connections with like-minded farmers helped consolidate this knowledge. The emergence of the Winona and Stoney Creek Grape Growers Club by 1878, which attracted many growers living in nearby Saltfleet Township, signalled general interest in networking with other farmers.[30] Participants shared information about how to grow good crops in local conditions, and how to cope with problems they all faced, like the pests and diseases to which fruit was subject. The sort of expertise farmers acquired through time spent in this relatively stable region was consolidated through engagement with other farmers encountering similar conditions.

By the time E.D. Smith was finding success on his new farm, he had established himself as something of a practical expert in the fruit growing community. By July 1884, Smith was contributing a regular horticultural column to the *Canadian Stock Raisers' Journal*. Smith's columns represented an intriguing combination of the general and the specific: the best general farming advice, he argued in various ways at different points in time, was to pay attention to the specific conditions of the land in question. "To make farming remunerative," Smith made clear in a typical assertion, farmers must employ those techniques and crops "which are best adapted to our country and locality."[31] Through his time in the Niagara scarp region, Smith participated in the construction of certain knowledge within the local area, and he recommended that those in other areas seek to establish such knowledge for themselves. Through both his actions and his words, Smith argued this was the way to farming success.

By the early twentieth century, in Manitoba, those who lived in lowland areas were confronting their own barriers to agricultural progress. Frequent flooding meant many in the lowlands were profoundly dissatisfied with surface water management. Their public protests led the province to commission a government investigation of flooding in the region. The Manitoba Drainage Commission was formed by an Order-in-Council of 17 January 1919.[32] J.G.

Sullivan, a civil engineer much experienced with railroads, was appointed chair. The Sullivan Report, returned in 1921, included the recommendation that the costs of drainage be spread among all those in the involved watershed. Manitoba's watersheds stretched from the lowlands into the highlands above the escarpment, encompassing both those in dire need of drainage and those who perceived themselves to have no involvement whatsoever in surface water problems down below. Previously, drainage costs had been borne solely by those in the lowlands who were included in drainage districts. The province hosted a series of public hearings on the report, at which representatives of the involved areas were invited to speak. Lowland residents were largely in favour of the report, finding in it expert confirmation that drainage to date had been inadequate. They were glad to see the recommendation that the highlands should help finance lowland drainage, convinced as they were that land change up above the escarpment had worsened surface water conditions down below.

Manitobans who lived in the highlands saw things differently. According to their spokespeople before the public hearings on the Sullivan Commission's findings, the report was profoundly unfair. Many saw a rough equivalence between the land clearing necessary on the highlands and the land drainage necessary in the lowlands.[33] As they had received no assistance with clearing, why should they share the burden of drainage? Highlanders' conviction was at least in part a question of self-interest, as they feared the financial consequences of involvement in drainage. As J.G. Sullivan put it, people concerned with their own livelihoods "are so apt to see [their] own quarter section and nothing else."[34]

Highlanders before the commission, however, chose to argue over the issue of correctness at least as much as over the question of fairness. In their view, the Sullivan Report was simply wrong, and through their testimony, they offered instead what they believed to be their own certain knowledge of the lands they inhabited. Reeve D.F. Stewart of Thompson Municipality argued that "any farmer who knows anything about farming knows that cultivated land will not run off water faster than wild land."[35] Representative Guy from the Municipality of Lorne argued that his municipality "does not turn down the water any faster than it did years before."[36] Reeve Arbez from the Municipality of Gray argued that many people in his municipality felt they sent no more water than previously, and in many instances "even much less."[37] Others elaborated further, complaining that the uplands currently suffered from water scarcity.[38] How, then, could runoff from these areas be contributing to flood problems lower down? Individual highland residents presented their case with confidence, and found further confirmation for their views in the testimony of others. For highlanders, the public hearings on the Sullivan Report provided an opportunity for confirmation and display of certain knowledge.

Given the propensity of highland representatives to argue from personal experience, time spent in the region became a powerful justification for their views. For instance, Reeve D.F. Stewart of Thompson Municipality included in his testimony the information that he had lived in the highlands for thirty-seven years, including twenty-two years of direct involvement in work on water flow.[39] South Norfolk Reeve J.R. Scott based his assertions in part on the opinions of the municipality's old settlers, those who had lived in the area for twenty-five years.[40] In the context of a relatively stable landscape, providing details regarding the amount of time spent in the region became a key means of signifying certain knowledge. In the minds of highland residents at least, having experienced local conditions over a long period offered access to a type of environmental knowledge that should be respected, even privileged.

Highlanders' confidence in their knowledge is illustrated in their willingness to confront the Sullivan Commission's experts. In a typical statement, a representative of Thompson and Stanley municipalities explained that, in his districts, "bridges over creeks have been cut down, and the culverts have been eliminated, and most important of all, there is an actual shortage of water in some districts," and he presented these assertions as evidence "to disprove the theory" advocated by the Sullivan Report.[41] As highland representatives saw it, the commission report parroted a general principle without reconciling it to the local environment. Similarly, George Armstrong of Pembina Municipality argued that the report should have produced "actual evidence," rather than offering the "general theory that because we have removed scrub and built road ditches" the uplands had changed water-flow patterns in ways detrimental to the lowlands.[42] In the concise summary of Reeve J.R. Scott of South Norfolk, highland residents were convinced "the principle is wrong."[43] Implicit in the testimony of highlanders was the argument that local knowledge should trump the interpretation of those who had broader expertise but less local experience. In time spent in the region and like-minded neighbours, highlanders found certain knowledge to line up against the qualifications of outside experts.

In this section, we examined how in the Ontario example, local knowledge was perceived by many as necessary to agricultural success, while in the Manitoba case, local knowledge was perceived by some as superior to external expertise. Similar in both Manitoba and Ontario scarp landscapes was the confidence residents evinced in their own perspective, confidence rooted in shared long-term experiences of the environmental processes playing out in the areas they inhabited. Significantly, certainty in local knowledge amounted to scepticism about broad principles or expert authority. This was confidence not only in the correctness of local knowledge, but also in its importance. Indeed, long-term residents amounted to experts in the local landscape, in a generally recognized way with regard to E.D. Smith and in their own minds at least with respect to Manitoba's highlanders. In the relatively stable landscapes of the Manitoba and Niagara escarpments, it was the passage of time and engagement with neighbours that produced certainty.

Conclusion

In the scarp landscapes of Manitoba and Ontario, settlers felt certain they knew their lands and their regions. This chapter examines the reasons for this. Three key factors—a relatively stable landscape, long-term residency, and shared experiences with like-minded neighbours—are linked to confidence in environmental knowledge. Notably, these findings confirm the inverse of the findings of some scholars concerned with the historical experience of uncertainty. In the work of Taylor and Valenčius, uncertainty is related to unfamiliarity with a landscape, often sparsely inhabited, to which settlers are relatively recently arrived. As changeable or unfamiliar environments contributed to uncertainty in the works of Valenčius and Taylor, so it seems the combination of long-term experience and like-minded neighbours in the context of a relatively stable landscape contributed to what locals felt was reliable knowledge in Manitoba and Ontario scarp landscapes. If Valenčius and Taylor studied the experiences of those recently arrived in new locations, this chapter studied the processes of coming to know particular locations. These processes are more complicated than the simple plowing of fields and erecting of dwellings that accompany early agricultural settlement. This chapter has attempted to consider the conditions, both those in non-human nature and those inherent to the human experience, that empower locals to espouse certain knowledge about a place.

Emphasizing the landscape distinguishes this study from some recent scholarship on uncertainty. Among scholars concerned with exposure to toxic chemicals or otherwise hazardous environmental conditions, uncertainty has been described as "a historical artifact," a product of a way of thinking or of clashes between different ways of thinking. Studying the relatively stable landscapes of the escarpments emphasizes how, in the two instances here under study at least, certainty seems less an artifact, a purely human creation, than a product of a number of factors, both human and environmental. In the context of relatively stable environments, it seemed possible to abstract environmental experiences beyond the present moment. Knowledge accumulated from past experience could be applied to the future. The historical experience of certain environmental knowledge is produced in relation to a particular type of environment, one that is undergoing a period of relative stability at the scales easily perceived by unaugmented human senses.

Interestingly, there are contemporary efforts under way to protect from change the escarpments considered here. Under the World Network of

Biosphere Reserves program of the United Nations Educational, Scientific and Cultural Organization (UNESCO), the Niagara Escarpment and a portion of the Manitoba Escarpment (the Riding Mountain area) have been recognized as biosphere reserves. This recognition is intended to protect these areas from rapid or drastic environmental change. This chapter illustrates an historical connection between the apparent stability of the landscape and the sort of knowledge people believe themselves to have about it. Understanding this connection might lead to a fuller appreciation of what is at stake in land management decisions that bear on how, and particularly on how quickly, landscapes are changed. Endeavours such as the Biosphere Reserve program, then, are at least potentially as much about people, and about people's knowledge, as they are about places.

Paying attention to the degree of confidence historical actors had in their environmental knowledge has the potential to enrich our understanding of the lived experiences of historical individuals and groups, and to enhance our capacity to understand their decisions and behaviours. Paying attention to what makes for certain or uncertain knowledge also has the potential to help us to better understand our own world. Given the persistence of uncertainty over anthropogenic climate change even in the face of overwhelming scientific consensus, the myriad ways uncertainty contributes to inaction on the part of both individuals and nation states, and the reality that inaction threatens our survival as a species, interrogating the conditions that produce the experiences of certain or uncertain knowledge is a task not just of historical interest, but also of contemporary importance.

Notes

1. Andrew Goudie, *Encyclopedia of Geomorphology*, vol. 1, *A to I* (London: Taylor & Francis, 2003), http://lib.myilibrary.com.login.ezproxy.library.ualberta.ca/Browse/open.asp?ID=46293&loc=340.
2. Richard White, *The Organic Machine: The Remaking of the Columbia River* (New York: Hill & Wang, 1995); Richard White, "'Are You an Environmentalist or Do You Work for a Living?': Work and Nature," in *Uncommon Ground: Rethinking the Human Place in Nature*, ed. William Cronon (New York: W.W. Norton, 1996), 171–85.
3. *The Organic Machine* and "'Are You an Environmentalist or Do You Work for a Living?'" are both widely cited. Notable examples of scholarship that relates to White's include Mark Fiege, *Irrigated Eden: The Making of an Agricultural Landscape in the American West* (Seattle: University of Washington Press, 1999); Linda Nash, "The Changing Experience of Nature: Historical Encounters with a Northwest River," *Journal of American History* 86, no. 4 (2000), 1600–29; Alison Calder, "Why Shoot the Gopher? Reading the Politics of a Prairie Icon," *American Review of Canadian Studies* 33, no. 3 (2003), 391–414; M.W. Klingle, "Spaces of Consumption in Environmental History," *History and Theory* 42 (2003), 94–110; Maureen G. Reed, "Uneven Environmental Management: A Canadian Comparative Political Ecology," *Environment and Planning A* 39 (2007), 320–38.

10 ■ Escarpments, Agriculture, and the Historical Experience of Certainty in Manitoba and Ontario 173

4 Conevery Bolton Valenčius, *The Health of the Country: How Americans Understood Themselves and Their Land* (New York: Basic Books, 2002); Alan Taylor, "'Wasty Ways': Stories of American Settlement," *Environmental History* 3, no. 3 (1998), 291–310.
5 Valenčius, *Health of the Country*, 263.
6 Joy Parr, *Sensing Changes: Technologies, Environments, and the Everyday, 1953–2003* (Vancouver: UBC Press, 2010).
7 Liza Piper, "Subterranean Bodies: Mining the Large Lakes of North-west Canada, 1921–1960," *Environment and History* 13, no. 2 (2007), 155–86; Liza Piper, *The Industrial Transformation of Subarctic Canada* (Vancouver: UBC Press, 2009).
8 Gregg Mitman, Michelle Murphy, and Christopher Sellers, "Introduction: A Cloud Over History," *Osiris* 19 (2004), 13.
9 Ibid., 13–14.
10 Abram Lee, diary, 9 Jan. 1873, Erland Lee Museum, Stoney Creek, Ontario.
11 *Canadian Climate Normals—Temperature and Precipitation, 1951–1980* (Ottawa: Environment Canada, 1981). This climate data pertain to a later period from that considered in this chapter. In the absence of more appropriate information, however, it might be used to suggest the relationship between the areas occupied by the two weather stations.
12 *Census of the Canadas 1851–52, Census of the Canadas 1860–61,* and *Census of Canada 1870–71*, Library and Archives Canada.
13 Ibid., and *Assessment Roll for Saltfleet Township, Wentworth County, 1890*.
14 Danny Blair, "The Climate of Manitoba," in *The Geography of Manitoba: Its Land and Its People*, ed. John Welsted, John Everitt, and Christoph Stadel (Winnipeg: University of Manitoba Press, 1997), 33.
15 W.J. Carlyle, "The Management of Environmental Problems on the Manitoba Escarpment," *Canadian Geographer* 24, no. 3 (1980), 255–69.
16 Shannon Stunden Bower, *Wet Prairie: People, Land, and Water in Agricultural Manitoba* (Vancouver: UBC Press, 2011), 78.
17 E.D. Smith, diaries, 1855, E.D. Smith Company Archives, Stoney Creek, Ontario, 80.
18 Damaris Smith, "Pioneer Wife," E.D. Smith Company Archives, 3.
19 Ibid.
20 Ibid., 10.
21 Sean Gouglas, "A Currant Affair: E.D. Smith and Agricultural Change in Nineteenth-Century Saltfleet Township, Ontario," *Agricultural History* 75, no. 4 (2001), 438–66.
22 E.D. Smith, diaries, 1874.
23 Ibid.
24 Ibid., 5 May 1878.
25 Ibid., 27 Nov. 1880.
26 Ibid., 20 Nov. 1881.
27 Ibid., 6 Feb. 1881.
28 Ibid., 5 Oct. 1883.
29 *Canadian Live-Stock and Farm Journal* 5, no. 3 (1881), 85.
30 *Canadian Stock Raisers' Journal* 1, no. 6 (1884), 110.
31 E.D. Smith, diaries, 1874.
32 Order-in-Council # 30724, Report of a Committee of the Executive Council, 17 Jan. 1919, GR 1530, Archives of Manitoba (hereafter AM).
33 Stunden Bower, *Wet Prairie*, 96.
34 Testimony of J.G. Sullivan, Drainage Committee Report, GR 174, G 8373, file 13, AM, 158.
35 Testimony of Reeve D.F. Stewart, Representative of the Rural Municipality of Thompson, ibid., 33–44.

36 Representative Guy, Representative of the Rural Municipality of Lorne, ibid., 45–46.
37 Representative Arbez, Representative of the Rural Municipality of Lorne, ibid., 51.
38 Drainage Committee Report, ibid., 29, 37.
39 Testimony of Reeve D.F. Stewart, ibid., 33–35.
40 Testimony of Reeve J.R. Scott, Representative of the Rural Municipality of South Norfolk, ibid., 56; see also 133, for further examples.
41 Testimony of Black, Representative of Thompson and Stanley Municipalities, ibid., 30.
42 Testimony of George Armstrong, Representative of Pembina Municipality, ibid., 9.
43 Testimony of Reeve J.R. Scott, ibid., 56.

CHAPTER 11

Whatever Else Climate Change Is Freedom

Frontier Mythologies, the Carbon Imaginary, and British Columbia Coastal Forestry Novels

Richard Pickard

Representatives of the British Columbia forest industry have had little to say about climate change. To date, they have promoted the idea that the forest sector can play an important role in the carbon economy with intensively managed forests as carbon sinks, wood-frame construction as long-term carbon storage, and carbon credits for using wood in place of other materials in the manufacturing and construction industries. One industry group has gone so far as to aim for "industry-wide carbon neutrality by 2015, without the purchase of carbon offset credits."[1] In BC specifically, the companies have not contested the idea that the mountain pine beetle infestation has been caused by climate change, even exacerbated by logging-generated even-age stands of trees and by nearly a century of fire abatement, and there is no one on record saying that climate change is not anthropogenic.[2] While this all sounds positive, in fact, precious little work has been done within the private sector to retool the industry for future production, or to talk about retooling the forest itself (so to speak) except to intensify its current configuration in the industrial supply chain by speeding the pace of harvest.[3] No one in the BC forest industry is talking about a climate-altered forest, in other words, but about ways for the industry to remain financially successful and unaltered by government regulation in a climate-changed future.

It is perhaps uncharitable, though, to blame industry for following so closely the unequivocal recommendation on forest practices in the 2007 report of the Intergovernmental Panel on Climate Change (IPCC): "In the long term, a sustainable forest management strategy aimed at maintaining or increasing forest carbon stocks, while producing an annual sustained yield of timber, fibre, or energy from the forest, will generate the largest sustained mitigation benefit."[4] With these words, the 2007 panel clearly authorized a modified version of forest business as usual in BC, even though it called for significant change in jurisdictions short on reforestation requirements, protected areas, or planning processes. Industry also cannot be blamed for letting government take a lead role in responding to climate change, which in BC has meant two primary initiatives: the multi-jurisdictional Assisted Migration Adaptation Trial (AMAT, led in BC by the Forest Service) that has experimented with planting and protecting seedlings outside their traditional range; and the Future Forests Ecosystems Initiative (FFEI), managed entirely by the BC government.[5] As the report on climate change commissioned by the Land Trust Alliance of British Columbia (LTA) puts it, climate change responses need to involve "appropriate stewardship of living carbon":[6] the LTA sees this as an obligation for all individuals and organizations; however, in BC, the government has tended to do the work that might otherwise fall to industry. While industry could be doing more, the government has been busy enough exploring possible futures that industry has felt little or no need to interrupt business as usual. As forestry biologists recognize, though, forest business cannot proceed as usual in BC, no matter what the IPCC might have to say about it: the forests themselves are no longer usual, and they are becoming less like their past selves every year.[7]

Computer models of the effects of climate change on forests have been clear for some time, in spite of the enormous difficulty of modelling and imagining climate change at a broad ecosystem level,[8] and the most recent research in this area has confirmed the models. In brief, with a temperature change of only a few degrees Celsius, or a precipitation shift of a few percentage points, trees at the edges of their ranges are at risk of dying off. Trees are themselves not capable of moving, but forests can and do migrate north and south at different rates, depending on environmental needs for successful germination and growth (i.e., shade versus sun, nitrogen content in the soil, and so on). However, climate change is demonstrably outpacing forests' capacity to migrate, both in the American Northeast as well as across the North American west coast,[9] including British Columbia. As Zhu, Woodall, and Clark conclude in their large-scale study, "the direct comparisons of seedlings and trees at range limits do not inspire confidence that tree populations are tracking contemporary climate change[, and the] majority of species in [their] analysis shows a pattern consistent with range contraction at both northern and southern range limits." The authors do acknowledge the numerous complicating factors for this complex multivariable research; nevertheless, they are confident that their results "predict migration rates far below those required to track contemporary climate change."[10]

Even before these recent publications, however, individual tree species in the region were clearly suffering as a result of climate change. Trees have been distinctly struggling in areas where their species have been dominant since glaciation ended, up to 14,000 years ago. Cedars are dying on BC's south coast, pines are dying in the southern and central interior regions, and Douglas fir is fading in many areas across the province where it currently thrives.[11] Climate change will lead to dramatic changes in forest composition across BC in the period 2006–85, such as a predicted 336 percent increase in the size of the coastal Douglas fir zone as it expands northward, a 79 percent decrease in the mountain Hemlock zone, and an 800 percent increase in grasslands.[12] Industry has said little about what this means for the industry's long-term viability. The glossy multimedia, multi-jurisdictional, and multistakeholder response *Tackle Climate Change—Use Wood*, for example, authored by Roxane Ward, offers only a general comment that climate change will "chang[e] the range and distribution of plants and animals," followed by a longer, more technical remark about future directions in forest management: "Long-term forest planning that considers climate change can minimize potential mismatches between species and future climatic and disturbance regimes.... Although many of the impacts of climate change are decades away, North American resource managers are using computer models to explore possible adaptation strategies to reduce the vulnerability of forests."[13] While the publication does describe methods of intensive forest management,[14] it does not go so far as to discuss the idea of managing species change. I have had difficulty finding other comments even as explicit as this from the BC forest industry about forest species range and composition shifts, and I have not, so far, found a more detailed comment.

This is not to say, of course, that comments on forest species range and distribution shifts cannot be found in relation to the BC forest industry—just not from inside the industry. A 2010 report authored by Ben Parfitt as part of the Climate Justice Project, a research alliance led by the Canadian Centre for Policy Alternatives and the University of British Columbia, for example, representing the best information currently available on climate change's impact on BC forests and forestry, was co-published by four labour unions, three environmental activist organizations, and a left-wing think tank.[15] Not one of these organizations represents industry interests, so even though this report provides valuable detail about the issues facing forestry in BC, the lack of industry perspective still limits its discursive effectiveness. After listing several key issues in trying to determine what to do with a severely stressed forest system (beetles, blights, fire threat, and so on), suggesting a coordinated and greatly expanded program of tree planting, Parfitt moves to the overarching issue for all forest management in British Columbia:

Complicating matters is that climate models predict "wholesale redistribution of trees in the next century" as local and regional temperatures change along with precipitation patterns. (How rapidly and where such turnovers occur will be critical factors in determining what scale of human intervention, if any, makes sense ...) ... Careful decisions will need to be made based on verified field observations and educated guesses about what genetic variations of trees are most resilient in the face of a changing climate.[16]

Parfitt cites three scientific articles in this passage—Aitken et al. in *Evolutionary Applications* (2008), Wang et al. in *Global Change Biology* (2006), and Yanchuk et al. in *Tree Genetics and Genomes* (2007)[17]—as well as Gayton's 2008 literature review for a forestry consulting firm that includes numerous technical and scholarly references.[18] Clearly, the industry does have research on which it could draw. Instead, the industry appears to have made the decision not to discuss this issue—or anything else that might suggest business practices be either slowed or fundamentally altered—focusing instead on climate issues that will allow for even more intensive forest harvesting. *Tackle Climate Change—Use Wood*, for example, stresses that old-growth forests take up carbon less efficiently than young forests; that old-growth forests are subject to occasional, uncontrolled releases of carbon through fire, disease outbreaks, and so on; and that enough carbon is sequestered in the wood of an average wood-framed house to offset the carbon footprint of the rest of its construction. All true, though debatably so in some circumstances: but can't the forest industry in BC be more forward thinking than this? Is it incapable of imagining a future with a forest industry that does not look much the way it does now? Can it not imagine anything other than the current forest composition, but far more intensively managed toward timber-harvesting objectives?

Many approaches are possible in a study of forest industry actions in relation to climate change and social evolution in British Columbia. We could read with suspicion the reluctance to invest much capital in the mills: are they planning to stay for the long term?; the resistance to paying municipal taxes: do they want to belong to local communities?; and the extended duration of shutdowns: what relationship do they want with their workforce? We could read with more care the continued reliance on trucking logs between watersheds for processing, or the exporting of raw logs to other countries, particularly China and the United States. In this chapter, though, we will look narrowly at the industry silence on the topic of helping BC's forests adapt to climate change, by thinking carefully about forest migration and composition, and about acting to facilitate their adaptation. We will do this in large part through commenting on the genealogy of the forest industry's discourse of independence, as it appears in two early twentieth-century BC novels about logging. The reverberations between literature and history, between criticism and historiography can generate rich and complex questions about how cultural development occurs in relation to environment and can perhaps even help us imagine a way forward. In undertaking such a project, the chapter participates in the revolutionary global project of cultural responses to anthropogenic climate change, sometimes referred to as the environmental humanities.

Frontier Forestry: Origins

Forestry companies have always been at risk in BC, and workers have always been endangered. Of course, risk has always been part of labour and corporate operations across all industrial sectors, not just in forestry, but as Gordon Hak outlines in a series of brief but detailed—and occasionally stomach-turning—case studies, loggers and mill workers before 1900 were subject to a staggering array of death-dealing circumstances.[19] The small companies, too, operated so near the break-even point that the slightest crisis might lead to insolvency: in the 1880s, companies so often failed to make their hoped-for profits, and consequently to pay their employees for completed work, that "government brought in legislation to deal specifically with [non-payment to] loggers." In response to this industrial model, argues Hak, "the state created a climate conducive to the capitalist exploitation of the forests."[20] BC has held onto its Crown land, so 90 percent of the province's forests remain publicly held in 2011, but this does not reflect a desire by government to run things. Instead, Crown land was meant to facilitate the independence of corporations as they worked flexibly to harvest the province's timber, a corporate structure that would in theory empower individual men to choose their own roles in the noble and important task of building a province. To borrow Gordon Hak's terms, by the 1940s the companies took hegemonic control of the discourse of independence, representing themselves (with material effects) as the region's great individualists.[21] As Hak notes, the companies were working from a "worldview that celebrated individualism, the rights of private property owners, and small government."[22] A later poetic restatement of this world view, from the perspective of workers in the woods, can be found in Howard White's *The Men There Were Then* (1983), particularly the "Accidents" sequence.[23]

Of course, embedded in this ideology were many assumptions about gender, race, and class that affected patterns of industrial development. Asian workers were barred by statute from some types of work; Native workers were regarded as categorically unreliable; and race-based pay grids were the norm along the coast.[24] As Hak wryly puts it, "In the woods, race was not an issue: loggers were largely white,"[25] although non-white workers were able to take on employment less ideologically valuable and less financially remunerative. Any conversation about freedom and independence must occur with awareness of these structural inequities, which so distinctly narrowed the eligibility for independent action. Still, the discourse of independence remained comparatively vital, and it is at this point that we turn to matters more commonly within the purview of cultural studies, by engaging with Frederick Jackson Turner's hoary mythology of the American frontier.

Turner's thesis, developed in the early years of the twentieth century, was that America's development had been entirely unique among the world's nations, because of the long, slow opening up of the western frontier. He argued persuasively and influentially that American ideals developed as they did because of a historically unique relationship to geography and resources. Always there had been more land available for the colonial taking, and always there were men engaged in the nation-building task of moving into new, allegedly unoccupied territories. By the 1890s, though, with the definitive settlement of the Pacific Northwest, the frontier was extinguished, and Turner argued as a result that America was going to develop differently from how it had to that point. Turner's thesis remains influential, with a large succession of texts deferring to and developing his ideas; in 2009, Penguin published a selection of Turner's writings in its Great Works series as *The Significance of the Frontier in American History*. A great boost in the acceptance of Turner's ideas occurred in the 1960s and 1970s with the publication of Roderick Nash's *Wilderness and the American Mind*, which extended and deepened Turner's claims. Nash revised and republished his study a few times during these years, much more often than is the usual academic practice. Because of their significant influence within both the Sierra Club and the Wilderness Society, Nash's and Turner's ideas have had a disproportionately high profile in relation to developing environmental thought in North America.

However, Turner's geographical determinism has also long been critiqued, most influentially by Richard White in *"It's Your Misfortune and None of My Own"* and, with Patricia Limerick, in *The Frontier in American Culture*.[26] Turner has, among other things, been blamed for encouraging ignorant isolationism, for minimizing unhelpfully the role of cities in American history, and for neutering critique against industrial capitalism by focusing the discipline's attention on the frontier when it should have been observing the concentration of wealth and power among certain families and corporations.[27] While Turner's thesis has often been cited in relation to Canadian culture of the same time, there is at best an awkward fit between American and Canadian political systems, patterns of colonialism and settlement, and relations with the European nations. As Barry Gough argues, the American frontier myth of independence was directly opposed to the colonial bureaucracy's hopes for British Columbia.[28] However, even with these caveats, there are good reasons for the ongoing consideration of Turner's ideas in relation to the history of BC forestry. In writing of the period before 1914, especially, Gordon Hak notes claims of independence to be of significant value for both company and worker: "small producers celebrated virtue and independence in a decentralized political system," and BC saw the development of "a logging culture that extolled the independence of loggers."²⁹ As Gough and Pritchard show, the American model of independence was opposed in BC by an ideology of colonial governance. Workers and companies in the BC forests at the time, though, took very seriously ideas like Turner's, and as a result, the concept of "independence" is descriptively useful. BC's forests really did represent at least an imaginary frontier for the men working in them, both the labourers and their managers, though not for the provincial and national governments with jurisdiction over them.

Independence: *Woodsmen of the West*

Martin Allerdale Grainger's 1908 *Woodsmen of the West* is a portrait of just the period addressed by Hak's *Turning Timber into Dollars* discussed above. Set largely in 1907, and organized around Mart, the narrator's experiences with a small operation's boss named Carter, *Woodsmen of the West* lines up closely enough with verifiable history that the BC Archives has considered it tantamount to primary source material.³⁰ In spite of its apparent transparency, though, the novel needs to be read as representation rather than as transcription: accurate the details may be, but Grainger is nonetheless telling a story about the BC forest industry, about the BC forests, and about the men who work there. In particular, he draws out the implications of the conflict between community practices and individual freedom in the context of environmental resource extraction in a frontier setting.

Grainger's Carter, a tyrannical owner-operator of a small logging show, represents a pivot point for most of the forces (both natural and cultural) governing the BC forest industry. Fiercely independent but demanding surrender from other men, Carter's success derives not just from his own abilities but from the gentle, congenial character of his business partner Bill Allen. Before the narrator ever meets Carter, blacksmith Dan Macdonnell (who had once been approached by Carter for a potential partnership) introduces Mart to Carter's failings: "Being partners with him means obeying him and being his slave; a man of any independence couldn't stay with him five minutes. Carter's as pig-headed as they make them; and wicked. Everything's got to be done his way; your way is wrong, and he won't even listen to what you're going to propose; and he'll go against your interests, and against his own, and wreck his whole business rather than admit himself in the wrong."³¹ The weaknesses of thorough independence are laid bare in Macdonnell's summary. As Macdonnell points out, it's not enough for Carter that he be proven right, because power is more important to him than success. The narrator, once Mart has met Carter, agrees with the blacksmith, recognizing that "Carter, of course, can only tolerate a man who seems subservient to his every whim; a man who will slave for him; who will submit, in moments of Carter's anger, to be talked to like a dog."[32] The language used to describe Carter's relations with other men (the narrator's own language, as well as that of other characters in the novel) insists that equality is not important to Carter. As hard as he works, Carter holds himself apart from the labourers who perform the same tasks. For Carter, his own independence is the only one with value to him, and his actions throughout the book have to do with imposing dependence (of one kind or another, legitimately or otherwise) on all those around him.

Mind you, Carter is no cartoon villain for Grainger. In spite of hearing from Macdonnell of Carter's "wicked" nature, and in spite of agreeing with Macdonnell, Mart nonetheless finds himself working relatively comfortably for Carter for an extended term. Further, Mart sees in Carter something remarkably positive: "among the clinkers and the base alloys that make up so much of Carter's soul, there is a piece of purest metal, of true human greatness, an inspiration and a happiness to see."[33] Grainger (or Mart) never goes on to define this "purest metal," but Mart has already praised Carter's ability to succeed on his own or with very small crews.[34] Mart is smitten with the idea of proving his masculinity in the woods (spending his days without pay, for example, splitting wood without charge but in public at the Port Browning Hotel), and since his actions toward proving himself tend toward asserting his own independence, it would be a surprise for Mart not to have at least some desire to see himself in Carter.

On the BC coast in the early twentieth century, including inside the forest industry, independence exists in tension with community, though more subtly than in later years. Carter can—and perhaps must—be praised for his self-sufficiency, but he must also be damned for his inability (or at least unwillingness) to connect with other people. By book's end, Mart tires of Carter's mistreatment: angered not by the physical or mortal danger resulting from hazardously flawed technology or from reckless management decisions, but at being insulted by having his abilities called into question. Upon quitting, and precipitously leaving Carter's camp, Mart and his new companions deliberately and visibly exclude Carter from the conventions of a coastal logging community. Spontaneously, these five men come together in opposition to Carter: they do not confront him physically, but their temporary community is defined by its collective opposition to Carter. They're loudly drunk, to stress that whisky is nearby but unavailable to him; they sing a song full of coded insults of Carter, loudly enough for him to hear; they display a freshly killed deer where Carter can see it, but they do not offer to share. This brief community is unsustainably non-productive, and readers should be troubled by its deliberate refusal to accept other members (the worker François, who is left behind in Carter's camp), but otherwise it begins to offer an alternative mode of community distinct from anything contemplated in Turner's frontier thesis.

As Hak observes, at just this time there was a kind of double view of the logger, and by extension a double view of the forest industry as a whole. In the collective actions that represented the first steps toward unionization, which did not take effect broadly for three more decades, Hak sees the outline of a question challenging enough to shape future development of the industry and the discourse around it: "Did real men embrace adverse conditions without whining, independently dealing with employers and moving to a different camp if conditions did not meet their standards, or did real men act collectively with their fellow workers in a protest against unacceptable situations?"[35] As the example of Mart quitting Carter's camp suggests, particularly his joining up with Kendall and the others, unionization is not the only way to signal community involvement. Still, their community is non-productive as well as impure, since a man indebted to Carter has been left under Carter's control, alone. The closure of this group against François, and possibly Carter as well, contravenes egregiously the local standards of community, so it can be no role model either.

Importantly, crises in *Woodsmen of the West* lead to highly individualist responses: Mart flees Carter, as we have seen; Bill heads for Vancouver to drum up investors to combat local effects of the 1907 international financial crisis; men enter headlong and willingly into disasters of all kinds, even though they know how poorly equipped they are and how rigged the game is against them. A bank's unwillingness to lend capital, a log too heavy to be moved, an axe wound suffered on the job: the only viable option is to do your best, on your own. The government is next to invisible in *Woodsmen of the West*, and communities are transient at best, so in their minds, characters see themselves to be separately responsible for reaching or maintaining self-sufficiency.

Collectivity: *Timber*

Just a little over thirty years later, Roderick Haig-Brown could argue that it was time to embrace unionization and formalize employee–employer relations. Even though *Timber* (1942) draws much of its narrative force from the romantic relationship between Johnny Holt and Julie Morris, always in the background is Alec Crawford's drive toward unionization.[36] Further, while the official community represented by unionization remained in the future, *Timber* sees the role of government in loggers' lives to be clearer than it had been in the world of *Woodsmen of the West*: Haig-Brown's novel begins with a coroner's inquest after a man has died on Johnny Holt's crew, and it ends with a second inquest following the death of Alec Crawford. Between these two official processes, characters act freely, come to know themselves, and fight the companies, but in fact a powerful conservatism informs the novel's narrative arc.

In Grainger's novel, Carter's 1907 logging show is small enough for three men to operate, organized around a frail old donkey engine that yards logs along the ground and dumps them into the sea. In Haig-Brown's novel, the B. and A. show has multiple camps, several miles of railway, and two separate locomotives; it is also a high-lead show, as was common in the 1930s, with logs being transported aerially by cable (rather than along the ground) from where they are cut to where they are loaded on rail cars. By the 1930s, a large proportion of the labour needed in 1907 had been transferred to machines, and logging firms produced far more timber, far more rapidly than they had previously been able to produce. As Richard Rajala points out, though, technological change in the woods was not driven by changing harvest sites or markets: "Timber capital sought domination over nature not as an end in itself, but to secure control over the activities of those they employed."[37] The companies resisted unionization in the 1930s, especially with fresh memories of the Industrial Workers of the World or "Wobblies" in the years around the First World War, because they feared a loss of control over their employees. They wanted to control the pace and type of change occurring in the camps, socially and economically as well as technologically. As Haig-Brown and others recognized, technological change had placed far different demands on men working in the woods, and consequently the need to protect them had only intensified.

And yet, in *Timber* as in life, loggers themselves proved difficult to organize: "You guys make me sick," Alec snaps at one point. "So goddamned independent you'd drown before you'd catch hold of a chunk of wood to save yourselves."[38] Mill workers, at least initially, were even harder to organize since many of them lived in stable communities largely developed and supported by the companies (in arrangements recognized through very heavy municipal taxation into the twenty-first century), but for quite a different reason. Loggers resisted unionization not out of respect for the company, as mill workers did, but out of a desire to preserve their independence in order to resist the company. Rajala describes the consequences of worker freedom on forestry companies: "Anything but desirable from the operator's standpoint, the independence of these workers was disruptive, and would prove to be increasingly so as technological change enhanced the benefits of labor stability."[39] The effects of collective resistance would have been clear to the workers whose individual action was already disruptive, but organizing remained difficult anyway, a paradox that the characters in *Timber* discuss explicitly: "'If it's needed so badly why is it difficult, Alec?' [Dal] asked. 'I know the boys are independent, but what else?'" Alec's reply mentions several issues, from job-specific wage scales to frequent crew changes driven by a camp's changing labour needs, ending lamely with "Lots of other things like that."[40] As Howell Harris notes, too, unions should perhaps be seen as "a weak institution in a powerfully organized capitalist society,"[41] and one of Alec's "other things" was simply the management structure of a large union. The companies couldn't be trusted, but the unions were viewed as similarly corporate, with Eastern and/or American head offices controlling the actions of local offices: "Vancouver would run it. And the East would run Vancouver and some outfit down in the States would have the last word about everything."[42] Haig-Brown recognizes here precisely what recent labour historians have confirmed through their research, namely, that loggers' independence meant that neither the companies nor the unions could count on them, and that the loggers themselves were comfortable with their intransigence, and proud of it as well.

Haig-Brown's novelistic analysis, though, also connects with the geographical determinism of Jackson's frontier thesis: "this country's too near pioneering. A man likes to think he's his own boss and can work how he wants and where he wants,"[43] notes Alec, who distinguishes between an individualism needed during the building of a nation and the collective action needed to maintain the nation.[44] Rather than a colonial politics or a politics of national development, Haig-Brown's characters are working from a self-consciously class-based frontier politics. This is especially true for Alec, who explains one morning as the men watch the sun rise during their ride to the job site: "This country ought to grow good people, with things like that to look at every day.... Look at the people in these little European countries, always fighting and raising hell. I figure a lot of it's because they haven't got room enough to look around and see things. Canadians and Americans aren't like that and maybe they're better for being able to see things and get out into a big country."[45] Once he is blacklisted, Alec considers joining the Mackenzie-Papineau Battalion in order to fight in the Spanish Civil War, so the frontier perspective is far from stable even in the one character in *Timber* most aware of it. Still, Alec decides to stay in Canada, and eventually goes through a kind of conversion experience under the influence of the BC landscape that seems intended by Haig-Brown to bear out the accuracy of his philosophy.

After Alec and Johnny are blacklisted for their involvement with a unionization push at the B. and A., and following a dishearteningly long stint in Vancouver looking for work, they find employment on a small haywire outfit at Yellow River, on Vancouver Island. Johnny and Julie, now married with a small child named after Alec, continue to build an increasingly middle-class life together, but Alec, quite simply, is stuck in this camp. Too educated for his role at Yellow River, impatient with the low-quality tools and cheap management style, frustrated at seeing Johnny living happily with the woman he himself once loved, and frustrated perhaps most of all at the decline of the powerful homosocial bond he has long shared with Johnny, Alec goes on a lengthy hike that ends up resolving his existential crisis. Climbing steep cliffs, dreaming of Julie (but envying Johnny for more than just marrying Julie), he comes to a position that one might read as an evolved individualism: "I am myself, he thought, Alec Crawford, and I know my own job the way damned few men know it.... One more week with this cheapskate outfit so I won't be walking out on them, then I'll hit for town."[46] He emerges from his wilderness conversion newly confident in himself, refocused on his own career, and yet sensitive to others. Not only is he concerned that his departure not unduly hurt Johnny and Julie, but also he wants to limit the negative effects of his departure on the company. In essence, Alec emerges with a new perspective that adds a healthy dose of independence to the communal approach he has been persistently but unsuccessfully advocating through the novel.

Except that within a week Alec is dead, crushed by a chunk of a falling spar tree, a tree that the camp rigger has apparently been fired for refusing to rig up because he considered the tree unsafe. In the longer story of regulating the coastal timber industry, rather than taking on a management role to oversee the growth of unionization, Alec ironically performs exactly the same involuntary role as does Charlie Davies, whose death (largely from his own negligence) sparked the novel's opening inquest. As he tries early in the novel to calm the man who had controlled the log that killed Charlie, Alec says, "Charlie would have got it sooner or later.... It's the fault of all the crazy guys that want to show how fast and easy they can do things."[47] Even if a union had been in place to ensure safer working conditions, Alec goes on to admit, "Maybe Charlie would have got it anyways." Ed Nelson saw Charlie as "naturally reckless," while Johnny saw him as always "putting on a show to please himself," and the last words spoken to him were "Watch those logs, Charlie, you crazy son of a bitch."[48] Throughout the novel, Alec has been depicted as the one character best able to recognize future outcomes of current conditions: he has a preternatural ability to lay out rail lines and spar trees to overcome the limitations of terrain, for example, and he inhabits wild land with great ease (as in the hike mentioned above). The opposition between Alec Crawford and Charlie Davies is striking, and yet they play the same victim's role in the increased regulation of the forest industry. Alec's recognized skills, because they are the skills of just one man, offer him no protection against death in the woods due to industry shortcuts.

At each inquest, the jurors are free to ask questions. At each, one juror takes on the job of asking numerous questions, "just trying to understand the thing properly."[49] Whereas the questioner in the first inquest is a known quantity, a local storekeeper named Edwards, the questioner at Alec's is unnamed and previously unknown to Johnny: "a bull-headed little guy, bitter as hell,"[50] with fierce ideas about worker safety and corporate responsibility. The contrast between these men is accompanied by a shift in the inquest process itself. At the first, when Edwards complains aloud that jurors should be chosen who already know the logging business, the coroner disagrees: "The Crown wants the opinion of a group of intelligent citizens, not experts."[51] This first coroner's rebuttal emphasizes governance and collective decision making. At the second inquest, on quite the other hand, the coroner's verdict includes a recommendation that "all jurors on logging inquests ought to be made up of guys that know logging":[52] no longer "a group of intelligent citizens," but simply "guys," and with an accompanying erasure of government ("The Crown") in favour of an agentless recommendation. Johnny sees that the companies are tightening their control over the industry's regulation, or at least trying to tighten their control, but he doesn't object. Instead, he testifies honestly and in full at the inquest, managing against all his desires to avoid physically assaulting anyone, and then he returns to work.

At book's end, Johnny is hired back and promoted by B. and A., and assured that as long as production goes well, the new manager will not interfere with any collective action occurring under Johnny's watch, not even unionization. This manager was a former logger himself who had been the supervisor at the B. and A. camps during Johnny's first term there; his health-related resignation earlier in the novel, due to ulcers, had triggered the chain of events culminating in Johnny and Alec quitting. He had been trying to locate both Alec and Johnny to hire them back, failing to find them in time before Alec's death at Yellow River. As Rajala notes, BC's forest companies eventually came to see that non-radical unionization would lead to a more stable workforce, so the acceptance of unionization should be seen as one element in domesticating BC's wild loggers. Timber production with "the new technology placed a high premium on the ability of a crew to function as an organic production team,"[53] and unionization could—and at times did—contribute to the building of such teams. As Rajala states elsewhere, his own perspective is that "corporate resource exploitation takes place within the context of, in fact is dependent upon, exploitation of workers."[54] Clearly, this would have been the perspective of Alec Crawford and Johnny Holt as well; it's just that they wanted workplace protection during the exploitation, not exemption from exploitation.

The loggers' independence needed to come under the companies' disciplinary control, the same way nature had come under increased technological control, and for the same reasons. Once the companies judged that their hard-line anti-unionization stance had failed to achieve its management goal of disciplinary control, it turned out that unionization could achieve it instead, perhaps to the surprise of the radicals who convinced reluctant loggers to join unions. In the culture more broadly, the companies managed to inherit the loggers' reputation for independence, riding the decades-long dominance of free-market discourse and ideology, even if this claimed independence hasn't kept the companies from accepting (and soliciting) government support. In

Grainger's novel, the small outfit's boss is punished for his independence and his failure to live by community standards; in Haig-Brown's, workers die for and from their independence, but the community is a function of the company rather than of the workers. Again, as Hak suggests, the companies took hegemonic control of the discourse of independence and represented themselves as the region's great individualists.[55] In relation to the 1940s push toward such social benefits as health insurance, Hak remarks that the companies resisted such a push out of a "worldview that celebrated individualism, the rights of private property owners, and small government."[56] This same world view, I would suggest, largely endures today, and has impoverished the industry's response to climate change.

Living with Change

The forest biology of BC under a regime of global climate change is clear, leading to a situation succinctly described by Ben Parfitt:

> That certain forest ecosystems may also be prone to unravelling or changing in the face of climate change has prompted an increasing number of scientists to focus on how certain trees will fare in the face of higher temperatures and dramatically altered precipitation patterns. Some tree types will adapt better, and continue to occupy their ecological niches. Others will "migrate"—move south to north or from lower to higher altitudes. And still others will be extirpated, or face localized extinctions.[57]

Tree species throughout British Columbia are already struggling in their traditional ranges. In the even-age stands left behind after large-scale logging, and as a consequence of the warmer winters and shorter cold snaps marking the new climate regime, insects are taking a heavier toll than at any point in the past. Some areas have too little moisture to sustain existing dominant species, and other areas have too much for other species. Decades of fire suppression have meant that blazes are predictably unpredictable and increasingly intense, especially in forests where trees are drying out from reduced precipitation and higher soil temperatures.

The province's logging companies have, by and large, declined to comment publicly on these issues. Instead, they focus on selling reforestation and wood-frame construction as methods of locking up carbon and hence slowing the carbon emissions partly responsible for anthropogenic climate change. They continue to ask, as for so long they have asked, for a more streamlined regulatory model that would simply let them get on with their work in a global marketplace: lower tariffs and taxes, fewer export restrictions, and less oversight. Nowhere in their public statements is a recognition of how BC's forest composition is changing as a result of climate change, as Daniel Botkin was warning as long ago as 1990 that it would, in forests across the planet rather than just in BC: "A seedling of a species suited to the climate at the time of its planting may find itself in a climate too warm for the seeds it produces by the time it reaches maturity several decades in the future."[58] Nowhere is there an explanation of how BC's forest industry might respond to the compositional changes under way.[59] Richard Rajala points out to us that "timber capital's entrenched tendency to 'mine' the Douglas fir forest" is a tendency supported by the province's "reluctance to forge a rational integration of the scientific and policy-making processes":[60] and we should remember, of course, that mining is the one mode of resource extraction that is transparently non-renewable.[61] Though Rajala is here speaking primarily of the period before 1930, the BC forest industry has always operated in an environment where freedom has been prized. Pre-1910 logging bosses, pre-1940 loggers, and post-1990 international companies prize independence against their own long-term interests, so committed are they to the idea that if only they could be free to work on their own, things would come right.

The predicament of BC's forests now, and of those working in them, is that this independence offers no response to climate change and its effects. This new crisis cannot be solved by throwing new technology at it, or by more thoroughly controlling the employees. Dan Macdonnell's words about Carter in Martin Allerdale Grainger's *Woodsmen of the West* should be recalled, chillingly: "he'll go against your interests, and his own, and wreck his whole business rather than admit himself in the wrong."[62] The industry needs to participate in the conversation, and it needs to participate deeply. Its current silence is unacceptable. If the forests are at risk of changing so broadly as to alter radically their species composition, it is past the time for half-measures like wood-frame construction's mild defence against medium-term carbon emissions. Or as Alec Crawford complains in Roderick Haig-Brown's *Timber*, "So goddamned independent [they'd] drown before [they'd] catch hold of a chunk of wood to save [them]selves."[63]

A solution?

Well, we could do worse in BC than to drop the knee-jerk self-defence and empire building, and instead sit down together, all of us, to see what we can figure out. As Don Gayton asserts, because BC's environment is so complex geologically and biologically, "we must demand more of ourselves as British Columbians, and not be content with minimal responses to the challenge of climate change."[64] Ben Parfitt's 2010 report takes us part of the way there, maybe, but not without the participation of industry, government, and First Nations. We cannot, like a 1910 small logging outfit, get ahead through some form or another of exploitation that we haven't yet applied; we cannot, like a 1930s logger, quit and move on because the situation doesn't suit us; we cannot, like every large logging company in BC since at least 1910, ask for looser regulation to let us all see what we can accomplish individually. None of these old independences has any hope of working.

And maybe nothing will work. We need to keep that likelihood in mind.

But it may be that our only chance is to attempt to trade in our congenital but inherited independence, whose roots in BC culture can be seen in the long-ago logging novels of Martin Allerdale Grainger and Roderick Haig-Brown, in exchange for some form of enduring community yet to be represented in literature of British Columbia forests and forestry. Nothing else, let me repeat, has any hope of working.

Notes

Everyone working on questions related to BC forests and forestry in BC needs to acknowledge the work of Gordon Hak at Vancouver Island University and Rick Rajala at the University of Victoria, whose careers as environmental historians and historians of labour have been an inspiration to me. More personally, I wish also to thank Alexa Brosseau and Joanna Dawson for their research assistance; Fraser Hannah, Sean Kheraj, and David Brownstein for their comments; and my colleagues Nicole Shukin and Nick Bradley for their encouragement.

1 Forest Practices Association of Canada (FPAC), "Carbon Neutral Pledge," *FPAC.ca: The Voice of Canada's Wood, Pulp and Paper Producers*, n.d., http://www.fpac.ca/index.php/en/carbon-neutral-pledge.
2 Don Gayton, *Impacts of Climate Change on British Columbia's Biodiversity: A Literature Review* (Kamloops, BC: Forrex, 2008), 3; Andrew Nikiforuk, *Empire of the Beetle* (Vancouver: Greystone, 2011), 61–62, 133–35.
3 While BC's Council of Forest Industries (COFI) released in 2011 a teaching resource on climate change (*Climate Change, Our Forests, Our Future*, http://www.cofi.org/wp-content/uploads/2011/12/Tackle_Climate_Change_Secondary.pdf), this material focuses almost exclusively on the potential role of industrial forestry in mitigating climate change. Tellingly, COFI's teaching resource on sustainable forestry (*Sustainable Forest Management in British Columbia*, http://www.cofi.org/wp-content/uploads/2012/01/sustainable_forest_management_in_bc_02.pdf) does not mention climate change once, though it was released in 2012.
4 Quoted in Roxane Ward, *Tackle Climate Change—Use Wood* (Vancouver: BC Forestry Climate Change Working Group, 2009), 4.
5 See, e.g., Elizabeth M. Campbell et al., *Ecological Resilience and Complexity: A Theoretical Framework for Understanding and Managing British Columbia's Forest Ecosystems in a Changing Climate* (Victoria: BC Ministry of Forests and Range, 2009), or David Spittlehouse, *Climate Change, Impacts and Adaptation Scenarios: Climate Change and Forest and Range Management in British Columbia* (Victoria: BC Ministry of Forests and Range, 2008).
6 Sara J. Wilson and Richard J. Hebda, *Mitigating and Adapting to Climate Change through the Conservation of Nature* (Vancouver: Land Trust Alliance of BC, 2008), 4.
7 Gayton, *Impacts*.
8 Ronald P. Neilson et al., "Forecasting Regional to Global Plant Migration in Response to Climate Change," *BioScience* 55 (2005), 749–59.
9 Richard H. Waring, Nicholas C. Coops, and Steven W. Running, "Predicting Satellite-Derived Patterns of Large-Scale Disturbances in Forests of the Pacific Northwest Region in Response to Recent Climatic Variation," *Remote Sensing of Environment* (2011), online 3 Nov. 2011, doi: 10.1016/j.rse.2011.08.017.

10 Kai Zhu, Christopher W. Woodall, and James S. Clark, "Failure to Migrate: Lack of Tree Range Expansion in Response to Climate Change," *Global Change Biology* (2011), 7–10, doi: 10.1111/j.1365-2486.2011.02571.x. Zhu, Woodall, and Clark's research is based on "data on 92 species in more than 43,000 forest plots in 31 states." See Duke University, "Eastern U.S. Forests Not Keeping Pace with Climate Change, Large Study Finds," *ScienceDaily*, 31 Oct. 2011.
11 Gayton, *Impacts*; Wilson and Hebda, *Mitigating and Adapting*.
12 Wilson and Hebda, *Mitigating and Adapting*, 25; see also Gayton, *Impacts*, 6.
13 Ward, *Tackle Climate Change*, 11, 19.
14 Ibid., 54.
15 Ben Parfitt, *Managing BC's Forests for a Cooler Planet: Carbon Storage, Sustainable Jobs and Conservation* (Vancouver: CCPA-BC Office; BC Government and Service Employees' Union; Communications, Energy & Paperworkers Union; David Suzuki Foundation; Pulp, Paper and Woodworkers of Canada; Sierra Club BC; United Steelworkers District 3—Western Canada; and Western Canada Wilderness Committee, 2010).
16 Ibid., 27.
17 S.N. Aitken et al., "Adaptation, Migration or Extirpation: Climate Change Outcomes for Tree Populations," *Evolutionary Applications* 1, no. 1 (2008), 95–111; T. Wang et al., "Use of Response Functions in Selecting Lodgepole Pine Populations for Future Climates," *Global Change Biology* 12, no. 12 (2006), 2404–16; and A. Yanchuk et al., "Evaluation of Genetic Variation of Attack and Resistance in Lodgepole Pine in the Early Stages of a Mountain Pine Beetle Outbreak," *Tree Genetics and Genomes* (2008), 171–80.
18 Gayton, *Impacts*.
19 Gordon Hak, *Turning Trees into Dollars: The British Columbia Coastal Lumber Industry, 1858–1913* (Toronto: University of Toronto Press, 2000), 139–43. At most inquests into deaths in the forest industry, Hak notes, "employee negligence was deemed the cause of the accident," no matter how unsafe the working environment might have been (161).
20 Ibid., 138, 65.
21 Gordon Hak, *Capital and Labour in the British Columbia Forest Industry, 1934–1974* (Vancouver: UBC Press, 2007), 11–12.
22 Ibid., 134.
23 Howard White, *The Men There Were Then* (Vancouver: Arsenal Pulp Press, 1983).
24 Hak, *Turning*, 137–38, 153, 155.
25 Ibid., 134.
26 Richard White, *"It's Your Misfortune and None of My Own": A History of the American West* (Norman: University of Oklahoma Press, 1991); and Richard White and Patricia J. Limerick, *The Frontier in American Culture: An Exhibition at the Newberry Library, August 26, 1994–January 7, 1995*, ed. James R. Grossman (Berkeley: University of California Press, 1994).
27 George Rogers Taylor, ed., *The Turner Thesis: Concerning the Role of the Frontier in American History* (Lexington, MA: D.C. Heath, 1949).
28 Barry M. Gough, "The Character of the British Columbia Frontier," *BC Studies* 32 (1976/77), 38–39; see also Allan Pritchard, "The Shapes of History in British Columbia Writing," *BC Studies* 93 (1991), 48–69.
29 Hak, *Turning*, 86, 134.
30 Rupert Schieder, Introduction to *Woodsmen of the West*, by M. Allerdale Grainger (Toronto: McClelland & Stewart, 1964), ix.
31 M. Allerdale Grainger, *Woodsmen of the West* (Toronto: McClelland & Stewart, 1964), 49.
32 Ibid., 59.

33 Ibid., 60.
34 Ibid., 52–53.
35 Hak, *Turning*, 147–48.
36 Roderick Haig-Brown, *Timber* (Toronto: Wm. Collins, 1946 [1942]). This character is sometimes identified as Slim, sometimes as Alec. While the novel's naming conventions for him appear to follow at least a loose structural rationale, this chapter refers to him throughout as "Alec" for ease of reference.
37 Richard Rajala, "The Forest as Factory: Technological Change and Worker Control in the West Coast Logging Industry, 1880–1930," *Labour / Le Travail* 32 (Fall 1993), 79.
38 Haig-Brown, *Timber*, 17.
39 Richard Rajala, "A Dandy Bunch of Wobblies: Pacific Northwest Loggers and the Industrial Workers of the World, 1900–1930," *Labor History* 37, no. 2 (1996), 208.
40 Haig-Brown, *Timber*, 84–85.
41 Howell Harris quoted in Rajala, "Forest as Factory," 76.
42 Haig-Brown, *Timber*, 17.
43 Ibid., 83.
44 Ibid., 84.
45 Ibid., 99.
46 Ibid., 223.
47 Ibid., 13.
48 Ibid., 16, 8.
49 Ibid., 12.
50 Ibid., 244.
51 Ibid., 10.
52 Ibid., 244.
53 Rajala, "Forest as Factory," 103.
54 Richard Rajala, "Clearcutting the British Columbia Coast: Work, Environment and the State, 1880–1930," *Making Western Canada: Essays on European Colonization and Settlement*, ed. Catharine Cavanaugh and Jeremy Mouat (Toronto: Garamond, 1996), 112.
55 Hak, *Capital*, 11–12.
56 Ibid., 134.
57 Parfitt, *Managing*, 12.
58 Daniel Botkin, *Discordant Harmonies: A New Ecology for the Twenty-First Century* (Oxford: Oxford University Press, 1990), 194.
59 See Gayton, *Impacts*; Waring et al., "Predicting"; Wilson and Hebda, *Mitigating and Adapting*; and Zhu et al., "Failure."
60 Rajala, "Dandy," 126.
61 I am indebted for this comment to Sean Addie, who spoke on this topic at "Environmental Studying: An Interdisciplinary Eco-Colloquium," held at the University of Victoria, 11 Mar. 2011.
62 Grainger, *Woodsmen*, 49.
63 Haig-Brown, *Timber*, 17.
64 Gayton, *Impacts*, 17.

CHAPTER 12

Endangered Species, Endangered Spaces
Exploring the Grasslands of Trevor Herriot's *Grass, Sky, Song* and the Wetlands of Terry Tempest Williams's *Refuge*

Angela Waldie

> Birds have long been the measure of our fate, their extinction portending disaster for humankind. The caged canary in the coal mine performs the same service as the dove that in the story of the Golden Fleece was sent out by the Argonauts before they risked losing their ship in the narrow passage of the Clashing Rocks. The trial doves sent out by Noah in the Bible and Utnapishtim in Gilgamesh to determine whether the earth could support life after the Flood prefigure the actual birds of our own time that serve as measures of the ecological health of the entire world.
> — Lutwack, *Birds in Literature*

In his poem "In Aornis," Don McKay asks the reader to envision a land without birds, "Where each tangle in the foliage / is not a nest, where the wind / is ridden by machines."[1] He asks us to consider not the specific absence of a dodo or a passenger pigeon, but a much broader chasm encompassing all birds, and the unfathomable silence and stillness that would accompany their disappearance. In this landscape of absence, the most enduring images are recollections of what is missing, as the immediacy of what has been lost overshadows the monotonous and mechanistic landscape that remains. The felt absence of "a nest"—a word whose rich connotations evoke a time when it still had meaning—is more immediate than its pale future shadow: a mere tangle of foliage that cannot contain a bird. A land where only airplanes ride the wind evokes longing for skies filled with the intricate choreography of avian flight paths. Even the "unsung sun" that "comes up anyway" pales in comparison with a sunrise invoked by birdsong.[2]

Refuge by Terry Tempest Williams and *Grass, Sky, Song* by Trevor Herriot convey landscapes that are the antithesis of McKay's birdless land, where the nests of numerous species are evident to those with eyes discerning enough to see them, and where the wind is made visible by wings. By conveying the flight paths of birds, Williams and Herriot each extend the scope of their home skies; and by evoking birdsong, they deepen the dimensions of dawn and dusk. By interweaving these animated skies with personal accounts, both offer memoirs that illustrate connections between human health and the health of our natural surroundings. Describing the nuances of bird species while detailing threats to these species and the landscapes they inhabit, they explore issues of habitat loss and environmental toxicity. Through compelling accounts of imperilled species, they advocate for awareness and attention to circumvent the possibility of a birdless land.

In the prologue to *Refuge: An Unnatural History of Family and Place*, Williams locates herself on the floor of her study surrounded by journals that contain the genesis of her memoir. As she opens them, "feathers fall from their pages, sand cracks their spines, and sprigs of sage pressed between passages of pain heighten [her] sense of smell."[3] By foregrounding physical and sensory aspects of the natural world, she establishes the significance of her surroundings to the story she is about to tell, as she invites the reader to join her in a realm where language, landscape, and self are inextricably linked. Memory, in Williams's writing, cannot be separated from the place it occurred, and place is shaped through an accumulation of geological, ecological, cultural, and personal memories.

Williams wrote *Refuge* in response to two devastating events in her life: her mother's death from ovarian cancer and the flooding of the Bear River Migratory Bird Refuge, Williams's spiritual home. Although this pairing is related chronologically but not causally, she emphasizes an enduring correspondence between humans and the landscapes in which we live. "I could not separate the Bird Refuge from my family," she writes. "Devastation respects no boundaries."[4] As the subtitle of her memoir, *An Unnatural History of Family and Place*, suggests, she continually connects her family with the Great Basin landscape, demonstrating how place can inform and nurture our relationships. By juxtaposing her mother's cancer—a disease perceived to be "unnatural"—with the flooding of Great Salt Lake caused by natural oscillations of the water cycle, Williams invites us to consider how environmental and cultural contexts influence our experiences of disease, grief, and healing. As she blurs the distinction between nature and culture, she also encourages us to question the arbitrary line we draw between the human and more-than-human realms, and to acknowledge a connection between the health of the natural world and human health.

Rooted genealogically, personally, and politically in Utah, Williams demonstrates a loyalty to place rare in contemporary North American society.[5] As Lucy Lippard writes in *The Lure of the Local*, "Few of us in contemporary North American society know our place ... and even if we can locate ourselves, we haven't necessarily examined our place in, or our actual relationship to, that place."[6] In contrast to the contemporary tendency toward residential mobility, Williams emphasizes the importance of developing an intimate and enduring connection to one's home landscape. In a 1996 interview with Ona Siporin, she states, "It may be that the most radical action you can commit is to stay home."[7] She explains that by developing a commitment to community and place, one is more likely to develop a corresponding commitment to the surrounding natural environment. "It is in that context," she asserts, "that we begin to learn the names of things."[8] This attention to the specifics of place has been essential to her work as a writer, as well as to her role as curator of education and later as naturalist-in-residence for the Utah Museum of Natural History.[9]

Knowing the names of things, for Williams, extends beyond an intimacy with Utah history and landmarks to include an interest in its geological, archaeological, and cultural history and, most notably, in its ecology. Although trained as a naturalist, Williams articulates a vision of interrelatedness that extends beyond the traditional bounds of ecology.[10] Her ecology incorporates a symbiosis of natural history, family history, myth, and spirituality, as she reveals both her family and the birds of the Great Basin to be integral to her sense of belonging. Her paternal grandmother, Mimi, inspired her interest in the language and lore of birds. When Williams was five years old, Mimi gave her a copy of Roger Tory Peterson's expanded *Field Guide to Western Birds*, and on birdwatching excursions Mimi impressed upon her granddaughter the importance of learning the names of bird species, as well as their intersections with human history and culture. As John Tallmadge notes in "Beyond the Excursion: Initiatory Themes in Annie Dillard and Terry Tempest Williams," Williams's excursions in *Refuge* are generally social rather than solitary, and "[her] narrative, therefore, is always embedded in a context of social relations."[11] During these excursions, birds often ground Williams and her companions in the present moment, acting as emissaries of the present place and time, but also offering portals through which Williams embarks upon intimate meditations on family, illness, and the healing power of place.

Refuge achieves a blurring of form and genre that imbricates the "subjective" literary memoir with "objective" natural history, foregrounding the extent to which the health of individuals, families, and communities is connected to the health of the natural places in which they reside. As Cheryll

Glotfelty contends in "Flooding the Boundaries of Form," this memoir is in many respects structured like a natural history. Williams titles each chapter with the name of a bird and includes an appendix of "Birds Associated with Great Salt Lake," arranged according the phylogenetic order used in field guides. Her intricate knowledge of avian ecology, the guidance of other scientists with whom she collaborates, and her expertise as a naturalist allow her to guide her readers through the Great Basin landscape on walks, bird counts, and even archaeological expeditions. She invites her readers to enter each chapter of *Refuge* as we might enter a marsh, learning the names and natural histories of the birds and other species that reside there, observing details of their nesting and migratory habits, and listening to Williams's transliterations of their songs and calls.

In the chapter entitled "Burrowing Owls," for example, we learn much about the species *Athene cunicularia* that we might find in a field guide. Nesting on the ground, often in abandoned prairie dog burrows, these owls prey on small rodents, birds, and insects. The rattlesnake-like hiss of "*Tttss! Tttss! Tttss!*" one might hear in the vicinity of their nests are "the distress cries of the burrowing owl's young."[12] Rather than separating this information into field guide categories such as "description," "voice," "habitat," "nesting," and "range," Williams weaves these elements of natural history into her personal account of rootedness in the Great Basin, revealing how the owls have helped to enhance her understanding of home. "There are those birds you gauge your life by," she writes. "The burrowing owls five miles from the entrance to the Bear River Migratory Bird Refuge are mine."[13] Standing as "sentries" atop a mounded nest that "rises from the alkaline flats like a clay-covered fist," burrowing owls assert their place in the desert community.[14] Williams further illustrates the natural history of these birds by explaining that historically they nested in the abandoned burrows of prairie dog towns, which followed herds of bison across the plains. Burrowing owls, prairie dogs, and black-footed ferrets once flourished at the edges of a momentous migration, but with this migration that aerated the plains now silenced, they inhabit the margins of survival.

In spite of their endangered status, the burrowing owls at the entrance to the Bear River Refuge represent consistency for Williams, alerting her "to the regularities of the land" and helping her to define the seasons; "In spring, I find them nesting," she writes, "in summer they forage with their young, and by winter they abandon the Refuge for a place more comfortable."[15] The owls stand sentry to stability and the comfort of knowing that amid the changes Williams faces in her life, they will appear with the reliability of the seasons. Since discovering the burrowing owls' nests in 1960, she and Mimi have returned each year to "pay [their] respects" to the birds.[16] When Williams seeks to visit the owls in the autumn of 1983, however, she finds the mounds gone—gravelled over for the construction of the Canadian Goose Gun Club. Beginning *Refuge* with the disappearance of these birds she once believed to be constant, she foreshadows the change and uncertainty that will soon unsettle her home landscape.

Williams subtitles each chapter with the water level of Great Salt Lake, illustrating the birds' reliance on the marshlands of the Bear River Refuge. Located at the bottom of the Great Basin, the largest closed system in North America, Great Salt Lake has no outlet to the sea and therefore its water level fluctuates substantially in response to annual changes in precipitation. In the mid-1980s, as a result of higher-than-average rainfall, Great Salt Lake flooded, threatening the Bear River Refuge and infrastructure in the surrounding area. On a Sunday afternoon, Williams watches from her mother's hospital room as thousands of volunteers sandbag State Street, turning it into a river to prevent the flooding of the Latter Day Saints church office building. This transformation from pavement to a rushing river reflects the unsettling effects of illness on Diane Tempest and her family. What was once solid and certain becomes fluid and unstable. As Glotfelty suggests, the rising and falling lake level corresponds closely to the plot structure of *Refuge*, as though it were a novel rather than a memoir: the lake level rises as Diane Tempest's cancer worsens, peaks at her death, and falls during the grieving and healing period that follows.[17] By emphasizing the symbolic parallel between the rise of Great Salt Lake and the worsening of her mother's cancer, Williams develops the metaphor of landscape as refuge. The disappearance of a beloved landscape is akin to the loss of a loved one, as fluid uncertainty submerges a place that once nurtured life.

As a liminal space, wetlands illustrate the central question of *Refuge*: "How do you find refuge in change?"[18] While this question applies most immediately to Williams's uncertainty and lack of stability as she struggles with the loss of her mother, it also reflects on the broader impacts of environmental change in landscapes increasingly dominated by human settlement. Although the rise of the lake was a natural occurrence caused by a period of heavier than usual rainfall, its impact on resident and migratory bird populations is complicated by the infrastructure surrounding the lake. Prior to human settlement, the birds could have simply moved to higher ground, but highways and neighbourhoods now limit their ability to find suitable habitat on the shifting shoreline. This conflict is heightened by the ecological significance of the Bear River Migratory Bird Refuge. Supporting 208 species of birds, sixty-two of which are known to nest there, Great Salt Lake is "a crucial link in the chain of primary migratory, breeding, and wintering sites along the great shorebird flyways that extend from the arctic to the southern tip of South America."[19] By describing the wetlands of Great Salt Lake as "a fertile community where the hope of each day rides on the backs of migrating birds," Williams extends the ecological fertility of wetlands to a sense of hope and renewal that spans continents.[20]

In "Imagined Territory: The Writing of Wetlands," William Howarth traces shifting perceptions of this critical biome in literary depictions of wetlands. Although he does not include *Refuge* in his analysis, Williams's portrayal of the Bear River Refuge closely corresponds to the values he sees wetlands as representing, including "difficulty or uncertainty," "change," and "contingency or possibility."[21] Howarth contends that as they become increasingly rare, wetlands "are the landscape equivalent of extinct or endangered species."[22] Although the flooding of the Bear River Refuge is temporary, it serves as a synecdoche for the disappearance of wetlands throughout North America. Williams notes threats to individual species throughout *Refuge*, and her ecological focus on the temporary loss of the Bear River Refuge reflects numerous permanent losses of wetlands throughout North America.[23] "Nationwide, seventy-six endangered species are dependent upon wetlands," she writes. "Marshes across the country are disappearing without fanfare, leaving the earth devoid of birdsong."[24] By juxtaposing the ecological and sensory impacts of habitat loss, she reveals that if the Bear River Refuge were to permanently disappear, not only would the Great Basin lose a vital source of amphibian and avian refuge, but the quality of the air would change, becoming quieter, less animated, and less evocative.

The contrast between song-filled and silent skies is most evident to those who have learned to appreciate and distinguish the instruments in a symphony of birdsong, making works such as *Refuge* crucial to an understanding of the music of wetlands. In *Topophilia*, Yi-Fu Tuan suggests that our bond with place is greatly impacted by our sensory perceptions. Although auditory sensitivity is not particularly acute in humans, he nevertheless asserts, "Our experience of space is greatly extended by the auditory sense which provides information of the world beyond the visual field."[25] The landscape of *Refuge* is not merely a visual realm but one in which land and water meet in a space of auditory expansiveness. The various wetlands Williams visits throughout *Refuge* are awash with birdsong from dawn, when "each voice in the marsh awakens,"[26] to dusk, when "meadowlarks and yellow-headed blackbirds sing the shadows longer."[27] She often transliterates the birds' calls, filling the skies of her memoir with the "*Plee-ek! Plee-ek! Plee-ek!*"[28] of avocets, and the "*Ip-ip-ip! Ip-ip-ip!*" of black-necked stilts.[29] While these transliterations cannot begin to capture the birds' voices, the appearance of italicized phonetic calls alerts the reader to a landscape filled with the voices of other beings. "There are other languages being spoken by wind, water, and wings," Williams observes. "There are other lives to consider: avocets, stilts, and stones."[30]

While most chapters in *Refuge* echo with avian voices, others accentuate prolonged silences, portending pathology in both the family and place on which Williams relies for emotional sustenance. As Great Salt Lake floods, eclipsing habitat for numerous species, its shorelines and the Great Basin skies become increasingly silent. As the rising lake level corresponds to the worsening of Diane Tempest's cancer, so does a gradual silencing of the surrounding landscape. The chapter that details her mother's death is called "Sanderlings," and although Williams has conditioned her readers to expect the sounds of these birds, they are silent. They make only a fleeting appearance toward the end of the chapter, "wheeling over the waves of grief."[31] While this muted presence aligns with Williams's grieving process, it also reflects the skies above wetlands that no longer serve as sanctuary. As the quality of the land changes, so does the quality of air; and as terrestrial avian refuges become increasingly rare, the appearance of birds in flight will correspondingly diminish. If we fail to care for the health of wetlands and other avian habitat, permanent silences will replace temporary ones.

In *Grass, Sky, Song*, Trevor Herriot echoes the lyricism with which Williams conveys the wetlands of the Great Basin as he reflects upon the avian species that share his prairie grassland home. In a preface entitled "The Way Home," he introduces his deepening appreciation for his home landscape with the image of grassland birds returning to the northern prairie in late April. "As day breaks," he writes, "they will look at the land below and compare its patterns of colour, light, and shadow with patterns they carry in their genes or in their memories."[32] From the birds' annual return to a familiar landscape, Herriot segues to his own deeper engagement with place derived through his attentiveness to these birds. On a guided hike at Buffalo Pound Provincial Park in 1984, he was stirred by "the thought of creatures being endemic to the place I lived,"[33] and this discovery launched him on a quest to learn the names and natural histories of the grassland bird species that he has come to regard as "the presiding genius of the northern Great Plains."[34] His exploration led him to an appreciation of the species that inhabit the grasslands accompanied by a parallel awareness of their endangerment. By conveying a prairie inspirited with threatened species, Herriot offers the reader an animated but elusive landscape, as a prairie spring filled with horned larks, burrowing owls, and Sprague's pipits becomes an increasingly endangered season.

Like Williams, Herriot writes of a landscape central to his family history. His great-grandparents, James and Mary Herriot, settled near the village of Plum Creek, Manitoba, in 1881.[35] Their son, Herriot's grandfather, moved west to farm the Great Sand Hills; but later, when during the Great Depression his crops failed to return even the seeds he had planted, he moved north to bush country. Although Herriot and his family live in Regina, their search for a connection to native grasslands led them to share in the purchase of a 160-acre parcel of land on a tributary of the Qu'Appelle River, where they spend much of their leisure time. By narrating his struggle to find land in a place that still supports Sprague's pipits skylarking on a summer morning, Herriot reveals both his own dedication to grassland birds and the increasing challenge these birds face when seeking to find viable habitat.

The grassland landscape has always been a shifting one, altered by fire and the movements of buffalo herds. The recent shift to an intensely cultivated agrarian landscape, however, has been detrimental to grassland bird populations. As Bridget Stutchbury explains in *Silence of the Songbirds*, "The world under the wings of our migratory songbirds is not the same landscape that their ancestors flew over."[36] The shift from native grass to cultivated fields has made it increasingly difficult for migrating birds to find patches of prairie suitable for them to breed and raise young. Following cultivation, birds relegated to field edges may persist where grass and brush offer a semblance of their former native grassland habitat. Yet with the industrialization of agriculture, pressure to farm even these grassy margins has resulted in the increasing loss of these remnants of habitat, forcing the birds off the edges of the land they once populated into endangerment and immanent extinction.

The distinction between altered and unaltered grassland is a visually subtle one. To the eye accustomed to seeing grasslands as a patchwork of cultivated fields, conceiving the prairies as a clear-cut far more extensive than those we condemn in forested areas remains a perceptual challenge. As Herriot explains, however, "of this grassland that makes up the midriff of North America, somewhere between 75 and 99 percent has been ploughed under, depending on the kind of prairie under consideration."[37] Although it might appear as though these lands could be easily reseeded with native grasses, the ecological complexity of native grassland takes years to reproduce, just as it takes decades or centuries to re-establish the diversity of an old growth forest. In *Grassland*, Richard Manning characterizes grasslands as the "most degraded and most misunderstood" of biomes in the United States.[38] In both the United States and Canada, legislators have been slow to protect grasslands as national, provincial, or state parks, as these landscapes are not commensurate with an inherited view of "wilderness" that prioritizes scenic mountain vistas over nature's more subtle biomes. Even as the conservation movement has acquired the wisdom of preserving remnants of all ecosystems regardless of their perceived aesthetic value, sanctuaries such as Grasslands National Park have met with resistance by those who believe such lands can be adequately preserved as ranchland.[39]

In *Grassland*, Manning writes, "This story of grass ranges like the prairie's hills, and like the hills, is best taken from the small to the large, from the specific to the general, and from the material to the spirit."[40] *Grass, Sky, Song* invokes this shifting landscape of specificity and generalization, materiality and spirituality, as Herriot portrays the landscape through alternately broad and focused lenses, frequently blending physical and spiritual perceptions of his prairie home. He shifts from general to particular by introducing species in geographical, historical, and often anecdotal contexts, as well as providing the reader with information relating to the natural history of each species. Like Williams, he emphasizes the importance of learning the name of each species, "not so much in the possession it afford[s] as in its capacity to call things forth from generality into a particularity that allow[s] for admiration, familiarity, even wonder."[41] By emphasizing the names and natural histories of the species that inhabit his account of the grasslands, he inspires not only wonder, but also a capacity for concern achieved by juxtaposing the vitality of species with an account of their decline.

While the chapters of his memoir reflect the centrality of grassland birds to Herriot and many of his fellow prairie dwellers, the profiles of grassland bird species that follow each chapter often reflect species clinging to the margins of survival. In the profile of the western meadowlark that follows the chapter "Birds of Promise," for example, Herriot emphasizes threats to the western meadowlark in a textual structure that mimics the increasing marginality of the meadowlark, as even its ability to thrive on the edges of agricultural landscapes is threatened. "If more farmers do not take steps to preserve or restore grassy margins on their land," he contends, "the meadowlark singing from the fence line will one day fade from the working landscape to become merely another piece of settler nostalgia on the Great Plains."[42] As Herriot argues for the preservation and restoration of grassy margins, the meadowlark's persistence at the borders of his text emphasizes the significance of margins, both textual and geographical. Such margins can only contribute to the preservation of the meadowlark, however, if they serve as a space from which meadowlarks can reoccupy the prairie landscape.

Herriot's search for endangered birds is often a collaborative journey, and he acknowledges the contributions of many field guides to his understanding of grasslands. While "some were books," he writes, "the best were the ones who guided by walking along with me in the field, showing me the sounds and shapes of things, answering questions, and sharing their delight."[43] Some of the many "field guides" who inform *Grass, Sky, Song* are biologist Stephen Davis and poet Don McKay. On a day spent birdwatching near St Peter's Abbey, Herriot and McKay share not only observations of numerous grassland species, but also the challenges of translating avian voices into text. Of the Sprague's pipit's flightsong, Herriot writes, "Don called it a 'skirling sound.' More like the air speaking than a bird, he said. That sounds right—this thin downward spiral of a song, ethereal yet so adapted to its ambience that it seems to be coming from the sun or the sky itself."[44] As he transitions seamlessly from McKay's onomatopoeic "skirling sound" to his own description of a "thin downward spiral of song," Herriot offers a glimpse not only of the challenge of conveying the voices of birds, but also of the benefits of collaboration. As McKay and Herriot seek to translate the untranslatable, they offer a layered description of a unique and otherworldly sound falling from sky to ground.

Throughout *Grass, Sky, Song*, Herriot encourages a respectful curiosity about avian species that reflects McKay's belief in the importance of poetic attention. This attention, which McKay describes in the essay "Baler Twine" as "a sort of readiness, a species of longing which is without the desire to possess,"[45] Herriot depicts as "a leaning toward the other without wanting to possess it or turn it into forms of knowledge, a way of listening that might over time deepen our sense of what it means to be in a place."[46] Although Herriot contends that this is "different from scientific attention," he nevertheless acknowledges that scientific research can be conducted with a spirit of reverence in spite of its invasive and demystifying tendencies.[47]

Another collaboration featured in *Grass, Sky, Song* is Herriot's 2005 journey with ornithologists Stuart and Mary Houston to mark the 125th anniversary of John Macoun's survey of the Canadian prairies. Herriot portrays this journey as an imbrication of past, present, and future—as he retraces a journey from the past, details a present in which ecological and human communities have been damaged by continued mistreatment of the landscape, and questions whether endangered species and human and ecological communities can be healed in the future. Recruited by Sanford Fleming, chief engineer for the proposed Canadian Pacific Railway, Macoun participated in five surveying expeditions in the Northwest between 1872 and 1881.[48] These expeditions corresponded to a period of particularly high rainfall, so he reported that much of the Northwest was ideally suited to agriculture.[49] As 2005 was a wet year, Herriot and the Houstons experienced the prairie much as Macoun would have in terms of climate. However, they traversed a land extensively altered as a result, in part, of Macoun's agricultural recommendations. The cultivation of native prairie, followed by a move toward increasingly mechanical and chemically intensive farming practices, has resulted in a prairie far less resonant with birdsong than that which Macoun would have encountered.[50]

The loss of grassland bird habitat has resulted not only from the segmentation of the prairie into farmland, but also from an insidious cause Herriot discovers when he learns that his wife, Karen, has been diagnosed with breast cancer. Joining a growing number of works incorporating toxic discourse, *Grass, Sky, Song* suggests a parallel between the indiscriminate use of agricultural pesticides and increased rates of cancer in rural areas. In *Writing for an Endangered World*, Lawrence Buell defines toxic discourse as "expressed anxiety arising from perceived threat of environmental hazard due to chemical modification by human agency."[51] An activist discourse aimed at consciousness-raising, it foregrounds the consequences of chemical toxicity to the natural environment and human health, stressing the inherent link between our health and the health of the places we inhabit. As Karen's diagnosis with breast cancer prompts Herriot to explore a link between pesticides and human and avian health, he segues from effects to possible causes. While in the waiting room of a cancer clinic during one of Karen's treatments, he overhears two farmers discussing the latest insecticide they have been using on their land as their wives, with their backs turned, discuss their recent cancer treatments and the prevalence of cancer among farm women. The irony of these exchanges reflects our society's resistance to exploring the effects of the myriad chemicals used in food production, even as they threaten our health and the health of our loved ones.

Seeking connections between the endangerment of grassland birds and the health of humans who inhabit the grassland landscape, Herriot discovers a trail of research and registration policies that tend to veil, rather than illuminate, the effects of pesticides on grassland inhabitants. His research into the chemicals to which Karen may have been exposed leads him to carbuforan, a pesticide used throughout the prairies to control grasshopper outbreaks in the 1980s. He illustrates a trend whereby pesticides are easily registered, often requiring only safety studies conducted by the manufacturer, and thereby leaving the burden of deregulation to underfunded scientists and environmental groups. Granular carbuforan, for example, was registered in the United States and Canada "despite the manufacturer's own supervised field trials in which at least forty-five species of birds died."[52] It was not restricted until the 1990s, after causing more than a decade of immeasurable damage to human health and the health of grassland bird species.[53] In the meantime, it was central to an extensive spraying campaign that extended beyond privately owned farmland to public lands, such as grasslands, ditches, road allowances, and irrigation canals.

In Part III of *Grass, Sky, Song*, entitled "Pastures Unsung," Herriot unveils a web of economically driven policies that hamper the development and funding of research examining the effects of pesticides. Although tasking pesticide companies with the responsibility to demonstrate the safety of their products is unlikely to yield unbiased results, this practice persists. Meanwhile, organizations that prioritize marketable research often overlook studies into the adverse effects of pesticides and the development of safer alternatives. Research on the effects of pesticides is limited not only by funding, but also by the challenge of demonstrating cause and effect in ecosystems influenced by a myriad of variables.[54] As Herriot writes, "the actual, multi-dimensional mystery in which birds live, rear their young, and die is nothing like a laboratory.... Ecosystems are borderless, chaotic, and ultimately unfathomable."[55] By emphasizing the "unfathomable" nature of ecosystems, he points toward the need to implement measures of toxicity that extend beyond scientific proof.

While he supports and celebrates scientific research throughout *Grass, Sky, Song*, often paying tribute to the careful research scientists have contributed to efforts to restore grassland bird populations, he also acknowledges the shortcomings of a regulatory system that demands evidence of toxicity, rather than evidence of a lack of toxicity: "Beneath the aspirations of biology, of all science, there is the power of a deeper way of knowing, a knowledge that has intuitive and moral dimensions too important to abandon merely because science is unable to provide incontrovertible evidence. Each of us knows that it is not good to kill creatures wantonly. Each of us knows that it is not good to pour poison on the land that feeds us. Any child can tell you these truths."[56] Syntax mimics sense in this passage, as Herriot subordinates the "aspirations of biology" to a knowledge founded in "intuitive and moral dimensions." While the syntactic layering of the first sentence mirrors the complex nature of scientific experimentation, which resists certainty, the simplicity of the final sentences reflects the straightforward "truths" it conveys.

Agricultural policy in North America has overwhelmingly followed the path of legislating economic privilege, while entrenching no obligation to care for the long-term health of the land or its occupants. As Herriot reveals, the obligation to care for the land rests not solely with the landowner, but with all who benefit from a system of chemical-intensive and inexpensive food production. "Like most North Americans," he observes, "I depend on industrialized systems to grow, process, package, and deliver much of my food."[57] If we are to forge changes in agricultural policy, we must acknowledge our complicity as consumers in encouraging the production of cheap and plentiful food. Just as it is difficult to assess blame or victimhood to the farmers and their wives in the cancer clinic, it is difficult to assess blame as the economic benefactors, and thus the indirect perpetrators, of chemical-intensive farming practices.

Both Herriot and Williams acknowledge the lack of absolute "proof" that the cancer suffered by their loved ones resulted from environmental toxicity. While legislation demands proof, often allowing the application of chemicals to continue until research provides incontrovertible evidence of their risk, narratives of toxic discourse validate experience. "In contemporary toxic discourse," Buell asserts, "victims are permitted to reverse roles and claim authority."[58] Having witnessed the suffering of loved ones, Herriot and Williams have gained the authority to advocate for caution in the absence of proof. Given the length of time it takes for cancer to manifest itself, it is often difficult to identify its origins, but the prevalence of cancer in agricultural areas and among those exposed to nuclear fallout and radiation implies a compelling connection between toxicity in the landscape and long-term impacts on the health of humans and other species.

In the prologue to *Refuge*, Williams writes, "Most of the women in my family are dead. Cancer. At thirty-four, I became the matriarch of my family."[59] Throughout *Refuge*, she makes no overt accusation as to the cause of the reproductive cancers devastating her family; however, in a polemical epilogue entitled "The Clan of One-Breasted Women," she suggests that these high levels of cancer were caused by above-ground atomic testing conducted in Nevada between 27 January 1951 and 11 July 1962. From her father she learns that her recurring dream of a flash of light in the desert resulted from a nuclear explosion she witnessed as her family drove past Las Vegas at dawn on 7 September 1957, the day before her second birthday. The fourteen years after this date that it took for her mother's breast cancer to become manifest is commensurate with the amount of time cancer from radiation requires to develop, according to Howard L. Andrews, an authority on radioactive fallout at the U.S. National Institutes of Health.[60] Although Diane Tempest survived this initial cancer, she passed away from ovarian cancer three decades after seeing the nuclear blast. As Reg Saner asserts in "Technically Sweet," "Since latent cancer may take twenty years or more to announce itself, plutonium is the perfect industrial murder."[61] Williams reveals the irony that in testing a weapon in order to maintain military supremacy over other nations, the U.S. Atomic Energy Commission caused extensive and untold damage to the landscape and people of its own nation. Although invisible, radiation illustrates ecological connections between the land and its occupants that extend across time and space.

In *Grass, Sky, Song*, an Ursuline nun, who runs a holistic healing centre, suggests to Karen that "cancer is an invitation to wake up and come to wholeness."[62] Herriot contends that pathology of the landscape should yield a similar awareness, encouraging its human caretakers to consider the health of the earth, water, air, and species that sustain us both physically and spiritually. As *Refuge* documents the eventual retreat of the floodwaters and the return of species to the Bear River Refuge, *Grass, Sky, Song* concludes with glimpses of hope based on bird species showing tentative signs of recovery. In 2005, the Saskatchewan Breeding Bird Survey indicated a resurgence in the numbers of Sprague's pipits, a species whose endangerment Herriot details extensively in *Grass, Sky, Song*. While such improved numbers do not assure a recovery of the species, they support his assertion of the importance of research and habitat preservation to preserve grassland species.

Early in his memoir, Herriot compares the potential extinction of grassland birds to "the tide going out for the last time, a long, slow wave drawing back into a sea that exists only in memory."[63] While much of his memoir documents statistics that support the image of a retreating tide, the reappearance of birds in regions that their species have not occupied for some time reflects a regenerative force in nature that endures in spite of continued threats. Herriot provides an example of such persistence in his account of a burrowing owl that appears on the margins of a zero-till field north of Regina. Although the owl does not stay long, its reappearance in the landscape after a twenty-year absence encourages optimism. For the brief time it stayed, Herriot notes, "it may well have been the northernmost burrowing owl on the planet."[64] He also provides an account of a pair of burrowing owls that successfully raise their young in a field of durum wheat, illustrating "what can happen when farmers make room for grassland birds."[65] The protection the farmers offer to these birds, which includes a circle of unseeded land around their nests and perches constructed so the owls can see above the crop, represents a level of awareness and concern contrasting to the indifference of the gun club owners who paved over the owl burrows near the Bear River Refuge. Although in a different landscape, in a different nation, and relayed in a different text, Herriot's anecdote reveals the impact concerned landowners may have on the survival of a species. While the seasonal tides of migratory birds returning to the prairies and wetlands are weaker than in the past, birds continue to offer spring to those willing to listen for it. By echoing their calls, Herriot and Williams emphasize the importance of gaining an understanding of and appreciation for the avian inhabitants of threatened landscapes.

Notes

1. Don McKay, "In Aornis," in *Slip/Strike* (Toronto: McClelland & Stewart, 2006), 66, ll. 1–3.
2. Ibid., ll. 13–14.
3. Terry Tempest Williams, *Refuge: An Unnatural History of Family and Place* (New York: Vintage, 1992 [1991]), 3.
4. Ibid., 40.
5. Williams's ancestors immigrated to the Great Basin from Illinois in the 1850s, establishing a family lineage in Utah. Although born in Corona, California, on 8 September 1955, Williams was raised in Salt Lake City, Utah, surrounded by a large extended family. In *Refuge*, she reveals her close relationships with her parents, her brothers Steve, Dan, and Hank, her maternal grandparents Lettie Romney Dixon and Donald "Sanky" Dixon, and her paternal grandparents Kathryn Blackett Tempest and John Henry Tempest, Jr.
6. Lucy R. Lippard, *The Lure of the Local: Sense of Place in a Multicentered Society* (New York: New Press, 1997), 9.
7. Ona Siporin, "Terry Tempest Williams and Ona Siporin: A Conversation," *Western American Literature* 31, no. 2 (1996), 101.
8. Ibid.
9. In university, wishing to major in both literature and biology, Williams approached the English department at the University of Utah to ask if she could major in "environmental English" and the biology department to ask if she could major in "literary biology." Both refused, so she settled for a major in English and a minor in biology. During her undergraduate studies, Williams spent three summers at the Teton Science School in Jackson Hole, Wyoming, and she later collaborated on her first book, *The Secret Language of Snow* (1984), with Ted Major, founder of the Teton Science School. Katherine R. Chandler and Melissa Goldthwaite, "Beginning Words: Introduction and Invitation," in *Surveying the Literary Landscape of Terry Tempest Williams: New Critical Essays*, ed. Katherine R. Chandler and Melissa Goldthwaite (Salt Lake City:

University of Utah Press, 2003), xi. In graduate school, her interest in blending literature and ecology continued, as she completed a master's degree in environmental education. As an ecologist and storyteller, Williams brings to ecological study the power of story and to storytelling the specificity of ecological knowledge, demonstrating the ways in which these disciplines can inform and strengthen one another.

10 Derived from the Greek roots *oikos* ("house" or "dwelling") and *logia* ("the study of"), the term *oecologie* was introduced in 1866 by German zoologist Ernst Haeckel. In 1869, in a lecture at Jena, Haeckel defined *oecologie* as "the body of knowledge concerning the economy of nature—the investigation of the total relations of the animal both to its inorganic and to its organic environment; including, above all, its friendly and inimical relations with those animals and plants with which it comes directly or indirectly into contact—in a word, ecology is the study of all those complex interrelations referred to by Darwin as the conditions of the struggle for existence." Robert C. Stauffer, "Haeckel, Darwin, and Ecology," *Quarterly Review of Biology* 32, no. 3 (1957), 141.

11 John Tallmadge, "Beyond the Excursion: Initiatory Themes in Annie Dillard and Terry Tempest Williams," in *Reading the Earth: New Directions in the Study of Literature and Environment*, ed. Michael P. Branch, Rochelle Johnson, Daniel Patterson, and Scott Slovic (Moscow: University of Idaho Press, 2008), 202.

12 Williams, *Refuge*, 9.
13 Ibid., 8.
14 Ibid., 8–9.
15 Ibid., 8.
16 Ibid., 11.
17 Cheryll Glotfelty, "Flooding the Boundaries of Form: Terry Tempest Williams's *Unnatural History*," in *Change in the American West: Exploring the Human Dimension*, ed. Stephen Tchudi (Reno: University of Nevada Press, 1996), 160–61.
18 Williams, *Refuge*, 119.
19 Ibid., 263.
20 Ibid., 22.
21 William Howarth, "Imagined Territory: The Writing of Wetlands," *New Literary History* 30, no. 3 (1999), 521.
22 Ibid.
23 This approach reflects recent trends in conservation biology to focus on the preservation of ecosystems for the range of species they support, rather than simply on individual species.
24 Williams, *Refuge*, 112.
25 Yi-Fu Tuan, *Topophilia: A Study of Environmental Perception, Attitudes, and Values* (New York: Columbia University Press, 1974), 9.
26 Williams, *Refuge*, 151.
27 Ibid., 108–9.
28 Ibid., 15.
29 Ibid., 19.
30 Ibid., 29.
31 Ibid., 231.
32 Trevor Herriot, *Grass, Sky, Song: Promise and Peril in the World of Grassland Birds* (Toronto: HarperCollins, 2009), 1.
33 Ibid., 12.
34 Ibid., 13.
35 Plum Creek later became known as Souris, Manitoba.
36 Bridget Stutchbury, *Silence of the Songbirds: How We Are Losing the World's Songbirds and What We Can Do to Save Them* (Toronto: HarperCollins, 2007), 26.
37 Herriot, *Grass, Sky, Song*, 23.

38 Richard Manning, *Grassland: The History, Biology, Politics, and Promise of the American Prairie* (New York: Penguin, 1995), 3.
39 Grasslands National Park was first proposed in 1956 by the Saskatchewan Natural History Society as a means of preserving remnants of the grassland ecosystem. A final agreement to create the park was signed in 1988. The park has yet to reach its proposed size of 900 square kilometres, as land continues to be purchased from ranchers in the surrounding area willing to sell.
40 Manning, *Grassland*, 1.
41 Herriot, *Grass, Sky, Song*, 12.
42 Ibid., 42.
43 Ibid., 257.
44 Ibid., 85.
45 Don McKay, "Baler Twine," in *Vis à Vis: Field Notes on Poetry and Wilderness* (Wolfville, NS: Gaspereau Press, 2001), 26.
46 Herriot, *Grass, Sky, Song*, 218.
47 Ibid.
48 Trained as a botanist, Macoun captured a fleeting portrait of native grasslands that his recommendation for settlement would immediately begin to endanger.
49 Herriot's presence on the prairie is tangentially linked to Macoun's journey, as shortly after Macoun returned to Ottawa with a glowing report of the potential for agriculture on the prairies, Herriot's great-grandfather settled near Souris, Manitoba. When his son moved farther west to farm in the Great Sand Hills, he found that the area Macoun had characterized as agriculturally viable was far too arid to support crops in drier years. Herriot, *Grass, Sky, Song*, 211.
50 In addition to the extensive habitat loss birds suffered as a result of the conversion of native grassland to cropland, grassland bird populations are also negatively influenced by environmental toxicity, West Nile disease, energy extraction industries, and drought brought on by global climate change. Given the scope of grassland bird migration, Herriot notes, most species would be subject to some if not all of these forces at some point on their migratory journeys.
51 Lawrence Buell, *Writing for an Endangered World: Literature, Culture, and Environment in the U.S. and Beyond* (Cambridge, MA: Harvard University Press, 2001), 31.
52 Herriot, *Grass, Sky, Song*, 159.
53 As Stutchbury explains, "Granular carbofuran is a notorious bird killer, yet its use was virtually eliminated in the United States only after a twenty-year struggle by environmental groups and scientists." Stutchbury, *Silence of the Songbirds*, 107.
54 The complex and uncontrolled nature of ecosystems tends to favour research detailing the lethal rather than the sublethal effects of pesticides. Lethal effects refer to the immediate deaths caused by a pesticide application, while sublethal effects refer to deaths caused by impacts that manifest themselves over a long term. Although pesticides may not kill a bird directly, exposure to pesticides may cause disorientation or illness, preventing it from finding adequate food, avoiding predators, or successfully nesting, feeding, and protecting its young. This legacy of sublethal effects must be further explored in order to determine the true impact of toxicity in the grassland landscape.
55 Herriot, *Grass, Sky, Song*, 177.
56 Ibid., 181.
57 Ibid., 164.
58 Buell, *Writing for an Endangered World*, 44.
59 Williams, *Refuge*, 3. Another version of this statement appears in "Testimony," a record of Williams's testimony before the Subcommittee on Fisheries and Wildlife Conservation and the Environment concerning the Pacific Yew Act of 1991: "My name is Terry Tempest Williams. I come to you as a woman concerned about health. I am thirty-six years old. I am the matriarch of my family. Nine women in my family

have had mastectomies. Seven are dead. My mother passed away from ovarian cancer in 1987. We have had subsequent deaths in 1988, 1989, and 1990. I am aware of the intimate, painful struggle of women, families, and cancer." Williams, "Testimony," in *An Unspoken Hunger: Stories from the Field* (New York: Vintage, 1995), 125. Williams testified before this subcommittee to advocate for the careful management of Pacific yew trees, which contain taxol—used as a treatment for ovarian and breast cancer.
60 Williams, *Refuge*, 286.
61 Reg Saner, "Technically Sweet," in *The Four-Cornered Falcon: Essays on the Interior West and the Natural Scene* (Baltimore: Johns Hopkins University Press, 1993), 76.
62 Herriot, *Grass, Sky, Song*, 186.
63 Ibid., 68.
64 Ibid., 228.
65 Ibid.

CHAPTER 13

What Should We Sacrifice for Bitumen?
Literature Interrupts Oil Capital's Utopian Imaginings

Jon Gordon

There is a narrative of progress built into the political and economic culture of Canada. The surmounting of obstacles has been, and remains, prominent in narratives of settlement and development, which can be located within a larger narrative of Western rational, scientific, and technological progress. As Karl Clark, the University of Alberta scientist who pioneered the hot water extraction process for separating oil from sand, wrote in 1949, "The seamed cliffs on the Athabasca carry a smirk on their faces, and no one who visits the Northland can miss its meaning. It says plainly: 'You are doing big things with everything else in this North Country, but I have still got you beat. When are you going to do something about it?'"[1] Nature poses a challenge; human ingenuity holds the answer. When Clark wrote those words, something was being done about it: attempts to commodify bitumen had been ongoing for nearly seventy-five years. It was only toward the end of his career that those efforts started to show success. Such efforts, though, require the sacrifice of various forms of human and non-human community, sacrifices justified in the name of progress. Pieter Vermeulen shows how both Benedict Anderson's and Jean-Luc Nancy's understandings of community find common ground "where they reflect on the role of death and loss in the constitution of community."[2] It is in the community's response to loss that it justifies its existence. Perhaps we can see this type of response most clearly in something like social contract theory as articulated by Thomas Hobbes, where the state arises when individuals agree to a mechanism for mediating conflicts between their interests: if I am wronged, if my property is stolen or I am murdered, the community, the state, or the sovereign will prove its legitimacy in how it responds. However, an extension of this contractual function operates through a symbolic process of abstraction, which seeks to enable mourning for members of the community who die in its service. Part of industry's challenge in developing bitumen is creating methods for managing the suffering produced as a result of human and non-human sacrifices required to realize that development. The rhetorical battle over bitumen development (among various political parties, federal and provincial governments, industry, academics, NGOs, environmentalists, and so on) can be understood, in part, by considering who is attempting to manage the losses produced by bitumen development and who is trying to mobilize them to change the status quo.

Literature is one site of response, considered by both Anderson and Nancy, which can justify the forms of loss that enable community or counter them by imagining other ways of responding. This chapter will think about a few of the forms of loss caused by oil capital in Alberta, consider how government and industry responses attempt to put those losses in the service of liberal capital, and explore the possibility for literature to defy those attempts.

Ducks and the Nation State

Perhaps the most widely publicized loss to bitumen production was the deaths of sixteen hundred ducks on Syncrude's Aurora tailings ponds in April 2008. Even though there have been subsequent losses caused by bitumen development, these ducks continue to haunt the industry and government in ways that the injury and death of workers and cancer rates in First Nations communities downstream do not.³ Despite the continuing recurrence of this scene of loss, there has not been a response to the duck deaths that has successfully changed the social licence for development. Partly this is due, I will argue, to the primary role of the nation state as mediator and diffuser of tensions, even when it has ceased to be the dominant imagined community at work. As Len Findlay has put it, "transnational corporations and the United States both need nation-states as satellites and proxies to talk (down) to, and to help develop the deregulating political and fiscal instruments necessary for further 'globalization' of the market economy; at the same time, citizens need nation-states as sites of resistance to such political and economic hegemony exercised by unelected elites and insiders."⁴ In addition to serving as an actual site of resistance, the nation state also serves as a perceived site of resistance, a focus for citizen action, in situations where its role is limited.

Where the potential for effective resistance does not exist, or such change would be counter to dominant interests, the nation state provides a focus for symbolic action. Vermeulen continues, "The nation-state relies on the readiness of its subjects to die for their community, and in order to promote such self-sacrifice, it forecloses the horror and sadness of dying by integrating death with an overarching framework in which it acquires public value and meaning."⁵ This logic of death and self-sacrifice seems to operate at levels of imagined community other than the nation state as well: as just one example, despite the deaths of and injuries to workers in bitumen operations, extraction continues at an increasing rate; workers continue to put themselves at risk in the service of capital. To understand how the logic of self-sacrifice underwrites bitumen development we need to rethink Benedict Anderson's foundational theorizing of the imagined community. We can see through the public relations work of the bitumen industry that, despite negative press, it has managed to retain its social licence to operate. That pollution generated by industry is threatening human and non-human lives is both denied and managed within the myth of progress. The illness and death of workers, Natives, and wildlife is both denied and shown to be compensated for by the benefits of development—energy security, profit, growth, jobs, funding for culture, education, and so forth. Further, as many argue, "environmental issues are largely local or regional in geographic scope while the economic benefits are provincial if not national."⁶ The decision to develop bitumen is a trade-off, and the global considerations trump the local. Therefore, even thoughtful commentators on bitumen development, like Vice-Chair of the Board of the Edmonton Economic Development Corporation Peter Silverstone and University of Alberta Business Professor Andrew Leach, often focus on greenhouse gas emissions rather than heavy metals, say, because one is a "global" problem while the other is merely local.⁷ An additional variant of this response is pragmatic relativism: Ezra Levant makes this point explicitly in *Ethical Oil*, where he argues—on what he claims are progressive grounds—that Alberta bitumen is better than oil from other nation states: better environmentally, better in its contribution to world peace, better in its treatment of minorities, and better in terms of economic justice.

When these forms of management fail, though, to foreclose the horror of particular losses, as with the death of the ducks or as with University of Alberta water ecologists Erin Kelly and David Schindler's revelation of water pollution caused by industry,⁸ promises of improvement fit into the Enlightenment narrative that the application of reason will achieve its liberating potential sometime in the future. Syncrude will improve its duck deterrent systems; the provincial government will convene a panel of scientists to compare the water data and make the appropriate changes. After the duck deaths, then Syncrude CEO Tom Katinas made a "promise to do better";⁹ this, as we will see, is the promise of modernity—the future will be better—any deaths endured today will have been worthwhile because of the kind of society that we will inhabit in the future. In the ensuing court case, in which Syncrude was charged under provincial environmental laws and the federal Migratory Birds Convention Act, the company was fined $3.2 million, which went to fund a University of Alberta research project into bird migration and deterrent systems, the Alberta Conservation Association, and a wildlife management program at Fort McMurray's Keyano College, despite their defence lawyer's claim that "If [...] Syncrude is guilty of this crime, the government is complicit and the industry is doomed" because "if by having a tailings pond we're guilty of this charge, we have to stop having tailings ponds": if that's the case, "we're done."[10] By implication, if industry is doomed, then so is the future that it is supposed to make possible. Of course, the guilty verdict and the fine have not doomed the industry or put an end to tailings ponds, despite the truth in the lawyer's claim about government complicity. If it is against the law to emit a deleterious substance into the environment, and the government has granted a licence to do precisely that, then the government is complicit and the verdict should require ending the use of tailings ponds. The verdict and the fine, rather, have precisely enabled a return to the status quo by providing the appearance of justice being done and a form of symbolic closure to the sadness expressed over the ducks' deaths. The form the fines took—with some of the money directed to a research project into bird migration and deterrent systems, the Alberta Conservation Association, and a wildlife management program at Fort McMurray's Keyano College[11]—can be read as an attempt at "integrating death with an overarching framework in which it acquires public value and meaning."[12] Like soldiers dying for their country whose sacrifices are said to make the lives of their fellow citizens safer, the deaths of these ducks will make the world safer for other ducks, will make the coexistence of ducks and bitumen mining possible.

However, as Imre Szeman puts it, "It is not that we can't name or describe, anticipate or chart the end of oil and the consequences for nature and humanity. It is rather that because these discourses are unable to mobilize or produce any response to a disaster we know is a direct result of the law of capitalism—limitless accumulation—it is easy to see that nature will end before capital."[13] The question, then, is why are these critical discourses unable to mobilize any effective response? Why does pointing out the flawed logic of an imagined community not lead to any collective reimagining?

*

21 February 2011: Libyan rebels ransack a Suncor Energy Ltd drilling platform in the Sahara Desert, forcing three Canadian workers to walk sixteen kilometres to another Suncor worksite.

*

Spinning Sacrifice

For Anderson, "the national imagining of community, when considered as a particular style, in the final analysis consists in a particular stylization of death";[14] thus, "imagining community differently," Vermeulen concludes, "also means finding another way to relate to the death of the other."[15] We might then understand the current debate surrounding bitumen development as a tension in what has been the dominant stylization of death, loss, and sacrifice required by development. Within the collective imaginary there continues to be a series of attempts to find other ways of relating to the death, loss, and sacrifice of the other, an other understood as both human and non-human. How do we make sense of the duck deaths? What story do we tell? Is it part of a larger ecological tragedy or a freak accident or, pace Levant, better than the alternative(s)? Could the story change the way we develop bitumen?

The way we imagine collectivities in time is crucial in this respect. This imagining goes beyond naming, describing, anticipating, or charting effects. For Anderson, national imagining takes place in "homogeneous, empty time," where "simultaneity is, as it were, transverse, cross-time, marked not by prefiguring and fulfillment, but by temporal coincidence, and measured by clock and calendar."[16] Thus, though "an American will never meet, or even know the names of more than a handful of his 240,000,000-odd fellow-Americans […] he has complete confidence in their steady, anonymous, simultaneous activity."[17] Today, though, while this form of imagining continues to bind members of national collectivities, another form of imagining is, perhaps, even more important, and has shifted the focus onto a future that makes up for the failures of the past. Therefore, while Anderson points out that "having to 'have already forgotten' tragedies of which one needs unceasingly to be 'reminded' turns out to be a characteristic device in the later construction of national genealogies,"[18] in moving beyond the nation as the key site of collective imagining, past tragedies are incorporated into the narrative as developmental stages. They need to be remembered as moments enabling progress. Anderson writes,

> English history books offer the diverting spectacle of a great Founding Father whom every schoolchild is taught to call William the Conqueror. The same child is not informed that William spoke no English, indeed could not have done so, since the English language did not exist in his epoch; nor is he told "Conqueror of what?" For the only intelligible modern answer would have to be "Conqueror of the English," which would turn the old Norman predator into a more successful precursor to Napoleon and Hitler. Hence "the Conqueror" operates as […] a kind of ellipsis […] to remind one of something which it is immediately obligatory to forget. Norman William and Saxon Harold thus meet on the battlefield of Hastings, if not as dancing partners, at least as brothers.[19]

However, while they might prompt critical reflection on the role of the nation state and make one question identifying with it, recalling these moments does not necessarily enable resistance to the narrative of modernity. This narrative can assimilate them as stages in the ineluctable march of abstract progress.

*

In his 1927 novel *Oil!* Upton Sinclair wrote of the dangers of working in the California oilfields, "of all the thousands of men who had worked there, seventy-three out of every hundred had been killed or seriously injured during the few years of the field's life! It was literally true that capitalist industry was a world war going on all the time, unheeded by the newspapers."[20] Undoubtedly the industry would today point to much better statistics about workplace injuries and fatalities. Progress is being made.

*

A *Fort McMurray Today* article, "Killed Workers Remembered," recounts the observation of an International Day of Mourning for workers killed and injured on the job, and quotes Kirk Bailey, vice-president of Suncor, as saying, "We can do better and we are going to commit ourselves to having a workplace that's absolutely free of injuries and I know we believe that at Suncor and I know the other industry players here in the region are equally committed to that goal."[21] This focus helps to defer potential opposition to workers' deaths, or bitumen extraction more generally, into the future (e.g., we understand we are not perfect, but we're working to improve, and we will not settle for less than perfection *at some unnamed future date*). Similarly, industry and its apologists like to trot out statistics such as the fact that "to produce one barrel of oil sands oil takes 38 percent less emissions now than it did in 1990."[22] Yes, worker deaths are tragic; yes, pollution is bad; but these sacrifices are necessary to secure the future of limitless energy free of negative effects.

*

How will the sacrifices of those working to stabilize Japan's damaged Fukushima Daiichi nuclear reactors be remembered?

*

Writing of tombs of unknown soldiers, Anderson states, "The cultural significance of [tombs of unknown soldiers] becomes even clearer if one tries to imagine, say, a Tomb of the Unknown Marxist or a cenotaph for fallen Liberals. Is a sense of absurdity unavoidable? The reason is that neither Marxism nor Liberalism are much concerned with death and immortality. If the nationalist imagining is so concerned, this suggests a strong affinity with religious imaginings."[23] While the monument in Fort McMurray's Howard Pew Park is not a tomb to an unknown soldier, Carol Christian's photo shows a ceremony strikingly similar to the ritual laying of wreaths at cenotaphs on Remembrance Day (see Figure 13.1). Perhaps the focus is not on death and immortality in the way Remembrance Day ceremonies are, not on the deaths that secured the immortality of the nation. But how is death being imagined here? Is the labour movement concerned with death and immortality, with religious imaginings?

Figure 13.1 An unidentified man lays flowers at a memorial for killed and injured workers in J. Howard Pew Memorial Park in Waterways (Fort McMurray), Regional Municipality of Wood Buffalo, Alberta. Photo: Carol Christian, *Fort McMurray Today* staff.

The International Day of Mourning was established by the Canadian Labour Congress in 1984 as a national day of mourning for workers killed or injured on the job. The quote from the representative of labour provides a striking contrast to that of the industry representative, Kirk Bailey. Christian quotes Perry Turton of the International Brotherhood of Electrical Workers as observing, "We often hear the numbers of people that have passed through the years ... while at work. What we won't hear is the names of those that have passed." He continues, "This is something that bothers me as it feels we are allowing industry to turn us all into statistics. We are not numbers. We are people with families, friends."[24] The focus on numbers rather than names abstracts from the particularity of individual suffering and death and shifts the discussion from mourning to progress, a shift that is also the focus of environmental public relations.

Robert Bott, writing for the Canadian Centre for Energy Information, puts the environmental issues this way: "Human ingenuity has already accomplished a great deal by making the oilsands economically competitive with conventional oil. Environmental and social challenges are being engaged. Continuous improvement in science, technology and management are helping to overcome the remaining challenges to meet society's expectations for sustainable development."[25] We might read this as an oblique acknowledgment that bitumen is not sustainable, but it maintains the promise of sustainability in the future. The "remaining challenges" will be "overcome" someday; meanwhile, resource industries, enabled by government and citizen complicity, sacrifice the environment and society in the name of economic competition. Downstream from Fort McMurray, Fort Chipewyan, the oldest continual settlement in Alberta, provides one example of sacrifice. For many residents, hunting, trapping, and fishing remain tied to an intimate knowledge of place and natural cycles. When production continues twenty-four hours a day, three hundred and sixty-five days a year, bitumen's industrial activity of extracting abstract capital from concrete places is incompatible with such a relationship to place. Not only is their *way of life* being sacrificed for capital, but, given the high rates of unusual cancers that are appearing in the community, *lives* are being sacrificed.

However, each apparent failure of science and technology to produce the promised perfection in our world creates an opportunity for reimagining that collective myth. Vermeulen states, "Nancy's notion of the inoperative community receives its critical momentum precisely by resisting that process of abstraction [by which individual deaths are seen as glorifying a collectivity]. Against such ideological abstractions of death, Nancy puts forward a notion of existence as 'unsacrificeable.'"[26] Similarly, we might argue that current deaths are not justified by the promise of fewer, or no deaths in the future. Those individual workers with families and friends, those members of the communities downstream from bitumen-mining operations cannot be sacrificed; more money, better schools, well-funded cultural institutions do not compensate for those deaths, nothing does.

*

Monday, 21 March 2011, on CBC's *The Current*, an interviewee in Hiroshima, Japan, states there are "many people exposed to radiation in the power plant site, but, on the other hand, many people get a lot of electricity from their sacrifice."[27]

*

While the imagined community denies the irreducible singularity of death by glorifying and monumentalizing it or ignoring it or projecting it into a utopian future or rationalizing it as being in the common good, Nancy considers death and existence as "always inappropriable and always there, each and every time present as inimitable."[28] This applies, I believe, not just to literal human death, but also to the ways in which other forms of loss are managed through abstraction. Ezra Levant, in the *National Post*, responding to the public outcry over the duck deaths on Syncrude's Aurora tailings pond in 2008, trivializes citizens' concerns and reinscribes a utilitarian view of nature as "common sense": "even the Prime Minister called the deaths a 'terrible tragedy.' I agree. I can't believe that 500 ducks were wasted that way. Imagine all the lost foie gras."[29] Initially, I took Levant's sarcastic point to be that there is no consequence of development that cannot be compensated for by profit, and the only tragedy is that the ducks' lives were lost without being commodified, or, perhaps also, that those who would eat foie gras cannot take umbrage at these deaths in the production of another commodity that they also consume. While this may be his ideological position, with his publication of *Ethical Oil* his argument has shifted to comparing the consequences of bitumen extraction with other forms of oil on what he calls "liberal"[30] grounds. If fewer people are killed by the bitumen industry in northern Alberta than in Iraq, for example, then bitumen should be extracted faster, in Levant's view, to lessen reliance on Iraqi oil, and, thereby, save lives. He states, to illustrate this point, that every barrel of oil from Sudan, for example, "has a thumb or an eyeball in it"; there is, he states, "no comparison between the oil sands, which killed a thousand ducks [...] and 300,000 dead in Darfur."[31] Of course, this ignores the fact that demand equals supply and that Iraqi, Sudanese, Saudi, Libyan, and Albertan oil will all find buyers, and, in fact, they are indistinguishable (i.e., fungible) in the world market. It is impossible for bitumen production to accelerate fast enough to reduce, let alone eliminate, demand for oil from elsewhere. More importantly, though, this kind of relativist calculus cannot account for Nancy's point about the unsacrificeable. Each death is full of horror and sadness, and this is true of duck deaths as well as human (as the published photographs and videos, and international reaction to the footage of the ducks as they were dying made evident). An ethical ranking of types of oil does not absolve us of culpability with the problems of even the best type.

I wouldn't go so far as to suggest that death is never justified; it doesn't take long to recognize the multitude of deaths that make my life and lifestyle possible (including the duck deaths). In *The Gift of Death*, Jacques Derrida states, "I am responsible to any one (that is to say to any other) only by failing in my responsibilities to all the others, to the ethical or political generality. And I can never justify this sacrifice, I must always hold my peace about it. Whether I want to or not, I can never justify the fact that I prefer or sacrifice any one (any other) to the other."[32] Levant does not hold his peace about this. He justifies it. Suncor Vice-President Kirk Bailey does not hold his peace about this. He justifies the death of a worker today—he sacrifices that worker—for the imagined life of a worker tomorrow, for the future utopia where "hunger and labour, disease and war" have been overcome.[33] Indeed, in writing this chapter I have been unable to hold my peace about it. We might even say of Derrida, as Blanchot did of Wittgenstein, "Wittgenstein's all too famous and all too often repeated precept, 'Whereof one cannot speak, there one must be silent'—given that by enunciating it he has not been able to impose silence on himself—does indicate that in the final analysis one has to talk in order to remain silent."[34] As I hope to suggest at the end of this chapter, although it can only do so temporarily, literature provides one means of talking in order to remain silent, enables us to dwell in our failure to all the other others, allows us to recognize and hold our peace about it.

Lament for the Nation's Petroculture

Indeed, the justification of others' sacrifice may be the narrative of modernity in general of which nationalism is only a reaction and mediation, the drawing of a border, a limit. Anderson's famous definition of a nation as "an imagined political community—and imagined as both inherently limited and sovereign"[35] is useful here. He clarifies the term "limited" by saying, "The nation is imagined as *limited* because even the largest of them, encompassing perhaps a billion living human beings, has finite, if elastic, boundaries, beyond which lie other nations. No nation imagines itself coterminous with mankind. The most messianic nationalists do not dream of a day when all the members of the human race will join their nation in the way that it was possible, in certain epochs, for, say, Christians to dream of a wholly Christian planet."[36] However, it is not impossible to imagine an entirely capitalist (or communist) planet. Indeed, to return to the Canadian context, in *Lament for a Nation*, George Grant argues against the progressivist view of history, claiming that the necessary and the good need not be equated, that because the death of one individual or group may be necessary it is not necessary to claim it is good: "They are identified," he writes, "because men assume in the age of progress that the broad movement of history is upward. Taken as a whole, what is bound to happen is bound also to be good. But this assumption is not self-evident. The fact that events happen does not imply that they are good. We understand this in the small events of personal life. We only forget it in the large events when we worship the future."[37]

*

Is the rapid declaration on the part of the government and individuals in Japan that they will rebuild their country, after the tsunami and nuclear meltdown, such an act of worshipping the future? The justification for today's suffering lies in the future?

*

Grant goes on, in the next chapter, to claim,

> The universal and homogeneous state is the pinnacle of political striving. "Universal" implies a world-wide state, which would eliminate the curse of war among nations; "homogeneous" means that all men [sic] would be equal, and war among classes would be eliminated. The masses and the philosophers have both agreed that this universal and egalitarian society is the goal of historical striving. It gives content to the rhetoric of both Communists and capitalists. This state will be achieved by means of modern science—a science that leads to the conquest of nature.[38]

The continual references to science and technology in government and industry documents about bitumen extraction echo Grant's point, as does the funding of two post-secondary programs with the Syncrude fine.[39] For Grant, "It is the very signature of modern man to deny reality to any conception of good that imposes limits on freedom."[40] Therefore, insofar as nations are limited, they are counter to the thrust of historical progress, and in the case of Canada, in Grant's view, fated to be swallowed up by the modern liberal capitalism of the American empire. For Grant, this event took place in 1963 with the "Defence Election," in which Lester Pearson defeated John Diefenbaker, allowing his minority government to arm Bomarc missiles on Canadian soil with nuclear weapons.

I would like to link the end of Canada as a nation to different but historically contemporaneous events. The publication of the Blair Report in 1950, which showed that bitumen could be extracted and upgraded at a profit (produced for $3.10 and sold for $3.50 a barrel),[41] led, in 1953, to the creation of the Great Canadian Oil Sands consortium, whose major investor was Sun Oil of Philadelphia. In *Lament for a Nation*, Grant asks if his readers can "imagine Canadians expropriating the oil properties and taking on international capitalism as Cardenas did [in Mexico] in the 1930s?"[42] While discussion of nationalized development of bitumen did occur in Canada at the time, Karl Clark expressed the dominant view about it in a letter of 12 June 1963: "The political opposition parties in Alberta are saying that the Social Credit government is selling the oil sands, or rather giving them to the Americans—that Alberta should be doing the developing itself. I think Premier Manning is very willing to let private enterprise take the risk of a first plant, at least."[43] This willingness to cede to American capital the right to extract Canadian bitumen, the right to abstract the particularities of the Athabasca region into money, has proven more significant than the arming of Bomarc missiles with nuclear warheads. The particular losses suffered in the Athabasca region may prove to be less widespread and catastrophic than those of a nuclear war would have been (and than widespread nuclear energy use could still be, as events in Japan are intimating again), but they still have a singular horror.

*

Laurie Essig writes, in *The Chronicle of Higher Education*, in response to the nuclear crisis in Japan,

> In both the U.S. and Japan, a certain sort of voice was heard saying "don't worry," "stop being hysterical," and "nuclear power is rational." For instance, over at *Slate*, William Saletan calls us "nuclear overreactors" and says that fear of nuclear power is not based in fact, but feeling. According to Saletan, the rational thing to do is study what went wrong and fix it since if Japan, the United States, or Europe retreats from nuclear power in the face of the current panic, the most likely alternative energy source is fossil fuel. And by any measure, fossil fuel is more dangerous. The sole fatal nuclear power accident of the last 40 years, Chernobyl, directly killed 31 people. By comparison [...] from 1969 to 2000, *more than 20,000 people died in severe accidents in the oil supply chain*. More than 15,000 people died in severe accidents in the coal supply chain—11,000 in China alone. The rate of direct fatalities per unit of energy production is 18 times worse for oil than it is for nuclear power.[44]

Stacking up all the deaths in the oil and coal supply chains with the thirty-one directly killed by Chernobyl is questionable to say the least: Were any uranium miners killed? What about the long-term effects of radiation exposure from the Chernobyl meltdown? With 80 to 90 percent of the world's energy coming from fossil fuels, how would nuclear stack up in terms of safety if it were providing that percentage? What if Alberta builds nuclear plants to supply power to bitumen facilities? However, the two forms of energy do not operate in isolation. Weighing the relative costs and benefits is a way of justifying the status quo, of managing the horrors of the sacrifices caused by our lifestyles.

Clark wrote in a letter dated 14 July 1960 that at hearings in Calgary regarding the Great Canadian Oil Sands (GCOS) application for its first large-scale extraction plant (45,000 barrels a day), Bob Hardy testified with "quite a story about how we were going to have to stack tailings up 160 feet and were in all likelihood going to start slides that would engulf everything in catastrophic ruin. That picture lost its horror when it was pointed out that our tailings were going back into the mined-out pit enclosed on all sides."[45] However, the horror returned for Clark in 1965 when he visited the GCOS site. His daughter Mary writes, "It affected him deeply to see the landscape of his beloved Athabasca country scarred as the construction gangs began stripping away the overburden for the operation,"[46] and, shortly before his death, he confided that "he had no wish to return again to the scene."[47] I can easily imagine Clark lamenting the loss of a particular place, dying with the irony that his life's work had helped it happen. Grant notes, "The practical men who call themselves conservatives must commit themselves to a science that leads to the conquest of nature. This science produces such a dynamic society that it is impossible to conserve anything for long."[48] I don't know if Clark called himself a conservative or considered himself one in any sense. He voted Liberal federally and against the Social Credit dynasty provincially—not, he claimed, in a letter of 12 June 1963, "that [he] want[ed] to see the Social Credit government defeated but rather that [he] would like to see the government have a decent opposition in the Legislature."[49] Regardless, at least at the end of his life, the scene described by his daughter shows a man who wanted to conserve something.

Bitumen Conserving Bison

The management of the form of ecological loss that affected Clark so deeply near the end of his life is undertaken in many ways, but is perhaps seen most clearly in the representation of bison. They are monumentalized in Syncrude's Wood Bison Gateway, which marks the entrance to the Syncrude site, and Wood Bison Viewpoint, which concludes both the Suncor and Syncrude tours, by offering tourists a chance to see wood bison grazing on reclaimed land (see Figure 13.2).

In its 2007 Sustainability Report, Syncrude describes how it "established a small bison herd on reclaimed land in 1993."[50] Since then the herd has expanded and the company has for several years "assisted the Fort McKay First Nation in arranging a traditional bison harvest"[51] (see Figure 13.3). As Nicole Shukin argues in writing about her tour of Syncrude's operations, Syncrude "pursues what Mark Seltzer calls a *'logic of equivalence'* between the space of industrial extraction and that of cultural preservation."[52] It is

Figure 13.2 Syncrude's Wood Bison Gateway, which marks the entrance to the Wood Bison Trail System, which includes Gateway Hill, the first reclaimed site to be certified by the Alberta Government. Syncrude Image Library.

Figure 13.3 Bison grazing on reclaimed land at Syncrude's Beaver Creek Wood Bison Ranch. Syncrude Image Library.

because of oil sands mining that the Fort McKay First Nation can conduct a "traditional bison harvest" with twelve animals that were "donated to the community from the Beaver Creek Wood Bison Ranch."[53] As the word "donated" in the quotation suggests, however, the bison are the property of Syncrude until the corporation decides to give them to the community.[54] Thus, the bison exist within the realm of capitalist accumulation rather than as part of the wilderness commons they previously inhabited. Whatever community is possible between humans and bison is closely managed and controlled.

The "bison harvest," a phrase that shifts the traditional Native economy into the agricultural realm, occurs only because of the benevolence of Syncrude's donation. We may ask why a "harvest" and not a "hunt"? Hunting is unpredictable and requires wild animals. Harvests are supposed to be more predictable in terms of what they yield. Wild ducks, for instance, do not yield foie gras. Syncrude, as "the nation's largest industrial employer of Aboriginal people,"[55] is responsible for the preservation of First Nations culture in the area, but works to determine which aspects of the culture will survive and to ensure they are compatible with industrial development. Members of the Fort McKay First Nation actually need to book in as tourists to visit the bison.[56] Because of the associations of First Nations culture and bison in the cultural imaginary, which Syncrude has strategically exploited, their use of bison as a monument and as a "mascot" for the industry[57] establishes the idea that bison and Native culture can be preserved and recovered under the direction and control of transnational capital. Nothing need be sacrificed to extract bitumen. Again, like the death of workers or ducks, other forms of loss will be compensated for or superseded in the future. "Liberalism is," Grant writes, "the faith that can understand progress as an extension into the unlimited possibility of the future."[58] This view can be promoted or criticized through literature, and I would like to look at two examples, one that promotes and one that criticizes.

Pro and Contra Bituminous Literature

In 1913, Sidney Ells, senior engineer for the Government of Canada's Department of Mines, began his decades-long work to extract oil from the bituminous sands of northern Alberta. To this end, he undertook surveys, dug test pits and quarries, drilled core samples, and experimented with separation techniques. In addition to writing technical reports, though, he also wrote his memoirs, a collection of short stories and poems entitled *Northland Trails* and his *Recollections*. In his introduction to this latter volume, Ells writes that in the process of developing bitumen, "Every line and phrase of Kipling's 'If' were [...] lived."[59] William Dillingham argues that Kipling's world view, expressed in "If," emphasizes "the role of the hero as one who by an exercise of the will and through self-discipline and inspiration defies the meaninglessness of the universe and the chaos of life by superimposing order—in the form of one's 'work' or 'craft'—upon disorder."[60] Ells positions himself, then, at the beginning of his book, as one of the heroes worthy to be called a "man" by Kipling's speaker. However, that vision of imposing order on self and otherness, which leads to a belief that one possesses the "Earth and everything that's in it"[61] depends on "the fact that the admired spontaneity of freedom is made feasible by the conquering of the spontaneity of nature."[62] While Ells was instrumental in achieving our contemporary technological freedoms in

Canada, he was, like Clark, also aware, to some extent, of an order beyond that imposed by human conquest.

In addition to his reference to Kipling, Ells's book begins with a poetic epigraph and concludes with a poetic epilogue, both of his own creation. The first describes the call of nature to men of various enterprises—"the uplands" to the "shepherd," "the plain" to the "herdsman," "the sea" to the "sailor-man," and so on—but, in addition to these vocations,

> Now the rock ribbed northland – empire of vale and hill –
> Echoes the thud of bursting charge, the clink of sledge on drill,
> For college don and tenderfoot and seasoned pioneer
> Have turned their faces northward – men of the new frontier![63]

Development efforts in the Athabasca region are direct extensions of North American westward expansion in these lines: an encounter of civilization with wilderness, mediated by the receding horizon of the frontier.

Ells's epilogue, by contrast, offers a romantic conception of nature. Imagining his grave marker he writes,

> I asked not stately man-made shaft of stone,
> Within some crowded city of the dead,
> One of a mighty host, – and yet alone,
> While restless feet hurry above my head.
>
> Out on a wind swept ridge then let me lie,
> A rugged twisted pine my marker rude,
> Where owls' deep call and loons' sad wavering cry,
> Alone will break my peaceful solitude.
>
> Yet not alone beneath my tree I'll lie,
> For all about me furry things will play,
> While stately antlered monarchs wander by,
> Friends of the long, long trails of yesterday.[64]

This poem makes no mention of the industrial activity of the first poem and suggests that nature will remain unaffected by bitumen extraction after the speaker's death, allowing friendship with the animal world to continue into the afterlife. His texts, like Syncrude's Sustainability Report and strategic use of bison, thus appear as attempts to balance opposing values: the globalizing forces of liberal, rational, technological progress that promise freedom and equality, and the local, particular, traditional forces that claim virtues of communion and continuity with nature.

Attempts to strike this balance, at least rhetorically, continue in the oil sands work being carried out today. The industry works very hard to show that the "thud of bursting charge" and "peaceful solitude" are not mutually exclusive, that the industrialization will not compromise communion with nature, as seen in Syncrude's representation of the bison.[65] Ezra Levant makes this argument as well: "Forests will still grow and critters will still frolic on the land [...] And even the [land] that is mined will be reclaimed once the oil is pumped out [...] the first oil sands mine reclamation projects have already been certified. They're gorgeous hiking trails now, with forests and pristine lakes."[66] Never mind that the single, tiny reclaimed site Levant is referring to was not mined, but only ever used for storing overburden (all of the earth and rock that overlay bitumen deposits), and is nothing like its previous topography. Sacrifice is not necessary to secure the future in this narrative; when the necessary sacrifices are unavoidably apparent, they are justified with reference to that ideal future.

By contrast, Mari-Lou Rowley's poem "In the Tar Sands, Going Down" suggests an alternative to modernity by dwelling in the state of our loss, by remembering our vulnerability, which an imagining of a future utopian community tries to forget. She writes to oil as a "luscious baby," a "perfector of defects" providing "pipelines across continents and so many new / cars, jobs, cans of Dream Whip."[67] As the poem progresses, the speaker has sex while bitumen extraction pollutes the world. While "fondling under leaves" the "sky [is] mortally dazed / under clouds weeping acid."[68] She will continue to satiate her desires while the "boreal blistering"[69] goes on around her, "until," left with "the forestless birds," "the fishless rivers," "the songless, barren face of the earth," "we go down."[70] The "going down" from earlier in the poem, the "knees to ground / head to groin"[71] act of fellatio directed toward escapist orgasmic release from the tragic destruction going on around the copulating humans, in the end includes them (and us) in the exhausted collapse (they and we) were attempting to escape through the pleasures made possible by oil extraction.

Though Rowley's poem may be debunked and dismissed as just as ideological as Ells's, it is still significant that Rowley's poem critiques one option for life under liberal modernity while Ells's glorifies another. Grant writes, "The vaunted freedom of the individual to choose becomes either the necessity of finding one's role in the public engineering or the necessity of retreating into the privacy of pleasure."[72] Ells found his role in the public engineering, but tried to manage its consequences by imagining a future free from them: we do not see the individual tragedy of the speaker's death, only the pleasant consciousness of his afterlife. Rowley's poem follows our hedonistic retreat into private pleasure through to its grim conclusion. It shows how we are implicated in the public engineering of the oil economy, with the duck deaths, with genocide in Darfur, however far we try to retreat into the privacy of pleasure. Our pleasures, our escapes, our bodies are supported by an oil economy that is inexorably bringing us down at the expense of the beautiful otherness of the boreal forest that is sacrificed to extract bitumen, and the infinity of other otherness in Nigeria, Iraq, Sudan, Libya, Venezuela, Iran, Syria, Russia, and so on.

Vermeulen writes, "Both Nancy's and Anderson's work allow us to conceive of literature as a crucial medium for the imagining of a different community—a community that, unlike nationalism, does not rely on the abstraction and anonymization of death. Literature, that is, can mediate loss in such a way that the singularity of death is neither simply foreclosed by nationalist ideology nor fastened on to in a way that makes a determinate imagining of community impossible."[73] This raises a number of questions: is this still true of literature (and in what sense are we understanding that term)? If so, who will write the story of the duck deaths? Who will write the story of the people of Fort Chipewyan? These are not unfamiliar events—the stories have already been told by journalists, pundits, politicians, and activists but not, I think, in a way that has avoided fastening onto them a determinate imagining of community, not in a way that refuses to justify the sacrifice of one other to another, that holds its peace.

*

Essig tells us, "The hubris that is 'rational science' is built on those forms of knowledge that claim to be outside the bodies producing that knowledge [...] They are 'objective' and the rest of us are 'biased.' If only nuclear physicists were required [...] to consider the limits of knowledge, the very human messiness of its production, and the deadly consequences of dismissing your critics as hysterical."[74] But it's not just nuclear physicists who ignore the limits of knowledge or the messiness of production or who dismiss their critics as hysterical. These traits are paradigmatic of liberalism.

*

Robert Kroetsch, shortly before his death in an automobile accident, wrote in the first issue of *Eighteen Bridges*, "Story is often about difficult or impossible choices. It challenges our very being. Pure entertainment makes the other guy the goat; it makes us, as mere voyeurs, feel comfortable and privileged. And safe."[75] In the case of bitumen, and modernity more generally, entertainment makes us feel safe by getting us to live, to imagine our relationships to others and other others, in the future. As in Ells's poems, after our deaths everything will be fine. Stories, by contrast, require us to dwell in the present. We need more stories about bitumen and less entertainment, less of "the oil sands porn" that Levant says "sells magazines and gets donations"[76] but also less of the self-satisfied nationalist apologetics that his book provides.

Notes

1 Karl Clark, "The Athabasca Tar Sands." Karl Clark Papers, 76-93-16, University of Alberta Archives (hereafter UAA).
2 Pieter Vermeulen, "Community and Literary Experience in (Between) Benedict Anderson and Jean-Luc Nancy," *Mosaic* 42, no. 4 (2009), 96.
3 Carol Christian, "Killed Workers Remembered," *Fort McMurray Today*, http://www.fortmcmurraytoday.com/ArticleDisplay.aspx?archive=true&e=1546066; and Hanneke Brooymans, "Fort Chipewyan Wants Answers about Cancer Rates," *Edmonton Journal*, 5 Sept. 2010, A1.
4 Len Findlay, "Is Canada a Postcolonial Country?" in *Is Canada Postcolonial? Unsettling Canadian Literature*, ed. Laura F.E. Moss (Waterloo, ON: Wilfrid Laurier University Press, 2003), 298.
5 Vermeulen, "Community," 97.
6 Colin Babiuk, *Oil Sands and the Earth: Framing the Environmental Message in the Print News Media* (Charleston, NC: VDM Verlag Dr Muller, 2008), 13.
7 Peter Silverstone, *World's Greenest Oil: Turning the Oil Sands from Black to Green* (Edmonton: PHS Holdings, 2010), 3; and Andrew Leach, "Don't Let EU Define Our Emissions Policy," *Edmonton Journal*, 2 Mar. 2011, A15.
8 Hanneke Brooymans, "Oilsands Boosts Toxic Metals in Athabasca Watershed: Study," *Edmonton Journal*, 31 Aug. 2010. And Erin N. Kelly, Jeffrey W. Short, David W. Schindler, et al., "Oil Sands Development Contributes Polycyclic Aromatic Compounds to the Athabasca River and Its Tributaries," *Proceedings of the National Academy of Sciences USA* 106, no. 52 (2009), 22346–51.
9 Tom Katinas, "An Apology from Syncrude—and a Promise to Do Better," 2 May 2008, http://www.syncrude.ca/users/news_view.asp?FolderID=5690&NewsID=121.
10 Darcy Henton, "If Syncrude Convicted, Oilsands 'Doomed': Defence," *Edmonton Journal*, 29 Apr. 2010, A1.
11 Josh Wingrove, "Syncrude to Pay $3M for Duck Deaths," *Globe and Mail*, 22 Oct. 2010, http://www.theglobeandmail.com/report-on-business/industry-news/energy-and-resources/syncrude-to-pay-3m-for-duck-deaths/article1769027/.
12 Vermeulen, "Community," 97.
13 Imre Szeman, "System Failure," *South Atlantic Quarterly* 106, no. 4 (2007), 820.
14 Vermeulen, "Community," 99.
15 Ibid., 104.
16 Benedict Anderson, *Imagined Communities: Reflections on the Origin and Spread of Nationalism* (New York: Verso, 2000 [1991]), 24.
17 Ibid., 26.
18 Ibid., 201.
19 Ibid.
20 Upton Sinclair, *Oil!* (New York: Penguin, 2007 [1927]), 526.
21 Christian, "Killed Workers Remembered."
22 Ezra Levant, *Ethical Oil: The Case for Canada's Oil Sands* (Toronto: McClelland & Stewart, 2010), 6.
23 Anderson, *Imagined Communities*, 10.
24 Christian, "Killed Workers Remembered."
25 Robert Bott, *Canada's Oil Sands*, 2nd ed., ed. David M. Carson and Tami Hutchinson (Calgary: Canadian Centre for Energy Information, 2007), 39.
26 Vermeulen, "Community," 97.
27 David Gutnick, "Hiroshima," *The Current*, CBC Radio, 21 Mar. 2011, http://www.cbc.ca/thecurrent/episode/2011/03/21/hiroshima-david-gutnick/.
28 Vermeulen, "Community," 97.

29 Ezra Levant, "Shed No Tears for Ft McMurray's Ducks," *National Post*, 5 May 2008, www.nationalpost.com/ (accessed 8 June 2008). While initial reports put the number of dead ducks at 500, this was subsequently increased to an estimated 1,600.
30 Ezra Levant and Andrew Nikiforuk, "Is Oil-Sands Oil the Most Ethical Oil on Earth?" *Q*, CBC Radio, 15 Sept. 2010, http://www.cbc.ca/q/blog/2010/09/15/is-oil-sands-oil-the-most-ethical-oil-on-earth/.
31 Ibid.
32 Jacques Derrida, *The Gift of Death*, translated by David Wills (Chicago: University Chicago Press, 1995), 70.
33 George Grant, *Technology and Justice* (Toronto: Anansi, 1986), 15.
34 Maurice Blanchot, *The Unavowable Community*, translated by Pierre Joris (New York: Station Hill Press, 1988), 56.
35 Anderson, *Imagined Communities*, 6.
36 Ibid., 7; original emphasis.
37 George Grant, *Lament for a Nation: The Defeat of Canadian Nationalism* (Montreal and Kingston: McGill-Queen's University Press, 2005 [1965]), 37.
38 Grant, *Lament*, 52.
39 Funding a program in Wildlife Management at Keyano College and funding a University of Alberta study on bird migration patterns and the effectiveness of bird deterrent systems suggests that through improved scientific study and, through the application of that study, improved technological control of nature, our historical striving will be, if not realized, at least advanced.
40 Grant, *Lament*, 55.
41 Karl Clark, "Commercial Development Feasible for Alberta's Bituminous Sands," Research Council of Alberta, Clark Papers, 76-93-205, UAA.
42 Grant, *Lament*, 45.
43 Karl Clark, "Letters," Clark Papers, 84-43, UAA.
44 Laurie Essig, "Your Inner 'Hysteric' May Just Be Right," *Chronicle of Higher Education*, 17 Mar. 2011, http://chronicle.com/blogs/brainstorm/interpreting-nuclear-disaster/33297.
45 Clark, "Letters."
46 Mary Clark Sheppard, *Oil Sands Scientist: The Letters of Karl A. Clark 1920–1949* (Edmonton: University of Alberta Press, 1990), 89.
47 Ibid., 90.
48 Grant, *Lament*, 65.
49 Clark, "Letters."
50 Syncrude Canada Ltd, "2007 Sustainability Report," 35, http://sustainability.syncrude.ca/sustainability2007.
51 Ibid.
52 Nicole Shukin, "Animal Capital: The Politics of Rendering," PhD dissertation, University of Alberta, 2005, 139. Original emphasis.
53 Syncrude, "2007 Sustainability Report," 35.
54 Shukin, "Animal Capital," 195.
55 Ibid., 134.
56 Ibid., 195.
57 Ibid., 142.
58 Grant, *Lament*, 56.
59 S.C. Ells, *Recollections of the Development of the Athabasca Oil Sands* (Ottawa: Department of Mines and Technical Surveys, 1962), 4.
60 William B. Dillingham, *Rudyard Kipling: Hell and Heroism* (Gordonsville, VA: Palgrave Macmillan, 2005), 187.
61 Rudyard Kipling, "If," http://www.swarthmore.edu/~apreset1/docs/if.html (accessed 16 June 2009).

62 George Grant, *Technology and Empire* (Toronto: Anansi, 1969), 31.
63 Ells, *Recollections*, n.p.
64 Ibid.
65 For another example, see Canadian Association of Petroleum Producers, "Steve Gaudet, B.Sc., Syncrude Canada," 11 June 2010, http://www.youtube.com/watch?v=iXtHFPGXjlY (accessed 28 Aug. 2012).
66 Levant, *Ethical Oil*, 4.
67 Mari-Lou Rowley, "In the Tar Sands, Going Down," in *Regreen: New Canadian Ecological Poetry*, ed. Madhur Anand and Adam Dickinson (Sudbury, ON: Your Scrivener Press, 2009), 63.
68 Ibid., 64.
69 Ibid.
70 Ibid., 66.
71 Ibid., 64.
72 Grant, *Lament*, 56.
73 Vermeulen, "Community," 107.
74 Essig, "Your Inner 'Hysteric.'"
75 Robert Kroetsch, "Is This a Real Story or Did You Make It Up?" *Eighteen Bridges: Stories That Connect* 1 (Fall 2010), 19.
76 Levant, *Ethical Oil*, 10.

INTERLUDE

Symphony for a Head of Wheat Burning in the Dark
Harold Rhenisch

Instructions for Carving a Mask for the Winter Ceremony

Fires burn. Constellations spin slowly through the rooms.
In immense height, darkness gloves all men.

There are no discussions about the meaning of time –
how to limit it. There is only certainty.

I catch a bird in my hand: the bird speaks me
with a movement of its wings;

I catch myself in my mind:
I open my hands and set myself free.

Birds scatter up, black against the sun –
small specks receding into distance.

In our absence – all the proof we need: the world endures;
it is passing; we are on a journey – that is the journey.

The first grammar and the last: men do not fly off
above marsh cinquefoil. Their presence does.

Snow falling into still water: the flakes, for a moment, weighted,
crystalline, float among trout, that drink the dark from the dark.

One day, eagles are chasing each other through the trees.
The next day a man neglects to bury a dead horse.

The rain drums on the skylights of a deaf man's house.
Periodically, he goes to the door to see who wants to come in.

We have wrapped our banana trees in Salvation Army sheets.
11 a.m. "All staff go to the back room to pray." The poor wait.

The First People are sitting on the grey rocks of the breakwater.
We have forgotten we are remembering something that used to exist.

The mountains are being taken down and loaded on rusty ships.
At the next dock, rich women are painting their yachts. I go back and forth.

Luther said his God was an impenetrable fortress.
Bach set it to music. We buy it from amazon.de., on plastic.

Despite quantum mechanics, there is only one earth.
Although there is only one earth, there is quantum mechanics.

Look up. Along the river, pink salmon used to stink to high heaven.
Linguists are now hired to decipher the languages of machines.

Salmon are now being raised in net pens.
In the earliest dramas, the devil had all the punch lines.

Local players put on a show with twenty acting parts.
Most of the players aren't acting, so the audience does.

With Fraser at the Meeting of Two Rivers

The ghosts of salmon swim through the air.
Each word a man writes is the ghost of a word.

The fish have disappeared into the heartwood of trees
where they are swimming among the stars.

The stars are gravel, glinting in the sun.
Yellow ripples of light float over our hands and faces,

as we stare up, our gills opening and closing
on opening and closing. We are the name of the wind,

whisper the trees to Simon Fraser, Esquire.
He knows it is his own mind speaking.

He sent it out the day before in a bullet
among the cottonwoods. Now it is returning.

In this country, an echo can last for centuries:
the leaves slowly bending before its blast and slowly

springing back before it is heard. As salmon swim to the moon,
their trembling breath echoes through the water.

Already a film of oil spreads across the ocean.
Already the smog fills the air from the cities' ah.

The process of emptiness leads from conception to trial.
Sleep needs the mind. It burns it into dust.

I carry away the past in my hands and scatter it to the winds.
The winds blow it back.

The Initiate after Eleusis

The sky has permeated the soil.
We walk on the songs of crickets.

When I have been a window, I have been a bird.
I have held a lake in my mouth.

The sun burns in the breast of each sparrow in the grass;
in each sparrow the same sun, cool on a spring morning.

As to why the sparrows have not invented speech,
well, things are their names; well, it is spring.

Because the grass lives in silence,
the grass amplifies sound and sends it on. And on. And on.

I have built a house of fieldstones that remember the sea.
Every year the plow turns them over one by one to face the stars.

Adam opened a doorway into the earth.
The earth walked through it into the light, and was gone.

The First Day of the World

On the first day of the world the trees turned into birds
and flew across the burning sea. Everything became everything else:

windows became mule deer; angels became the grass in a field;
ploughshares became oxen; the moon became a field of driving snow.

The legacy remains: a world in which all things
became that which they are not.

A red horse climbs up the ladder of a blue field,
its head bowed, pulling a heavy weight.

I have made wine out of the breath of starlings.
As it poured out of the press, it was a cloud.

Symphonies and theorems of light
build up and collapse and build up again out of drifted sand.

There are fish swimming in the sun, silver,
turning over and under like a breath,

receding from it, and returning,
closer and closer to an approximation of sound:

animals, trees, people, and stones.
A whole blue wind made out of air is striking the whole coast all at once.

A trout coalesces out of the waves. A man walks through the furrows,
with stalks of wheat for hair. His hands are sparrows,

that splash off and wheel on a cold current of the wind.
Their shadows move together over the broken soil.

The man has a wasp nest for a head.
He lifts it off and pushes it up into the sky. It floats.

The Birth of Light

Milkweed pods are bursting along the dike.
Poppy seeds rattle in their bone shakers.

September. Within the sparrow: the twig.
October. Within the twig: the wind.

Silence is a means to feel the approach of the machine.
To know the wine, I ate the vineyard soil: salt.

A river sings along the mountains under my feet.
The dead walk among us with their necklaces of pearls.

Endless columns of refugees push through the trees.
The city of the past is on fire.

People dressed in black are coughing on dust.
Great towers of cloud pour into the air.

The dead are choking, starved and confused,
stumbling into ranch yards, glaring accusingly.

They hold out their hats for food and water.
They bring their little children to the creeks and they pass through.

Iamblichus said that god is present in stone truly carved.
The hands of lovers grip each other and then release.

The logging trucks pour down from the plateau forests.
The dead watch from the ditches along both sides of the road.

Night and day, black leaves catch the moonlight.
Waves break into light on the edge of the land where we are alive.

PART 3
MATERIAL EXPRESSIONS

CHAPTER 14

Propositions from Under Mill Creek Bridge
A Practice of Reading

Christine Stewart

Edmonton's Mill Creek Bridge, Mill Creek, and Mill Creek Ravine meet at 82nd Avenue between 95A and 93 streets. The bridge deck runs east to west at the same height as the tree canopy. Supported by wide concrete girders the bridge is impressive and arching; its wide deck warms under the teleology of traffic. Dreaming of the other side, the bridge promises that we are moving in the right direction, that we will arrive, that the automobile is subject and citizen, that oil is capital.

Two ravens live on the southeast corner. They watch the traffic and chase the pigeons. I have seen them in the trees in the early morning darkness of January when the clock on Whyte Avenue reads 6:45 a.m. and 35 degrees below zero. Their breath steams with the traffic. The couple fledges each spring. In the late spring of 2008, I watched a teenaged raven spiral above the bridge and then drop into the trees below. I crouched, knees bent, head tilted, watching.

Down below, under the bridge at Mill Creek, the world shifts. Beneath the bridge's promise, the underbridge promises nothing and offers everything: dust, stream, sleepers, wire, mud, thistle, and the warm steam of new human shit that blasts from the nearby grate. Small morning fires send smoke into the mist and materialize the early slanted light. In the evenings, conned by the sirens from the road above, coyotes call. Here everything counts for nothing. What lies under the bridge is of little consequence to those above. What is present marks an absence, but that absence marks the presence of complex relations, contiguous, accumulating, radiating outwards, endless, and compositive.

How might we read this bridge, its birds and underbridge? How might we read these relations?[1]

How should we read the worlds that we encounter? How might we note the spaces and the marks that we do not usually look toward? What are the ethics of such a reading? What are the ethics of not reading?

"It is as if the text's strange will desires me, and it is for me to receive, to be inhabited by its alterity."[1]

How might we read with reception and release so that we may be inhabited by the world's alterities?

The Open: Man and Animal, Giorgio Agamben's consideration of biologist Jakob von Uexküll, contains possibilities for such a reading.[2] As Agamben notes, Uexküll reads the environment as a collection of spheres subjectively constituted by an animal species. Calling these spheres Umwelten, Uexküll claims that each Umwelt is an environment that a species of animal perceives according to its unique cognitive apparatus. As a result, each species pays attention to specific marks or "carriers of significance":[3] These marks correspond precisely to the animal's receptive organs and yet are characterized by blind reciprocity. That is, different species of animals exist side by side, deeply engaged with each other, and yet unaware of the other's Umwelt. As Agamben explains, the spider creates a web so perfectly designed to catch a fly that it is flylike, but otherwise, the spider is blind to the fly's world or Umwelt.[4] And this blindness is shared: the fly is also blind to the Umwelt of the spider.

Significantly for Agamben, Uexküll's framework radically shifts biology from evolutionary progress, genetics, and vertical thinking, toward spatial and "horizontal" biology.[5] Unlike the classical scientific model that arranges organisms from the lower to the higher—from blood fluke to human—Uexküll's model configures an infinite variety of worlds, uncommunicating and exclusive, but equally perfect and bound together, as if in a vast musical score.

What are the notes of this score, and what does it sound like?

Could we use Uexküll's metaphor to propose a horizontal and spatial way of reading, a form of listening that notes this gigantic score, that works contrary to ideas of progress and verticality?

In such a practice of reading, the researcher would be both ears and eyes. She would learn to recognize the notes of significance, the marks to which the animal responds.

To read toward this spatial and paratactic biology she would note her carriers of significance, according to her own Umwelt, listening for the musical unity that might occur in her moment of blind reciprocity.

Would this reader, confined to her Umwelt, be theriomorphic: having the form of an animal, bound to her purposeful readings, and yet bound to the text's strange will, bound to release, to be inhabited (blindly) by its alterity, thereby listening to its strange, participatory, discrepant symphony?

Would such a reader obstruct the mechanics of what Agamben calls the Western European anthropological machine?[6]

Would a theriomorphic reader exceed the process by which the European human is presupposed and determined always in its opposition to the animal?

A theriomorphic reader might note the marks before her in their arrangements according to her own Umwelt. She might listen for the voice of the other: utterly strange but materialized in the curve of the ear and in the dappled matter of the brain—heard in the room of the skull.

"*Recognizing interrelations means* hearing *how water speaks, in an unfamiliar syntax, along with moss and smog, mountains and mycorrizhal mat, musqueam and ktunaxa [k-too-nah-ha], lubicon, tahltan.*"[7]

Built in 1911, the original Mill Creek Bridge over Mill Creek Ravine at Whyte Avenue amalgamated the outlying communities of white settlers with Strathcona, and later with Edmonton. It ran diagonally from Whyte Avenue west over to 83rd Avenue on the east, two metres above the average creek level, and cost $40,000 to build. John Donnan and his milk wagon were the first across.[8]

Built by Burns and Dutton Construction in 1960 and 1961, the second Mill Creek Bridge was eight spans long. The precast girders were supplied by Con-Force Products of Edmonton. The wearing surface was made of high-density concrete supported on a cast, with a concrete deck of precast concrete girders, measuring 2.44 metres centre to centre.[9]

In 2008, the bridge was widened to make room for more cars and trucks. The present deck is 240 metres long with two lanes of arterial traffic and a sidewalk on each side. Look east along Whyte Avenue: the bridge merges seamlessly with the road.

To read theriomorphically would be to read with an ear turned and tuned to the musical unity of reciprocations and strangeness.

To read thus would be to note the siren and the coyote as nature and its end. As the population of the city increases so do the police cars and ambulances that frequently cross the bridge and move along the edges of the ravine. The number of coyotes also increases; urban scavengers, they are the only large wild animals left in the ravine.

In the ravine, the coyotes sing with the sirens at dawn and dusk—antiphonic and clear. In hearing, the other materializes, gathers, and is differentiated in the listening mind. In the mind's ear. Surrounded. Exceeded. Symphonic.

In the ravine, other spheres of existence proliferate: lush flora withstand the toxins from the stream, and the vehicular exhaust that spills from the road. The growth signals stubborn life, but also the absence of big grazing animals (bison, deer, elk) and the lack of uncontrolled and controlled fire (local

Indigenous communities traditionally used fire management to keep the firs and poplar populations down, in order to encourage the grazing animals). Now these grazing animals are gone, dense greenery shelters the tents pitched along the creek to the north and south of the bridge. The creek moves past the encampments. A bird sings. Smoke rises from a small fire. The scene is pastoral and apocalyptic. The tents mark dawn's light and the world's end, shelter, and poverty. The tents also mark the new condo development on the northeast corner: Trinity Point—"inner city living with a back to nature ambience," "ravine living at its best," plus the older orange-and-white condominiums to the southwest, designed by Mill Creek Companies—"building wealth through real estate." From the lawns of the southern condos you cannot quite see down to the concrete ledges of the underbridge that hold bedding—quilt, comforter, cardboard, garbage bags, sleeping bags, set here even after the temperature has dropped below 30 degrees below zero Celsius.

F. lives outside on the south side. He tells me that you can sleep outside safely at up to −20 degrees. Any colder and exposed skin will freeze in a matter of minutes. I remember this as I note that the sleepers of the underbridge often stay through extreme cold, fastened here by circumstance, fixed and marked by poverty, by the vulnerability of their bodies. Their conditions are the consequences of an economic system that cannot sustain itself, and yet the sleepers also exceed the logics of this system. How to read them by the remnants that they leave behind? How to read their resourcefulness and necessity? To breathe here is hard. To inhabit this space. To receive these material conditions, that which is willed to you by the sleeper who is also a reader, a listener. Through this materiality we encounter other ecologies, conditions of context and space. These conditions are signs of life, and they implicate the viewer.

To read horizontally, through Umwelten and their proximities, challenges the centrality of the human. It challenges the machinations that constitute the human as an endlessly negotiated binary of human and animal, of the animal and human—these negotiations that determined (and redetermine) who or what is most violent, most bloody. What are the politics and practical considerations of the separation of animal and human? Agamben asks this question, and it echoes here, under this bridge where the division is a motile border, shifting. What does it mean to read from a place of separation and caesura? That is, what does it mean to be human within this framework? And what if (as reading listeners), instead of participating in the division, we imagine the blind edges of our own worlds, whereby beings connect, by virtue of their own markers, in reciprocal unities, willed and unwilled, listening, hearing, embodying, embodied.

Each detail constitutes a different element every time it is in a different environment.

What if we read through intersections? Eyes reading back and blind to other Umwelten, yet listening for these deep zones of indifference, in the poverty of the world, against the anthropological machine, next to the ravens, by the creek, in the fixity of the body, in the fluidity of skin, in the world, under the bridge?

This reading would be a form of listening, centring and decentring the reader. Would it dehumanize the program so that the listening subject would proceed theriomorphically (animal-like), resting next to other species, linked and unknowing, yet listening?

In 2007, I get a ride to Edmonton from Vancouver. We follow the Thompson River north and then turn right. I've been on the coast for almost thirty years, and the land here feels locked and flat. I am a stranger. Estranged—*ostranenie*.[10] The sky is empty, bearing nothing but a searing light and weightless blue. No mountains. No sea. No thick clouds. No markers by which I can know where I am. It is as if I am seeing the world for the very first time. There is nothing to hold me in place. Frantic, I seek height, incline, and trees. I find them in the Mill Creek Ravine. There trees and steep banks place me. At dusk, I walk down into the shade and cool green; I breathe more easily. One evening my daughter walks with me. We follow the path along the creek that leads to the North Saskatchewan River. There is a bridge ahead. The sounds of traffic drop from the road above, and the path leads directly under the concrete supports. In the darkening space of the underbridge, the landscape is apocalyptic—the lush ravine ends in desolation. Garbage lines the deep ruts of erosion that run down to the small stream. To the left, a higher slope rises toward the bridge, toward more garbage, more dust, burned wood, torn wool, broken bottles, and a cement ledge that runs the width of the underbridge. There are men on the ledge. My daughter's voice fades into the high dark arch of the bridge. The men sit quietly. As if they have been waiting for us. As if we cannot see them. I move past them, quickly, toward the trees and low evening light, out from under the bridge. I say nothing, but will my daughter to follow me. She does. The men watch us move back into the light.

"Reading does change the world, but usually not in the way one might wish it to."[11]

When we first arrive in a flat city, it is as if we have no present. We can make sense of nothing in the here and now. We can only look to the past. Everywhere we can only see what has been, what has been lost. What the river was: the lost creek, the blasted ravine, the trees, the men under the bridge who are home, and homeless. Perception is grief. There is no house. No east, nor west. No north or south. What is a bridge?

A bridge is a marker. Something that locates. Something that contextualizes the subject and the stream. The bridge gathers the earth and ransacks the earth. It creates an entire Umwelt. Many Umwelten.

During my first year in Edmonton, from the fall of 2007 until the spring of 2008, I rode my bike under the bridge at Mill Creek every day to work. I was afraid of Edmonton and the underbridge—its dark ledges and waiting men. But I was more afraid of the flat earth and blank sky above the ravine. I needed the trees, the steep banks. C. told me that in Edmonton we have to face our fears. Hers was hockey. Mine was the underbridge. And so I continued to ride under the bridge every day, turning my face to the dark west ledge. I read the strange arching space carved out by the bridge above. I read path, ravine, desolate slope, and creek below: the graffiti, the stuff, and its dust. Eventually, as I rode, each day, watching, listening, the underbridge stopped being a liminal space. It began to run dendritic through the middle and around the edges of everything. The underbridge illuminated my own fear and uncertainty, and I realized that the underbridge was a condition: an impasse, a passage, a limbo, a meeting place of lives.

The landscape of the underbridge is unstable and incomplete, a landscape of incongruities, displacements, dispossessions and possessions, occupied by ever-shifting images, objects, and events, constantly in need of redefinition and renaming. Next to the huge cement piers, the trees elongate, tall and thin. The creek shrinks against the eroded banks.

I read the path under the bridge according to my own markers and their limits, and the paved path is marked with the objects that are thrown, every day, from above: rock, pumpkin, *Big Gulp*, broccoli, *Edmonton Journal*, smashed amber *Kirin*, metal ring, gladiola, broken CD, empty cigarette lighter, embroidered coat, ski jacket, large bone. I ride my bike around the broken glass, pumpkin, and junk.

I notice that what comes over the edge is different from the objects arranged alongside the ledges. On the ledge, where the men sit, the garbage becomes arrangement, comfort and bundle, or expressive scatter. I read the graffiti on the giant piers that hold the bridge. It changes quickly as somebody regularly erases the text with grey paint. One day, I find *freedoom*. I read this incorrectly for the first few days. I read freedom, but then I note the play and see the *doom*. I read *acid drop, OH NO! jesus, softly*, and *fuck you*. I look closely at the plants that grow against the line of extreme desert that runs under the bridge: globe thistle, something that looks like burdock, and *Chenopodium berlandieri* or pitseed goosefoot. I learn that goosefoot is both a weed and dinner, from the *Amaranthaceae* family, related to quinoa, beets, chard, and purslane. It has been cultivated and eaten in North and South America for thousands of years. It grows all over Edmonton. Now I weed it from gardens and eat it.

Attending to the underbridge, I read the plant life, graffiti, engineering bridge reports, scholarly and historical texts, geological reports, oral addresses, management reports, and any information available on the Web. I listen to people's stories. I listen to the traffic, the raven, the creek's gurgle and surge. Each site, each sound expresses a particular world, an Umwelt connected to the bridge and its underbridge.

The language of the bridge reflects the State and its subjects, its weight, burden, and regimes. Its infrastructure denotes life, death, construction, connectivity, resistance, safety, vulnerability, fixity, and movement. Underbridge states that the bridge won't save you. TRUCK LOAD DEADLOAD EFFECT: the weight of the concrete on the earth. GROSS VEHICLE WEIGHT—the consequences of the sheltering deck on the earth below. DECK JOINT. The sleepers that live there. LIVE LOAD EFFECT PROBABILITY OF FAILURE: the modern State functions and fails as a machine that subjectivates and desubjectivates the contemporary subject. LANE LOAD RESISTANCE: the State both breaks and makes the subject—POST-TENSIONED DEEP GIRDERS— also creating a space for something in-between.

The contemporary political terrain evoked by the underbridge "insures the destroyed subject's immediate resubjectivation by the State."[12] Supported by post-tensioned deep girders, the deck slab holds that road that holds the load. Underbridge rolls over and plays dead.

In March of 1987, the Engineering Department of the City of Edmonton reports on the load rating of the Mill Creek Bridge superstructure. The increase of the gross vehicle weight corresponds with increases in Edmonton's population, and makes it necessary to determine the Mill Creek Bridge's rated capacity under increased live loads. The photos taken to illustrate the existing conditions of the bridge show the apparently minor staining of the bridge girders and the extensive staining of the pier cap.[13]

In Photo 4 the extensive staining looks like bread mould—*cladosporium*. From Photo 3 you can see graffiti on the bridge piers. The graffiti reads *NO SCAB, PC SUCKS*, and *AFL*. You can see *KILL SCABS* just beyond the branches of a small tree. Then: *ANARCHY. NOW.*

The graffiti archives the six-month strike in 1986 by the United Food and Commercial Workers (UFCW) local at Gainers meat-packing plant. Police repression of strikers and support of scabs enraged the city's workforce. On 1 June 1986, eleven hundred workers at Gainers– members of UFCW Local 280P—went on strike. The strike lasted for six and a half months and saw the arrest of over four hundred union members and their supporters. It was one of the most acrimonious labour disputes in Alberta's history.[14]

In 1987, Edmonton's population is 576,249.

Later that year a tornado moves along the eastern edge of the city; twenty-seven people die.

Six thousand eight hundred years before, the volcanic eruption of Mount Mazama (in present-day Oregon) left a layer of ash in Edmonton's river valley. A white strip is still visible today, along the south bank of the North Saskatchewan River not far from where the Mill Creek now empties.

Once, the Mill Creek drained from South Cooking Lake. Now it is a storm-water management facility for the City of Edmonton. Mark more grief here. Read creek as phantom—a managed storm-water facility. Demolished. But also brown, often small, sometimes clear, with ducks, or frozen white, and then wild, with spring runoff.

In the summer of 2009, there is a burned-out armchair on the east side of the underbridge. A few weeks later, a large mural protests the arrival of Columbus and colonization. But nothing lasts long. I don't know who wipes away the art or takes away the armchairs. That autumn is bitterly cold, and I find an upside-down cooking pot on the dry slope. It is surrounded by a series of chalk markings. It is a text I cannot read. It is a carefully arranged ritual beside an exploded newspaper that opens to an ad for American Apparel: a young woman lounges in an armchair. The caption reads: STAY WARM.

In *Real Estate Weekly,* Lawrence Hertzog writes that William Bird started a flour mill at the mouth of the creek in 1870.[15] The mill only lasted three years as a result of the fluctuating water of the creek, but the name remains. In 1891, the railway wound its way along the creek and the area soon became an industrial and commercial site with brickyards, lumberyards, coal mines, an abattoir, acreages, and a few houses. In 1954, the railway was removed; in the late 1960s, the brickyards and the city's incinerator were closed, and the area transformed from what Hertzog describes as a "working class industrial [community]" into a "desirable river community." In 1971, the James MacDonald Bridge was built and the cloverleaf roadways covered over the last stretch of the creek. What Hertzog doesn't mention is that in the late 1800s William Bird was an important Métis businessman; or that city planners, who designed the cloverleaf in 1971, diverted Mill Creek from its original course. The creek now runs underground through a culvert, hidden from sun or wind, until it empties into the river, just across from 96th Avenue in Rossdale.

The original confluence of the North Saskatchewan and the Mill Creek Ravine was farther downriver, just east of Rafter's Landing. The stretch of sand, silt, and river rocks found there today is evidence of the old creek's flow. In the fall of 2011, in the bank, just above the bar, I find an old dump: bedspring, gutters, Orange Crush bottle, and Tonka Toy Truck. The truck and the Orange Crush bottle are circa 1970. The debris intertwines, in a complex configuration, with a standing willow tree that leans toward the river. The presence of the old junk embedded in the silt has protected the bank and the tree—just upstream the bank is badly eroded. Looking back toward the tree, I notice a four-door beaver-bank house just under the roots and garbage. Somehow, I hadn't seen these large round entrances before. The holes are festooned with and supported by bedsprings and gutters. I also notice that the sandy silt mound, in front of the doors and below the willow tree and the scraps of metal, is covered in twigs and logs, all marked by the beavers that live there. This looking extends before me, revealing a seemingly infinite series of relations and their strange dependencies.

The bridge, the altered creek, and the river the creek still runs down to meet represent the recent historic economic dynamics of Edmonton: the complex social and economic relations between Indigenous, European, and Métis families, the success of the fur trade, and the arrival of new European capitalists in the late 1800s: uninterested in the existing communities, but eager to buy and sell the land around Edmonton.

The bridge, like capitalism, runs counter to its own stability. The massive concrete piers that are built deep into the earth and rise up to the bridge's deck are embedded in the eroding bank of the creek. By its very presence, the bridge creates an environment of devastation. It fragments the landscape of the ravine. Consumption reflected in the trucks and condos above runs in deep contrast to the beds on the wide concrete ledges that run the width of each far end of the underbridge.

Standing under the bridge, I look west into the seam between the road and the dark underside of the bridge. Sometimes I stop and climb the desert slope—close, but not too close, to where the sleepers stay at night. Dust rises around me. It clings like soft talcum, pinkish fur, or thick skin in shades of grey. The sleepers are always gone by day, but their remnants remain—sometimes clothes and blankets, beer cans, burned logs and fire rings, once a sofa. For a few months, toilet paper hangs in a parody of festive streamers from the underside of the bridge. Maybe it is not a parody. People sleep here; they eat and keep warm. And, I am trespassing. The logics of their necessities are laid out in unknowable but precise orders.

Sometimes I photograph the ledges, the rocks, and the dust. This feels like a violation of a public space that has been made private. What is public space here? What is private? This looking is a problem. But I am compelled to archive these marks. These Umwelten. I want to trace them. To note their intersecting relations, to listen past my own blindness, acknowledging the lives that stop here. To understand the logics, carefully organized and articulated here, of tentative domesticities and absences. I don't want to forget them. But my wanting absolves me of nothing.

N. asks me about all this looking, and how I feel about it. I tell her I feel trapped. I can't decide if it is worse to look than to look away. There is discomfort and desperation either way. Wasn't my initial fear of this desert place and the men who sat on the ledge also a violation, a trespass? Should I just stay on the paths set by the City fathers, the paved paths for the runners? Then what? If I wander off the path and find the places where the sleepers rest at night, do I become another system of surveillance? Years earlier, in Vancouver, M. told me how hard it is to stay in the shelters. They aren't safe. For a man a ledge might be safer. Not necessarily for a woman. Where do the women sleep? Does this looking reveal what should remain hidden? The life of the human is raw here. Unsheltered. Encompassing, implicating.

Here, history marks particular Umwelten and it marks absences. 3 October 2009: Wikipedia reports that the Yukon & Pacific rail line ran from the CP Line at about 67th Avenue across 99th Street and down into Mill Creek Ravine (1891). It went north and out across the Low Level Bridge. This article also notes that the creek was named after the gristmill established by William Bird in the early 1870s, on river lot 19, at Rafter's Landing. Like Hertzog, this article leaves out the fact that Bird was Métis. But it does note that, in the 1970s, the City of Edmonton converted what was left of the rail line into a bicycle and walking path extending from one end of the ravine to the other and that, in 1988, the Edmonton Historical Board erected a plaque in Mill Creek Park commemorating the Edmonton Yukon & Pacific Railway.

One Sunday evening in the early winter of 2007, just before dark, just north of the underbridge, just before the creek freezes solid, a man crouches by the cold greasy creek washing his food in the water. The water is so cold. Submerge your hands in it even for a second and they will ache. Pull them out, let the blood come back—the pain sears like fire, like ice.

"With respect to bacteriological characteristics, all samples exceed the Canadian Drinking Water Standards. It is doubtful that the creek will be used for drinking, as its muddy appearance implies that consumption is at one's own risk. If water is not potable, then it can be generally concluded that it is not suitable for fish."[16]

In 2008, local government statistics state that there were 3,079 homeless in Edmonton. According to the same statistics, the majority of homeless people in Edmonton are Indigenous men between the ages of 31 and 54.[17]

In the fall and early winter of 2007, the population of people living in the ravine is high. The financial boom creates jobs and increases homelessness. Lack of rental controls in Alberta and a high demand for housing sends the rents up overnight. Many people lose their homes. A tent city is established in downtown Edmonton. By July, it houses over two hundred people.[18] The municipal government shuts it down in mid-September.[19] Many people moved from the tent city into Mill Creek Ravine. The shelters are crowded and unsafe. People in the condos complain about the campers in the ravine. From the underbridge, economic success, the rapid growth of industry is disastrous.

Under the bridge, people from the houses and the condominiums above the ravine walk their dogs past dust, bones, glass, quilts, and carefully stored shoes. From the underbridge, on either side of the desert thistle, shopping cart, mattress, sofa, plastic, rose, shit, grass lean toward the creek. The underbridge changes the people who walk under it, who sleep under it, and the ravens who live nearby. They contract into a fragile slightness against the vast concrete abutments. They become alterity and unknowable and next to. The creek is slight under the high arching space: brown and slick around the shopping carts, beds, bicycles, and creek grass. Last Tuesday there was a metal walker in the creek's rushing, greasy foam.

The progress that resulted in the railway that was built into the ravine in 1891 and the building of the first Mill Creek Bridge on Whyte Avenue in 1912 also brought disaster to certain Umwelten, a homelessness. As the white settler community expanded its infrastructures for new forms of trade and increased contact between other white communities in nearby settlements, a narrative developed that increasingly simplified the complex history of the area. The Métis and Indigenous families that had played such an intrinsic role in the history and development of the city were shifted to the background. Markers shifted in and out of visibility. Not until recently has the complex history of the area been explored.[20]

The beautiful ravine and the underbridge that runs through it like a gash express this past. The surrounded woods speak to a floral and faunal history. The reader cannot read this wounded space without wanting to know what else has been displaced: the sleepers, the sofa, the coyotes. What histories might we read here in these absences? Because of the complex actualities of this place, reading from this underbridge becomes a praxis of vast circumference.

According to archaeological records, the Wisconsinan glacier withdrew about 12,000 years ago, leaving an emerging North Saskatchewan River Valley that attracted moose, deer, elk, and bison.[21] River animals—beaver, muskrat, otters, and birds—were drawn to the rich plant life, as well as the fish: goldeneye, mountain whitefish, walleye, northern pike. Noting this abundance, wolves, coyotes, bears (grizzly and black), and lynx also entered the area.[22]

When you stand on the slope from the far side of the underbridge in the concrete dark, with the empty Chef Boyardee cans, ski jackets, graffiti, condom packages, and look across the creek, you can watch the joggers pushing wide baby buggies on the paved path. I've noticed that when I don't go back to the underbridge, when I just keep writing and stop wandering, the underbridge turns into something else, an embellished abstraction. Each time I return I have to face something I have forgotten: the cold, physical pain, boredom, desolation, my own lack of self, my own sense of loss and shame. It is uncomfortable. No matter what I do now. Agamben writes that shame is the "hidden structure of all subjectivity and consciousness, simultaneously embedded in the bio-political sphere and extricating itself from it;"[23] that is, because shame is "produced in the absolute concomitance of subjectification and desubjectification, self-loss and self-possession, servitude and sovereignty,[24] Agamben believes that it can move us toward a community that resists the bio-political sovereignty that reduces life. I run back down the slope hoping no one has seen me. I don't know who I mean by no one. I don't know what I mean by one. I don't know what I mean by I.

A unitary world does not exist either in time nor space. The Umwelt is the environment world that is constituted by a series of elements called "carriers of significance" or "marks," to which the animal is bound.[25]

According to contemporary archaeologists, Cree, Blackfoot, and Assiniboine communities followed the large mammals to the river valley; the quartzite and chert brought by the preglacial rivers flowing from the Rocky Mountains (1.6 million years ago) provided material for stone tools.[26] Seventy-one percent of the stone tools found in the Edmonton region are made of quartzite, and they provide evidence that Indigenous communities have lived in the river valley for the past 11,000 years.[27] According to Calvin Bruneau, Chief of the Papaschase Cree, the Indigenous people of this area have been here since time began.[28]

In Cree, Edmonton is *amiskwaciwaskahikan* (Beaver Hills House), and *amahkoyis* (Big House) in Blackfoot.

According to Papaschase scholar Dr Dwayne Donald, in the last half of the eighteenth century, the Blackfoot and Cree formed an extensive trade and military alliance with each other in the western Canadian plains.[29] When the Hudson's Bay Company established trading posts at Edmonton House and Rocky Mountain House in 1799, both were within Blackfoot territory. The establishment of the new trading centres shifted the delicate power dynamics between the Cree and the Blackfoot.

In 1870, there was a Cree attack on the Blackfoot just south of the North Saskatchewan River, where the university is now.[30]

Archaeological digs on the Rossdale flats have found evidence that the flats were a historic meeting and trading place for Indigenous people for thousands of years. These flats are just across the river from the new mouth of Mill Creek, west of its original point of entry. Goods from as far away as South America have been found there, and a tradition of trade and cultural integration continued throughout history—up to and including the arrival of European settlers in the area. In *The Rossdale Historical Land Use Study*, Harold Kalman and colleagues report that "there can be little doubt that early people preferred living near the North Saskatchewan River, or along the edges of major creeks.... This pattern of land use appears to have persisted for thousands of years, which forces one to question the traditional Eurocentric portrayal of places like Edmonton as a 'wilderness' prior to European settlement."[31]

But time feels thin above the ravine. Away from the eroding banks and underbridge, amnesia seems to circulate in a widening surface of malls and big-box stores. At the LRT Station at Grandin (just west of the Legislature Building, northeast of the river and northeast of Mill Creek) there is a mural that "celebrates" the role of Bishop Grandin in the history of Edmonton and St Albert. The panel depicts the large shining head of Bishop Grandin, next to a Grey Nun holding an Indigenous baby. A substantial building that resembles a residential school rises above a small community of Indigenous people and their tents. The scene is strangely pastoral. A plaque on the platform claims that Grandin "arrived in Canada in 1854 [and that] upon his arrival the West was a vast wilderness."[32]

In her essay on colonial conflict in the Rossdale Flats,[33] Heather Zwicker writes, "histories of Edmonton are scant (and some of them unreadably racist)."[34]

Within Uexküll's system of horizontal biology, a theriomorphic reader would not be in animal form in opposition to the human, but as well as the human. Thus she would work within her own Umwelt by virtue of her own marks and carriers of signification, as a listening reader with whom the historical and aggressive oppositional positionings between human and animal have been alleviated—the traditional hierarchies will have been dismantled, forms will exist next to each other, on a horizontal plane of encounter, in proximities that reveal deep connections and blind reciprocities: participatory discrepancies.

The mid-1870s smallpox epidemic killed more than half of the Indigenous people in Alberta. That, combined with the disappearance of the bison in the 1880s, placed the local Indigenous and Métis communities in crisis. The Chief of the Papaschase negotiated a reserve for his Cree band through Treaty Six on the southern side of the North Saskatchewan River in what is now south Edmonton. A federal surveyor calculated that the 241 members of the Papaschase band were entitled to 124 square kilometres on the south side of the river.[35]

Every minimal detail constitutes a carrier of significance, and constitutes a different element depending on the environment within which it is observed. Every detail is both familiar and radically other. To read this space suspends ideas of knowing and rests *next to*, in relationships of proximity, by way of non-agonistic perception. It offers a way of acknowledging others without reducing them to objects of knowledge: "to refuse appropriation (you are like me) or empathy (I am like you) to refuse to deny or suppress the alterity of others."[36]

Implicated in Grandin's narrative is my own unknown shape, blind, and unthinkable: a different ledge of grief and cold. That shape—that I that is mine—is more illusory than the creek that runs through the ravine as a conduit for city storm management. This self is spectral, managed, and regulated to feed on a bogus and damaging colonial narrative. What markers are these? What Umwelt? What now? I am consigned to something that cannot be assumed. What does it mean to be here (where I am) and what (I am) when to be there (here) is to deny what was (what is)? What does it mean to exist in this place as a subject under this bridge in this city, this country? Reading from the underbridge is not without casualty. Reading these Umwelten, the European subject is radically displaced and dissolved.

Influential white settlers like Frank Oliver, owner and editor of the local newspaper *The Edmonton Bulletin* from 1880 to 1923 (and later Minister of the Interior and Superintendent General of Indian Affairs from 1905 to 1911), pressured the government to remove the Papaschase band from their reserve, and advised against the creation of any Cree reserves close to Edmonton.[37] In his newspaper, Oliver created and sustained a narrative that degraded the Papaschase people and raised the value of the white settlers to the level of great founding fathers.

In 1888, when three men from the Papaschase band signed a surrender document on 19 November, at a meeting called with four days' notice by the government agent, the Papaschase people lost their entire reserve in south Edmonton.[38]

On 3 April 2008, in *Canada* v. *Lameman*, the Supreme Court of Canada ruled that the descendants of the Papaschase Indian Band had no legal right to pursue their land claim.[39]

According to Shirley Gladu, one of nine council members of the new Papaschase IR136 First Nation, traditional Papaschase lands extend north to the banks of the North Saskatchewan River, south to the Leduc area, as far west as the town of Devon, and east to the town of Tofield.[40] This includes Mill Creek, the bridge, and the ravine.

Underbridge gathers complex ecologies of subjects and non-subjects: of sleepers, coyotes, fescue, painters, waxwing, creek. Underbridge archives the marks of its Umwelten: *oh no!*, *fuck you*, *kill scabs*. Underbridge challenges bridge's span and gleam. Reading from its darkness, horizontally, against the anthropological machine, against the State. The underbridge challenges and extends the creek, the ravine, Edmonton, its history, the dream of the empire, its citizens, its subjects, lost bison, notions and nouns of being, Indigenous land claims, the ecologies of the human and the non-human, of fixity and mobility.

In 2000, at the University of Alberta, Indigenous scholar and lawyer Sharon Venne gave an address to a largely non-Indigenous audience about Treaty Six (1876):[41] "If you are walking around the University you are walking on Indigenous Land," she said.[42] She asked, "What are your treaty rights? Every non-Indigenous person should know their treaty rights ... without the treaty no one other than Indigenous people has the right to live here."[43]

Venne's presentation reflects what was told to her by her Cree Elders.[44] It is a history and a world view to which she has listened carefully. Venne's clearly articulated Umwelt challenges the Eurocentric version of the Treaty process and European history: "Our territories were not *terra nullius*."[45] Venne argues that there were complex systems of governance, culture, and spirituality that existed prior to the Crown and Canada, and that continue to exist. Venne's Umwelt denies the wilderness metaphor essential to the Bishop Grandin narrative, and exposes the workings of the European anthropological machine—one that designates the human in opposition to the animal. Venne's challenge evokes Agamben's discussion of the impact of the Eurocentric anthropological machine, its Umwelt and terrible proximity, but cannot be reduced to it.

The reader notes, listening. The bridge is not necessarily a solution. The creek is not necessarily in the way. Time is not necessarily Crown time. History is not necessarily European history. The human is not where or what we imagined. Working within the oral tradition, Venne speaks as reader *and*, importantly, as listener. Listening, she articulates markers of what has been, still is, and what leans now, with power and possibility into the future.

The site under Mill Creek Bridge is specific to Edmonton, to the stream and ravine. It is fixed in space, and yet it is "embedded in diverse political and social implication[s],"[46] that exist beyond Edmonton and across North America.

The Umwelten of this site, the rocky dust of the underbridge, among the trees of the ravine, also exceed Uexküll's designations. Here not only can we attend theriomorphically (animal-like). We must also attend this place phytomorphically (plantlike) and geomorphically (rocklike).

To read thus (theriomorphically, phytomorphically, geomorphically) would be to inhabit the sites of kinships in our selves, rather than to inhabit the human as site of separation. It would be to note our markers; to listen to the vast intersecting composition to which we are bound and blind. It would be to note that we are constitutive and oblivious to the extent of our relations, to the limits of our Umwelten.

Reading/listening under the bridge, we are compelled by the material world, to see and hear its surfaces, to reconcile ourselves to the specific world within which we move, but also to refract the human, the animal, the plant, the rock, to shift the machinations of the anthropological machine, and to locate an objective space that emerges as an environment created by our receptive organs, but where radically diverse elements are (noted as) intimately connected.

The underbridge is the interior site of a border that harbours and reveals what is censored, repressed, a place where consciousness shifts and processes information differently, where new data are taken into account, where new sense is made. I dream the underbridge. I read the underbridge dreaming. It rests between the sleeping and the woken, between the sleeper and the jogger, between having and not having, life and its absence. It is a place: a nonplace. It is a dynamic and it is a border that borders nothing: that is central to everything.

To note that strange border is to read with an ear turned and tuned to the musical unity of reciprocations and strangeness.

It is to hear the sounds of this space and its dust. Underbridge dust. Solid particles with diameters less than 500 micrometres. Postglacial oil molecule, pollen, skin, particles, till, concrete, meteor, dog, and bird. This pink/grey stuff. Light motes of sound rise and drift: minute, particulate, compositive, fragmented. Dust hangs suspended above trees, bridge, and creek's brown water. It covers the landscape with new definitions and dissolutions. It rises and shifts—articulate. It marks the underbridge as in between, as Umwelt among Umwelten. A suspension of suspension, a zone that marks the dream of the State and its ruin. Reading here, dust rising here. A morphic eye and ear. Underbridge is both text and symphony. It desires its reader and her ears—and the reader is inhabited, willed, unwilling, drawn by its familiarity, blind, and listening to its unknowable alterities. And this attending poses unanswerable questions. This reading hears elements in a composition, a vast and dense musicality that accentuates the discrepancies of worlds, their materiality, their horizontal, extreme, and necessary relations.

Notes

1. Lisa Robertson,"Lastingness," *Open Letter* 13, no. 3 (2007), 47; emphasis added.
2. Giorgio Agamben, *The Open: Man and Animal* (Stanford, CA: Stanford University Press, 2004), 39–47.
3. Ibid., 40.
4. Ibid., 43.
5. Ibid., 40.
6. Ibid., 33–38.
7. Rita Wong, *Poetic Statements KSW Positions Colloquium* (Vancouver, BC: KSW, VIVO, 2008), n.p.; emphasis added.
8. Tom Monto, *Old Strathcona before the Depression* (Edmonton: Crang Press, 2008), 13.
9. Con Create USL LP, "Whyte Avenue (Over Mill Creek) Bridge Rehabilitation Contract No. 0834 Edmonton, AB," 2008, http://www.uslltd.com/251.t1/images/usl.portfolio/Portfolio_48print.pdf.
10. Victor Shklovsky, "Art as Technique," *Russian Formalist Criticism: Four Essays*, translated by Lee T. Lemon and Marion J. Reis (Lincoln: University of Nebraska Press, 1965).
11. Robertson, "Lastingness," 47; emphasis added.
12. Ullrich Raulff, "Interview with Giorgio Agamben—Life, a Work of Art without an Author: The State of Exception, the Administration of Disorder and Private Life," *German Law Journal* 116 (2004), http://www.germanlawjournal.com/article.php?id=437.
13. Duthie, Newby, Weber, and Associates, *Increased Load Limit Program B082 Mill Creek Bridge* (Edmonton: Author, 1987).
14. Alberta Federation of Labour, "Poster Project," *Project 2012*, http://www.project2012.ca/default.asp?mode=posters&id=11.
15. Lawrence Hertzog, "It's Our Heritage," *Real Estate Weekly* 24, no. 22 (2006).
16. Edmonton (Alberta), Parks and Recreation Department, *Mill Creek Ravine. Appendices: Prepared for Edmonton Parks and Recreation* (Edmonton: Butler Krebes, 1974), 47.
17. According to the *Homeless in Canada—Resources* site, these numbers dropped in 2010. In October 2010, the number of homeless persons in Edmonton was recorded at 2,421. *Homeless in Canada—Resources*, http://intraspec.ca/homelessCanada.php#Edmonton.
18. "Province Won't Tear Down Edmonton's Tent City," *CBC News*, 18 July 2007, http://www.cbc.ca/canada/edmonton/story/2007/07/18/tent-city.html.
19. "Edmonton Closes Tent City for the Homeless," *CBC News*, 15 Sept. 2007, http://www.cbc.ca/canada/edmonton/story/2007/09/15/tent-city.html.
20. See, e.g., Linda Goyette's *Edmonton in Our Own Words* (Edmonton: University of Alberta Press, 2004).
21. The River Valley Alliance, *Alberta Capital Region River Valley Park*, http://www.rivervalley.ab.ca/media/uploads/rva-geological-history.pdf.
22. Harold Kalman et al., *The Rossdale Historical Land Use Study* (Edmonton: Commonwealth Historic Resource Management, 2004), 33.
23. Giorgio Agamben, *Remnants of Auschwitz: The Witness and the Archive* (New York: Zone, 2002), 128.

24 Ibid., 107.
25 Agamben, *The Open*, 40.
26 Kalman, *Rossdale*, 27, 36.
27 Ibid., 23.
28 Calvin Bruneau, presentation, University of Alberta, Dr Keavy Martin's class, 11 Mar. 2011.
29 Dwayne Donald, "Edmonton Pentimento: Re-reading History in the Case of the Papaschase Cree," *Journal of the Canadian Association for Curriculum Studies* 2, no. 1 (2004), 32, http://pi.library.yorku.ca/ojs/index.php/jcacs/article/viewFile/16868/15674.
30 Monto, *Old Strathcona*, 13.
31 Kalman, *Rossdale*, 37.
32 After two years of consultation, in March, 2014, the Grandin Mural was revised by the original artist Sylvie Nadeau. Two panels depicting Indigenous and Métis youth were added to the far sides of the original work. In addition, local Métis artist Aarron Paquette created a responding mural on the opposite side of the station, facing the original work. The original plaque that I describe here still remains. See http://www.edmonton.ca/city_government/news/2014/aboriginal-spirit-displayed-at-grandin-lrt-station.aspx.
33 Heather Zwicker, "Dead Indians, Power Conglomerates and the Upper Middle Class: Commemorating Colonial Conflict in Edmonton's Rossdale," presentation, Canadian Association of Cultural Studies Conference, Hamilton, Ontario, 2004, http://www.culturalstudies.ca/proceedings04/proceedings.html.
34 Ibid., 7. "There is virtually no architectural record: Edmonton tears down old buildings to make way for the new. There is no cultural record…. There is barely even an archaeological record…. And there is no literary record. Histories of Edmonton are scant (and some of them unreadably racist or poorly written)." Ibid.
35 Goyette, *Edmonton*, 114.
36 Lyn Hejinian, *Language of Inquiry* (Berkeley: University of California Press, 2000), 332.
37 "Frank Oliver and the Michel Band, Excerpt from the Report of the Royal Commission on Aboriginal Peoples (RCAP)," *Michel First Nation*, http://www.michelfirstnation.net/frank-oliver--the-michel-band.html.
38 Sharon Venne, "Treaties Made in Good Faith," in *Natives & Settlers, Now & Then: Historical Issues and Current Perspectives on Treaties and Land Claims in Canada*, ed. Paul W. DePasquale (Edmonton: University of Alberta Press, 2007), 2.
39 "Papachase Land Claim Resolved," Centre for Constitutional Studies, 2012, http://www.law.ualberta.ca/centres/ccs/rulings/papchaselandclaim.php.
40 Richard Cairney, "Confronted," *SEE Magazine*, Feb. 1999.
41 Treaty Six is the agreement between the Crown and the Plains and Wood Cree at Fort Carlton, Fort Pitt, and Battle River. The Treaty covers most of the central area of the current provinces of Saskatchewan and Alberta. One Manitoba band would also sign on to the Treaty by adhesion in 1898. The Treaty signings began in August 1876.
42 Venne, "Treaties," 2.
43 Ibid., 5.
44 Ibid., 1.
45 Ibid., 6.
46 Stephen Bocking, "The Nature of Cities: Perspectives in Canadian Urban Environmental History," *Urban History Review* 34, no. 1 (2005), 3.

CHAPTER 15

Understory Enduring the Sixth Mass Extinction, ca. 2009-11

Rita Wong

Maybe with many years of listening and studying, I could learn the languages of black bear, deer, salal, salmonberry, cedar, and yew. These are just a few inhabitants of the unceded Coast Salish watershed where my drinking water comes from. Ficus Chan, watershed educator and forest guardian, tells me that a bird like the Swainson's thrush only sings when the salmonberries are edible. The song that links bird to berry to fish is overshadowed by city traffic but can be heard if I attend to it.

With Ficus as guide, we walk in the forest that holds and nourishes the Capilano watershed, seeing trees as medicines, living vessels, arboreal water pumps, realizing that the hemlocks and firs are making and cleaning the air that we breathe, feeling grateful for what is around us, in and out of us, part of a larger flow and rhythm. Having inhaled the forest, I mull over how it infuses me and the city with life through clean water and air. The question of how urban dwellers, like myself, perceive our relationship to forest ecosystems, near and far, is important because it makes some actions possible, others impossible. The understory is what gives life to the forest, and to us, who breathe the oxygen produced by the trees, drink the water slowly held and released by the trees. Composed of shedded biomass (leaves, twigs, petals, etc.) and nursing logs, whose decay offers nutrients for new growth, what might look like a mess is actually what makes you and me possible.

What we can't see is every bit as important as what we can. And we forget this at our peril. When I walk in Coast Salish forests, admiring the cedars, firs, and hemlocks, my gaze turns repeatedly up as I inhale the distinct green smells of tree upon tree upon tree. But beneath my feet there is steady activity I must imagine as well—roots going deep, deep into the earth, holding vast stores of water, acting as what Vandana Shiva has called "natural dams."[1] They are part of the ancient cycle that establishes groundwater and keeps the watershed capable of retaining and releasing the water I drink every day.

An urban creature, I stumble a lot in the forest as I slowly awaken to the significance and situated knowledges of the larger ecosystems that my species has steadily paved and clear-cut the past century. Though I love walking in watersheds, I have spent much more of my life reading, and much of what I am now learning about the environment is mediated through texts, through the writings of knowledge keepers like Vandana Shiva and many more. I grew up ignorant of forests, notwithstanding rare camping trips, and am only now coming to remedy that cultural impoverishment. Yes, I say *cultural* impoverishment, for the forest has a distinct culture, whether humans recognize it or attempt to destroy it. The forest's inhabitants and constituents form cultures every bit as valid as human ones, even if we do not know how to speak their languages. A mature culture is one that respects, rather than merely selfishly exploiting, the vast non-human realms that make its existence possible.

Even if I do not learn the languages of bear and deer, cedar, and salal—for who has the time and energy to do so in a human world that so constantly calls, distracts, and stresses on a systemic basis—can I at least respect their worlds, and leave them alone to live as they will? Many of us do just that, you might say, for we don't go traipsing into the watershed to gawk at the animals who survive and remain, and city folk may well go home after taking a few photos of the cute deer. But our impact as a species has nonetheless damaged the habitats and lives of so many creatures—salmon, tuna, wolves, bison, beavers, to name just a few—that as we lose the natural wealth and diversity that has enabled our quality of life, I fear what we have destroyed will in turn destroy us.

While the earth has gone through five previous mass extinctions, many scientists believe that we are now undergoing the only mass extinction that may be caused by the accumulated impact of one species—us. As a lifelong lover of reading, I weigh my sources carefully, and I find these scientists credible, not infallible, but credible enough to take their very educated best guesses seriously. As such, it is not helpful to get into arguments about blame, but we do urgently need to decide how to respond to the situation. This danger is too vast and intimidating for me to tackle alone or head on, but I hold it in the edges of my consciousness like a worrisome dilemma, always wondering how it can be collectively addressed, hoping for clues and allies as I proceed. Yes, I'm scared but there is no point in panicking. Keep breathing, keep going, I tell myself. What can't be accomplished alone, can be accomplished together if you stay present and alert.

One fact that that we need to understand is the crucial role of the ocean in supporting our lives. After interviewing scientists around the world, Alanna Mitchell has described what is happening to the ocean today: a deadly combination of increasing acidification, growth of dead zones, and warming. This, combined with how industrialized fishing has depleted global fish stocks by roughly 90 percent in the past fifty years, is drastically changing the ocean's chemistry. In *Sea Sick: The Global Ocean in Crisis*, Mitchell argues that if all life on land were to disappear tomorrow, the creatures in the sea would survive or be better off without us. But if life in the ocean disappears, so eventually will we. Half the oxygen we breathe is produced by phytoplankton, and the phytoplankton who rely on calcium in the ocean will find that with the acidification caused by increased carbon dioxide in the atmosphere, which the ocean has been absorbing, they may not be able to form the way they have for millions of years.[2] If we continue at current rates of carbon emissions, this could happen as soon as 2050.

I find myself faced with the challenge of understanding how our lives depend on something as mundane and taken-for-granted as phytoplankton being able to live in the ocean, and how basic chemical reactions mean that, en masse, our car exhaust affects the living conditions of plankton. The implications are terrifying. And yet, there is an odd poetry to this interconnection, a call to grow into the world more attentively rather than to retreat from it in ignorance and fear. However terrible the situation is, the response matters, if only for ourselves to define who we choose to be in this moment. In an emergency, there is nothing to lose by trying to stay calm, open, and curious, and much more to be lost in closing up in panic or depression.

While the scale of the problem is global, and responses will need to become global, I also find myself turning to what's local, what's tangible. Start with where you are, with the home that you can experience in your everyday life, while remembering that the places you can't directly experience are also home, also matter. From the condo-crazed city I call home, I've observed that the neoliberal myth of economic progress, while superficially appealing, is fatally flawed in how it privileges economy over ecology, and the rich over everyone else.

Working from a philosophy of power *with*, rather than power *over*, others offers what Chickasaw writer Linda Hogan has called "a different yield." She writes, "We are looking for a tongue that speaks with reverence for life, searching for an ecology of mind. Without it, we have no home, have no place of our own within the creation. It is not only the vocabulary of science that we desire. We also want a language of that different yield. A yield rich as the harvests of the earth, a yield that returns to us our own sacredness, to a self-love and respect that will carry out to others."³

As someone who has inherited a history of migration, capitalism, and colonization, a history defined by some people exploiting and exerting force over others, I feel it is very late, but not too late, to cultivate cooperation with others, human and non-human. Home, both precarious and vital, is to be carefully made by respecting the cultures grown from the land on which one lives, wherever that may be, as well as language itself, as it flows much like water does, in rhythms both ancient and contemporary, manufactured, intervened stops and starts that bear the weight of industrialization. The fact that I am writing this in English rather than Hənq̓əmin̓əm̓ or Skwxwú7mesh Snichim or Chinese, already attests to a historical devastation.

My everyday life occurs within a fabric consisting of catastrophes that range from colonization to global warming, both of which are normalized to become unseen or often bracketed in the back of people's minds because they do not feel capable of addressing them. I cannot change the violence of the history I've inherited, but I can take responsibility for how I respond to it. How might we better learn from and assert solidarity with what remains of Indigenous cultures that have respected land and water for millennia? Many Indigenous people such as Dorothy Christian, Josephine Mandamin, and groups like the Keepers of the Water—composed of First Nations, environmentalists, and concerned citizens—assert that water is sacred. Following their lead, I find that water leads me to everything from the *Tao Te Ching* to the geopolitics of pollution to ecological thinking, decolonization, and learning how to swim. I begin to learn just how permeable, how interdependent I am, as I ask: what happens if we view our home through the lens of water?

The land that the Chinese used to call Saltwater City (in contrast to New Westminster, which was Freshwater City) is marked by stal'əw (what Musqueam speakers like Victor Guerin call the Fraser River) to the south, the Burrard Inlet, Salish Sea, and Pacific Ocean to the west, three watersheds to the North—Capilano, Seymour, and Coquitlam—as well as over fifty streams that used to be full of salmon while under the care of its Indigenous peoples, the Musqueam, the Squamish, the Tsleil Waututh. Today the land is marked by the name Vancouver. The story underneath, as told by salmon whose homes have been destroyed and paved over by colonizing and dam-building humans, would reveal how the age-old wealth of the world that feeds both us and the salmon has been systematically destroyed in the past century, with one humble and important exception.

The Musqueam Ecosystem Conservation Society looks after the one original wild stream that still gets the rare salmon returning to it. Knowing that this stream is struggling for its life, they dedicatedly improve its odds for survival. One Saturday morning, I learn from them how to rip out invasive ivy, Himalayan blackberry, holly, and laurel to make space for the indigenous plants that are trying to survive along the creek's banks. It is sweaty and surprisingly satisfying work for a city girl who is finally starting to remedy her ignorance of what the city tends to pave over and obscure.

Another Saturday morning, I participate in a stream walk led by Willard Sparrow, who tells the story of how they saved the city of Vancouver $40,000 by asking the city to stop removing tree "litter" from the creek. In "tidying up," the city's park staff were starving the fish, as it were. Learning to see the beauty of what the trees do, rather than imposing a colonial aesthetics upon them, turns out to be economically beneficial as well. Sparrow encouraged us to reimagine our relationship with the creek and the park when he paused and said, "Enjoy 10 minutes of yourself," reminding us of how the ecosystem gives us back to ourselves, if we quiet down and pay attention.

Nor is Musqueam Creek alone. Various community efforts to bring paved over streams back to daylight are evident, as newcomers belatedly realize the immensity of what has been lost. The Still Creek Daylighting Project, Spanish Banks Creek Daylighting, and the Sanctuary in Hastings Park are a few of the signs that a growing number of people understand the importance of the original waterways as being valuable beyond economic measurements, as inherent to the integrity of the land, to which we can contribute, rather than ignore or exhaust. The streams were buried because, in many cases, newcomers used them as garbage dumps. What needed changing was people's perception and behaviour, not the streams themselves. The streams could teach us to coexist with resilience, if we are wise enough to listen.

At the corner where I live, at St George and 5th, I can hear a small buried stream, now sewer, gurgling its longing to return to light. Its song calls me to help organize neighbourhood conversations, a street party for the creek, proposals to depave it, revive it, re-story it. I expect it will take decades to remove even some of the concrete, but I'm prepared to try.

In the face of my looming fears around global warming and mass extinction, how can trying to revive a few buried or struggling creeks make even a dent in the threats that face us? I freely admit the inadequacy of this, yet what's at stake, I feel, is how we perceive and act on our relationships with the non-human life that ranges from urban streams as biodiverse habitat to microscopic copepods in living, wild water to salmon, lichen, plankton, and a whole host of creatures who we rely on without knowing. The details matter. Alone, they are not enough, but in conjunction with other efforts across watersheds and oceans, who knows what could become possible when people more fully respect both the ecosystems that they live in as well as the ones they cannot see?

And from another angle—that of water—even if people do not survive the ecological crisis that we are wreaking, water will continue to flow, perpetually poisoned, perpetually healing, in its own time. As books like Alan Weisman's *The World Without Us*[4] remind us, water and land will do just fine without us. While I might draw some comfort from that, I'd selfishly like humans to be around for the long term, and so I embark on my ecological education, starting but not ending with the urban streams beneath my feet.

As important, I realize, is to cultivate and coordinate a range of ways to respect the land. This concept has already been well defined by many cultures, as my colleague Maria Hupfield, quoting Doreen Jensen, shares with me: "LaxHösinsxw is a very important word in the Gitxsan language. It means honouring and respecting others, place and space. In this city, at this site and at this time in history, it is a word we might learn from. The place and space in which ... Vancouver ... stands, physically and metaphorically, is a contested one. Here, where a city has been incompletely exchanged for the forest and newcomers have incompletely replaced the aboriginal inhabitants, LaxHösinsxw may be key to the creative process and to our future."[5]

How does one translate such respect into one's daily life? It begins, for me, with a concerted effort to decolonize, to listen, to observe, and to recognize the knowledge and concerns of Indigenous thinkers like Jensen, Hupfield, Christian, Sparrow, Hogan, Guerin, and more. It is very late, but not too late, for decolonization and reindigenization to happen. As writers like Lee Maracle and Jeannette Armstrong remind us, in a slim volume called *Give Back: First Nations Perspectives on Cultural Practice*, respect involves reciprocity.

We have taken and taken from the watershed and its Indigenous peoples. What do we give back in return? What might be worthy gifts to offer? How might practising LaxHösinsxw wherever we are help to strengthen both the places we see as well as those we cannot?

What the city gives back right now is toxic, inadequate. Metro Vancouver releases its treated waste water into the stal'əw (and Victoria is even worse, dumping its raw sewage into the ocean, shamefully). Selfishly, I am grateful that my drinking water comes from the forests that I have walked with Ficus Chan, not from the long, winding stal'əw, which is also downstream of Prince George and its pulp mills. But what of my fellow creatures trying to live in and with the stal'əw?

Sludge sinks and scum floats in the Annacis Island, Lulu Island, and Langley waste-water treatment plants. Tons upon tons of sludge sink and scum floats in the Lions Gate and Iona Island sewage plants. Once the solids are removed, the water that remains is chlorinated before it's returned to the stal'əw, the Burrard Inlet, eventually the Pacific. These sewage plants were not designed to deal with medical waste—their engineers did not imagine Prozac, birth control pills, antibiotics, and much more, dissolved and floating in their smelly, watery depths.

Is estrogen to estuary what chlorine is to kidney? Even if a watershed could sustain and adapt to one attack, such as excess nitrogen fertilizers leaking from industrial agriculture to form dead zones at a river's mouth, how would it recuperate from a toxic stew of man-made chemicals that includes thousands of endocrine disruptors, brominated flame retardants, and much more? Ninety-one percent of Canadians test positive for bisphenol A.[6] We are flush with the poisoned gifts of Monsanto, Dow, and other corporate chemical pushers. Every bit of seafood tested in Singapore had bisphenol A in it.[7] I hear of an engineering team at Ryerson University that has designed methods to clean pharmaceuticals from waste water,[8] and it is heartening to know what humans can do when they put their minds to a meaningful purpose. Will our public infrastructure move quickly enough to measure and remove these drugs?

Tracking water's flow also inevitably leads me to oil. Last summer, on a journey along stal'əw, I was stunned to find a pipeline built underneath the river, not far from its headwaters, within a so-called provincial park. Because of the Alberta tar sands,[9] a pipeline expansion is being proposed by Enbridge, and it is being met with strong resistance by both First Nations and municipalities.

As British Columbia's longest river, stal'əw carries silt from the Rocky Mountains out to the Pacific Ocean, bit by bit, over millennia. The river that created Vancouver is endangered by a poorly planned megaproject that is enormous enough to be seen from outer space, poisoning three to four barrels of water for every barrel of oil it extracts from the earth. If Calgary, rather than Fort Chipewyan, were downstream of the tar sands' tailings reservoirs, would those toxic reservoirs have been built? The tar sands strikes me as not only the biggest reason why Canada failed to meet its Kyoto Protocol targets, but also one of Canada's largest examples of environmental racism. If there weren't a revolving door between corporations and government, would the tar sands have been allowed to proceed? Ironically, the other structure in northern

Alberta that is visible from outer space is the world's largest beaver dam, found in Wood Buffalo National Park. Where the beavers have made the land habitable for many other creatures, humans have done the exact opposite.

In approaching my watershed, I find myself not only trying to think like a slowly eroding mountain range and an endangered, life-giving river, but also like a chaotic ecosystem, facing the political onslaught of human greed and hubris. Because what made this disaster was political and economic, a large part of the solution will also have to be political and economic. The flow of water, of pollutants, of information, circulates in so many directions, I feel at times bewildered. And humbled.

From beginning to end, petroleum compounds ecological damage. Not only does it increase Canada's greenhouse gas emissions at the front end, but plastic is also manufactured from petroleum, carrying its own deadly footprint.

I live in a world where it is now not unusual to find over a dozen pieces of plastic in a dead seabird's stomach if you are in the middle of the ocean. Chris Jordan, who takes photographs of dead albatrosses full of plastic debris on Midway Atoll, points out that these photos are not merely of objects "out there," but rather, they are self-portraits for the humans implicated in them.[10] The plastic is the material trace and implication of our own actions, the evidence of how we are related to those we may not see, regardless of intentions, ignorance, or denial.

The Great Pacific Garbage Patch, which may more aptly be called a plastic stew somewhere between twice the size of Texas and the size of Canada, is one of eleven gyres in the world, ocean currents sucking plastic trash from the land into the hearts of the ocean. While much of Vancouver's plastic litter probably stays within a slow-moving tidal basin along our coast, according to my colleague Duane Elverum we do receive some of the trash that circulates in a clockwise gyre extending from San Francisco, past Hawaii, northwards through Asia, circling under Alaska, and then down along BC's coastline.

Tracking the flows of water (and all that they dissolve and carry) through the locus of Vancouver helps to articulate some of the many relationships that exist between the seen and unseen, relationships that might also be addressed through a poetics attentive to the art of the possible, the relationship between what has been done and what remains to be done. Édouard Glissant writes, "A poetics cannot guarantee us a concrete means of action. But a poetics, perhaps, does allow us to understand better our action in the world."[11]

Emergent, improvisational, chaotic—without guarantees or certainty, a poetics receptive enough to encompass turbulence and stillness, wide enough to revoke corporate personhood and invoke gratitude to phytoplankton—open enough to articulate wonder at mycelium and embarrassment at human hubris—how might such a poetics sing and stutter the understory, cultivate a critical mass that enables a cultural shift for the planet that we are part of? Alongside my fear and frustration at so many industrial-scale disasters is the joy at still being alive, at being able to ask what needs to be opened in this moment, what possibilities, heretofore isolated from one another, might in concert offer some footholds from which to navigate and recompose, humbly, yet responsibly. A bit to my surprise, one foothold I happen to I stumble upon is Robin Blaser's *Holy Forest*.

Blaser writes, "Where the poesis reopens the real and follows its contents, the presuming discourse imposes form and closes it, leaving us at the mercy of our own limit.... It may be argued that the push of contemporary poetics toward locus, ground, and particularity is a remaking of where we are ... a remaking of the real is at stake. One needs only to notice how much of it is a common experience and also something regained, rather than an invention."[12]

Something else regained—how community might be reimagined—comes via Cherokee writer Daniel Heath Justice, who tells us:

> Indigenous nationhood should not ... be conflated with the nationalism that has given birth to industrialized nation-states. Nation-state nationalism is often dependent upon the erasure of kinship bonds in favor of a code of assimilative patriotism that places, and emphasizes, the militant history of the nation above the specific geographic, genealogical, and spiritual histories of peoples.

> Indigenous nationhood is more than simple political independence or the exercise of a distinctive cultural history; it's also an understanding of a common social interdependence within the community, the tribal web of kinship rights and responsibilities that link the People, the land, and the cosmos together in an ongoing and dynamic system of mutually affecting relationships.[13]

Or as Peter Warshall would phrase it, "When you turn on the faucet, picture the network that binds your body to the biosphere."[14] The web of relationships to the biosphere and the web of literature can both yield a respect for what is "outside" one's lived experience. On the one hand, there is no outside to the intended and unintended *effects* of global capitalism. The plankton are in our home with us, or we with them. On the other hand, the outside is what we do not actually experience directly in our daily lives; in that sense, the plankton remain outside my landlocked *Umwelt*. This does not mean they don't matter. The question: is how do I acknowledge or respect the relationships and the differences that exist? I start by realizing how contingent the "outside" is, and how reading and writing continually restructure our perception of it, as Robin Blaser notes:

> I know writing is in trouble these days—a cultural mongering tells me so. Yet, it seems to me that, in the midst of our cultural depletion, we've participated in a very great period of art. One that could change our experience of the outside of ourselves. It is that I meant one time in using the phrase "the practice of outside." And I believe that there is a larger audience for this art than ever before. That audience does not control mercantilism which controls public space and the forms of its devotions. I believe in necessary writers. And one of them, Avital Ronell, writes: "Resist the numbing banality of the they, the dictatorship of nonreading." And I find this is another necessary writer, Michel Serres, writing an imaginary dialogue: "Neither the world nor the market knows how to integrate suffering and happiness, nor the question of meaning, nor that of evil." And on a page of a third necessary writer, Giorgio Agamben, I'm reminded that "Dante classifies languages by their way of saying yes" and find this parenthetic notice: "(What is astonishing is not that something was able to be, but that it was able to not not-be)." And that is getting pretty close to whatever and whomever we mean by love.[15]

A poetics of love is simultaneously a politics of love. Having studied how people respond to disasters, Rebecca Solnit has observed that spontaneous mutual aid often comes to the forefront, materializing a wider, more generous sense of relationship than the narrow, privatized one that is normalized under late capitalism:

> The existing system is built on fear of each other and of scarcity, and it has created more scarcity and more to be afraid of. It is mitigated every day by altruism, mutual aid, and solidarity, by the acts of individuals and organizations who are motivated by hope and by love rather than by fear. They are akin to a shadow government—another system ready to do more were they voted into power. Disaster votes them in, in a sense, because in an emergency these skills and ties work while fear and divisiveness do not. Disaster reveals what else the world could be like—reveals the strength of that hope, that generosity, and that solidarity. It reveals mutual aid as a default operating principle and civil society as something waiting in the wings when it's absent from the stage.[16]

A poetics of love is simultaneously a poetics of relation, an aesthetics of the earth, not a romanticized natural earth in this era of late capitalism, but one marked by massive human intervention, as cogently articulated by Édouard Glissant:

An aesthetics of the earth? In the half-starved dust of Africas? In the mud of flooded Asias? In epidemics, masked forms of exploitation, flies buzz-bombing the skeleton skins of children? In the frozen silence of the Andes? In the rains uprooting favelas and shantytowns? In the scrub and scree of Bantu lands? In flowers encircling necks and ukuleles? In mud huts crowning gold mines? In city sewers? In haggard aboriginal wind? In red-light districts? In drunken indiscriminate consumption? In the noose? The cabin? Night with no candle?

Yes. But an aesthetics of disruption and intrusion. Finding the fever of passion for the ideas of 'environment' (which I call surroundings) and 'ecology,' both apparently such futile notions in these landscapes of desolation. Imagining the idea of love of the earth—so ridiculously inadequate or else frequently the basis for such sectarian intolerance—with all the strength of charcoal fires or sweet syrup.

Aesthetics of rupture and connection.
Because that is the crux of it, and almost everything is said in pointing out that under no circumstances could it ever be a question of transforming land into territory again. Territory is the basis for conquest. Territory requires that filiation be planted and legitimated. Territory is defined by its limits, and they must be expanded. A land henceforth has no limits. That is the reason it is worth defending against every form of alienation.[17]

Much of the damage that has been inflicted on the planet and its peoples is irreparable, tragic, unbearable. But much remains that is resilient, courageous, and creative, which keeps me alert for possibilities. Awake to rupture and potential connection.

In a world groaning and fevered under the burden of industrialization's poisoned gifts, I observe how the old is new, not naively but dynamically, persistently, *tidalectically*, to borrow a term from Kamau Braithwaite via Wayde Compton. Tidalectics is "a way of seeing history as a palimpsest, where generations overlap generations, and eras wash over eras like tides on a stretch of beach [...] Repetition, whether in the form of ancestor worship or the poem-histories of the *griot*, informs black ontologies more than does the Europeanist drive for perpetual innovation [...] In a European framework, the past is something to be gotten over [...] in tidalectics, we do not *improve upon* the past, but are ourselves *versions* of the past."[18] As Jared Diamond's *Collapse* points out, some societies have died out, while others have adapted successfully to their homes for thousands of years. Which version of the past are we?

As someone descended, however brokenly and imperfectly, from a tradition of poet-scholars, I seek what Myung Mi Kim has called "generosity as method," an "ability to read subtleties and nuances as to how you affect the systems around you, where they are intimate relationships or work or the poetics you explore, how can we attend to that whole *circuitry*."[19]

Despite systemic privatization under capitalism, water, air, language, and culture continue to form a circuitry that keeps alive the commons. As Vandana Shiva points out, existent cultures do self-organize for their common well-being, and the commons do "in fact apply the concept of ownership, not on an individual basis, but at the level of the group."[20]

An attentiveness to the circuitry of how we are materially interdependent helps to strengthen awareness of the often-overlooked commons in which we live—the commons that is our understory. Our governments and our structures of feeling are affected by the presence of both salmonberries and endocrine disrupters, the effects of both cedars and toxic waste reservoirs, all of which we are linked to through water. Whether or not our decision-making apparatuses account for this complexity, industrialization has made it so.

From plankton to people, how do we read and write spaciously and humbly enough to respect not only the familiar, but also the distant, the illegible, the unpredictable, the unknown? It may be in attending to the languages and dynamics of the understory—the watery commons that constitutes both our watersheds and our bodies—that we offer one another better gifts: peaceful kinship, meaningful lives, creative depths and breadths. Why settle for the ecological reductionism and social devastation of late capitalism when another world is still possible? What kinds of relations and changes do we, as readers of this planet, invite and nourish?

Notes

Earlier versions of this work appeared in Rita Wong, "Understory Enduring the Sixth Mass Extinction, ca. 2009," *Matrix* 84 (2009), 5, and in Rita Wong and Dorothy Christian, "Untapping Watershed Mind," in *Thinking with Water*, ed. Cecilia Chen, Janine MacLeod, and Astrida Neimanis (Montreal and Kingston: McGill-Queen's University Press, 2013), 232–53.

1 Vandana Shiva, *Water Wars: Privatization, Pollution, and Profit* (Toronto: Between the Lines, 2002), 3.
2 Alanna Mitchell, *Sea Sick: The Global Ocean in Crisis* (Toronto: McClelland & Stewart, 2009), 85.
3 Linda Hogan, "A Different Yield," in *Reclaiming Indigenous Voice and Vision*, ed. Marie Battiste (Vancouver: UBC Press, 2000), 122.
4 Alan Weisman, *The World Without Us* (Toronto: HarperCollins, 2007).
5 Doreen Jensen, "Metamorphosis," in *Topographies: Aspects of Recent B.C. Art*, Vancouver Art Gallery, 29 Sept. 1996–5 Jan. 1997, http://www.ccca.ca/c/writing/j/jensen/jen001t.html.
6 Solarina Ho, "Canada Tracks BPA Exposure, Finds in Most People," 16 Aug. 2010, http://www.reuters.com/article/2010/08/16/us-bisphenol-idUSTRE67F4NW20100816.

7   Chanbasha Basheer, Hian Kee Lee, and Koh Siang Tan, "Endocrine Disrupting Alkylphenols and Bisphenol A in Coastal Waters and Supermarket Seafood from Singapore," *Baseline/Marine Pollution Bulletin* 48 (2004), 1145–67.
8   "Ryerson University Students Design Innovative Wastewater Treatment for Removing Pharmaceuticals," 31 Mar. 2010, http://www.ryerson.ca/news/media/General_Public/20100331_rn_wastewat.html.
9   The tar sands is actually its historical name; it was known as this before industry rebranded it as the "oil" sands. What's there is bitumen, closer to "tar" than "oil," if truth be told.
10 Chris Jordan, *Midway*, http://www.midwayjourney.com/.
11 Édouard Glissant, *Poetics of Relation*, translated by Betsy Wing (Ann Arbor: University of Michigan Press, 1997), 199.
12 Robin Blaser, *The Holy Forest: Collected Poems of Robin Blaser*, ed. Miriam Nichols (Berkeley: University of California Press, 2006), xviii.
13 Daniel Heath Justice, "'Go Away, Water!': Kinship Criticism and the Decolonization Imperative," in *Reasoning Together: The Native Critics Collective*, ed. Craig Womack, Daniel Heath Justice, and Christopher B. Teuton (Norman: University of Oklahoma Press, 2008), 151.
14 Peter Warshall, *Septic Tank Practices* (Garden City, NY: Anchor, 1979), 45.
15 Blaser, *Holy Forest*, 405–6.
16 Rebecca Solnit, *A Paradise Built in Hell: The Extraordinary Communities That Arise in Disasters* (New York: Viking, 2009), 313.
17 Glissant, *Poetics of Relation*, 151.
18 Wayde Compton, *Blueprint: Black British Columbian Literature and Orature* (Vancouver: Arsenal Pulp Press, 2003), 17; original emphasis.
19 Myung Mi Kim, "Generosity as Method: An Interview with Myung Mi Kim," by Yedda Morrison, *Tripwire: A Journal of Poetics* 1 (Spring 1998), 78; original emphasis.
20 Shiva, *Water Wars*, 26.

CHAPTER 16

Seeding Coordinates, Planting Memories
Here, There, & Elsewhere in W.H. New's *Underwood Log*

Travis V. Mason

Asking Where

> *This place seems frail; the merest invention could make it disappear. How to know this land without vanquishing it.*
> —Tim Lilburn, "How to Be Here?"[1]

Questions about "here" abound in Canadian literary history, from Northrop Frye's famous "Where is here?" to Tim Lilburn's meditative "How to Be Here?" This demonstrated interest in geography vis-à-vis ontology and phenomenology, which Alan Morantz, in *Where Is Here? Canadian Maps and the Stories They Tell*, calls a "cartographic obsession," "has even affected Canadian novelists and artists who time and again probe themes of exploration and mapping as routes to self-identity."[2] Frye's and Lilburn's questions invite responses that attempt to get beyond the rhetorical and engage the spatial, temporal, and ideological concerns they evoke. Frye suggests that the problem of locating *here* has involved for Canadian writers, as W.H. New reminds in *Articulating West* (1972), an inevitable turn to the physical landscape,[3] which has contributed in myriad ways to the wilderness myth.[4] Place, in other words, takes on cultural meaning as soon as anything gets written about it, however uncertain the author's bearings might be and so long as incorrect bearings do not lead to real-world injury or death. Because Frye is talking primarily about settler writing when he poses his question, we might read "Where is here?" in the twenty-first century as "Where the hell are we?"—a translation that both cleaves "Where is here?" to the more common and strictly ontological "Who am I?" and points toward Lilburn's ethically inflected question, which is also attuned to the role poetry might play in learning how to be here. "How to be here?" for Lilburn is not only a philosophical conundrum. It is political and ecological insofar as he acknowledges the autochthonous presence of Canada's First Peoples alongside the unsettled presence of European descendants, and he wonders, "What is the source of [the] impulse to colonize the world psychically, bending otherness into human forms?"[5] The centrality of the human—human form, human consciousness, human culture—simultaneously informs these questions about place and complicates the search for answers, always and unavoidably anthropocentric in their drift. The emphasis on "here," as I posit below, also inevitably points us "there" and "elsewhere."

Like the temporal notion of "now"—which can neither be observed nor inhabited because of time's constancy—"here" poses geographical complications. A character in Timothy Taylor's *Stanley Park* (2001), an intertextual touchstone for New's work under examination here, embodies the geographical and ontological concerns identified by Frye, New, Lilburn, and Morantz. Siwash, a homeless man living in Vancouver's urban park, evades the ambiguities of such questions as Frye and Lilburn posit by claiming to know precisely where he is and by remaining fixed in place, inhabiting his square of land as unproblematically as most settler-colonials inhabit their homes on unceded territory. The apparent certainty regarding *where* Siwash stands renders moot the ethical question of *how* to be in place. Surrounded in his bunker by two-dimensional maps, Siwash ultimately eschews them because of their limitations and necessary distortions of a three-dimensional world. To mark his hyper-local position in space, he, ironically, uses a Global Positioning System (GPS); he spends his days counting the number of people who move through his space, reminding himself—convincing himself, rather—that he is "not in motion."[6] Another irony, of course, is that he is always in motion, as we all are. While knowing the coordinates of "here" does not necessarily encompass knowing how to be where we think we know we are—an ethical, ecological concern that complicates Frye's conclusive query[7]—that a character named Siwash doesn't feel the need to calculate his ethical relation to place implies a sense of belonging that challenges allochthonous inhabitants of place. But it also, given his name's historical and geological significance,[8] suggests that we might learn an ethics of belonging from a rock's slow movement through time and space.

W.H. New's book-length poem *Underwood Log* (2004) takes these fundamental Canadian questions about place, joins them with such texts as *Stanley Park*, and reiterates them historically, lyrically, and playfully. His sixth collection of poetry, *Underwood Log* comprises images and ideas about place that New has been contemplating for decades (and which address, indirectly, Frye's and Lilburn's questions).⁹ In this collection, New uses map coordinates in place of conventional titles to locate individual poems. New's situating, which locates in a seemingly precise way, does not suggest that the poem and the reader are meant to remain fixed in one place. Indeed, such stasis is anathema to the aesthetic and the poetics at work in *Underwood Log*, which in many ways resembles a traveller's log: the poem ranges all over the world, visiting India, China, Sri Lanka, New Zealand, Australia, Great Britain, Italy, Wales, the United States, and Canada. Despite this cosmopolitan aspect, though, the poem keeps returning to Canada's Pacific Northwest. This specificity is important because attention to place remains important in an increasingly international (and neoliberal) world. As Alison Calder argues, "literary analysis that attends to representations of specific places, or that connects itself to specific places, can help us to develop ideas about what is going on in the places where we live—what forces are acting on us, and how we might respond to them."[10] As a work of literature that comprises decades of New's critical thinking about place and writing, *Underwood Log* enacts a globalized poetics of travel intent on interrogating an ethics of place.[11]

I want to discuss three groups of poetic entries in *Underwood Log*, each of which indicates a different way of thinking "here." The first group I identify comprises entries that take place at the same geographical location, the most commonly recurring place/location among the scores of other log entries. (To be clear: I refer to *Underwood Log* as the poem, and the individual poems throughout as entries in the eponymous log.) The most recurrent coordinates in the poem are 49°20'N, 123°10'W, which is not far from where Siwash locates himself in Taylor's novel (49°18'32"N, 123°09'18"W), making it clear that the place to which the speaker of *Underwood Log* keeps returning is geographically (and imaginatively) close to the iconic, forested Stanley Park. This place is the poem's insistent Here. Yet attempting to locate Here in relation to Siwash's location radically unsettles any such insistence, particularly when using such resources as Google Maps and Google Earth. Compared with the coordinates Taylor uses, New's Here is north of Stanley Park, somewhere in West Vancouver, the municipality at the base of the Coast Mountains opposite Vancouver proper. But close attention to the poems New marks with the coordinates 49°20'N, 123°10'W suggests Here is in Kitsilano, a neighbourhood south and east of Stanley Park.[12] An early Here entry mentions "Timothy Taylor's book"[13] and describes the scene, mentioned above, with Siwash standing still in Stanley Park. The speaker implores his listener to think about Siwash, "one man standing // fixed on a single moment," and to contemplate the possibility that "whoever is always moving is // never here / or there."[14] Aligning this intertextual scene with the poem's Here, New effectively posits 49°20'N, 123°10'W as "a place to plant the feet" and resist the seemingly pointless expression *"neither- / here-nor-there."*[15] The illusion or memory of fixity, of a place to return, animates much of the poem's movement between Here and There.[16]

The poem's "there" comprises all of the other geographical locations in the poem, which are indicated by various coordinates, whether Creston, British Columbia; Everett, Washington, USA; or Brisbane, Australia. These entries tend toward travelogue and demonstrate a lifetime of movement across geopolitical borders and literary genres alike. Unlike the Here entries, New's There entries plant memories instead of feet in the various places they evoke; if Here represents geographical and psychic home, There proffers literal and literary touchstones that enable movement and return. There comprises a set of places in and from which to glean, to learn, to develop a repertoire accessible via memory and literature. The third group of entries are not labelled at all; they have not been geographically located. These are what I'm calling the Elsewhere entries, and they constitute a dialogue or, more accurately, a question and answer between old friends; it could also be read as an interior monologue. Elsewhere disrupts whatever stability exists between Here and There; beholden not to physical place but to the vagaries of mental space, Elsewhere thrills with generative potential and threatens to undo a sense of self accumulated through shared experience. Reading and rereading *Underwood Log*, the germ of an idea takes root. The multivalency of Here undermines any attempt to determine positional singularity. Call it heterogeneous geography or attention to place spliced with botany, ecology, history, memory, and literature. Here, the poem insists, is also There and, in slightly more imaginative (and potentially destabilizing) ways, Elsewhere, too.

All of which is to say that the poem's subject matter and formal strategies emerge from and help shape a critical engagement with environmental and postcolonial thinking. In keeping with the poem's title, trees play a prominent role, appearing, in name or in detail, in every entry. Attention to place concomitant with an acceptance of the impossibility of ever really knowing along with a recognition of movement, of change, and of the dynamic systems of which humans and others are a part inhabit *Underwood Log* in ways that invite meditation on ecological thought. The risks and rewards of getting lost weave in and out of these entries as Elsewhere insinuates itself into location and imagination. As the book's title suggests, this is a poem as much about the idea of forests shaping lives in Western intellectual and literary traditions as it is about books—whose pages, made from wood pulp, we leaf through—shaping the forest of the mind. This is a log, remember, written about and out of underwood, situated amid the giant conifers of the coastal Northwest and the myriad trees comprising memory and imagination. (Think, too, of how the title evokes another underwood of the forest, the nurse log, which waits on the forest floor for a bird or a squirrel to deposit some seeds and so nurse a seedling to life.)[17]

Trees and understory, fairy tales and nursery rhymes, street names and neighbourhoods converge in the poem to remind of the ways imagination, language, and story continually inform ideas, politics, and action. These convergences serve to remind how place is not static; Here is always multiple. As New himself writes in an issue of *BC Studies*, any "claim upon place, on what we severally call the reality of position is like an ice floe, always in motion, or potential motion, its fixity only ever illusory, its character imagined by the range of alternatives that it already anticipates and embodies."[18] In the same essay, he writes that "for a time at least, during a period of uncertainty about self (however long that lasts), the identity of 'here' remains susceptible to other people's empire, their desire for fixity and power."[19] It is not difficult to hear W.H. New the Commonwealth/postcolonial scholar interested in the imaginative and social power of borders and margins and edges, sites where colonial attempts to fix meaning and identity in place meet challenges from those who, despite their vulnerability, might just know better. Rather than reify instability in a way that neglects the real challenges faced by those displaced and uprooted by colonial forces, a postcolonial perspective acknowledges the dangers implied by uncertainty's slide into susceptibility. At the same time, New's Elsewhere unsettles Empire's reliance on a Here–There binary, introducing a category incompatible with colonial geography and cartography. The postcolonial thinking inherent in New's writing offers useful strategies for ecological thinking, as well, particularly if we consider *Underwood Log* as regionally situated yet globally inflected.[20]

Here

All of which brings us, to quote Cynthia Sugars, "back into the wilderness of 'Where is here?'"[21] In "Can the Canadian Speak: Lost in Postcolonial Space," Sugars identifies challenges faced by those Canadian critics who contemplate Frye's question from a postcolonial perspective; increased movement (cosmopolitanism) as a result of colonialism complicates the local/global (Native/cosmopolitan) dichotomy elucidated by A.J.M. Smith introducing *The Oxford Book of Canadian Verse* in 1943. If Sugars' focus on two critical questions—"What is distinctive about Canadian literature, and what is the connection between literature and nation?"[22]—vivifies decades of literary and cultural debate in Canada, New's Here entries frame such questions and their (potential) answers at once more locally and more globally. In other words, attuned to the problematical relation between the nation and the colonial history that enabled the nation to form (not to mention the relation between the west coast and the central regions of Canada), New arranges Here entries that rely less on national than on regional and local orientations. That slippage occurs between coordinates/entries, and that Here, There, and Elsewhere invoke one another (imaginatively if not geographically) indicate a poetic project that implicates the local in national and transnational imaginings and vice versa.

The poem's first Here entry illustrates nicely what I've been discussing:

think of those you read about, those
antique saints and halting scholars, wrestling

angels into alphabets, minuscules and
gleaming sins in rows –

chase them if you dare, down
animal tracks and city alleyways,

random as seed: they'll lead you into
storybook and wild mind, overseas and

inland – nothing will shade you from
ghost and glare except your willingness

to hear – you might think you're in command,
counting your way by milepost and mask,

but listen: what do you say,
and what do you apprehend –

> tree
> *shrub*
>
> grass
> *root*
>
> canopy
> *understory*

you have a name for everything
oh?[23]

The movement in this entry pulses with its insistence on multiplicity. The speaker challenges his listener to recall a lifetime of reading stories and the figures whose lives—fictive or historical—invite contemplation, perhaps even modelling behaviour for young and old. What we choose to follow—the plots and heroes, the twists and gestures—might be explained by any number of biographical realities: gender, class, geographical location, historical/temporal position. But predicting which of these choices will, to follow New's simile, germinate, take root, and grow is as impossible as knowing which seeds will find a suitable spot safe from squirrels or birds and take root.[24] Even the presumed order of storybook cannot guarantee "you're in command," the "wild mind" that imagines movement "overseas and / inland" unsettling your place Here. The ease with which the metaphorical slips into the actual implies a simultaneity, a both/andness, that invites introspection and outward consideration of the phenomenal world. The tension between what "you say" and what "you apprehend," according to New's speaker, reflects the relation between what "you read about" and where those books will "lead you"— namely, toward some version of forest: storybook or understory or Stanley Park or Siwash Rock. But Here is precisely where such instability can bear fruit, for no matter how wild or random one's storybook adventures, home awaits the reader on the other side. And home is another name for here.

There

> *No science of space (geometry, topology, etc.) can brook contradictions in the nature of space.*
> —Henri Lefebvre, *The Production of Space*[25]

New's There entries log moments, books, and places experienced at a location other than the home at Kitsilano's edge (or thereabouts). Those locations provide opportunities both for branching out into the wide world beyond Vancouver, British Columbia, and for transplanting versions of There, grafting onto the self all that will become memory. Though located geographically distant— to varying degrees—New's There entries welcome characters and readers alike into lyric forests, occasionally troubling the distinction between Here and There by locating entries close to Here or in places whose coordinates echo those of Here. A late entry, 23°20'N, 123°10'W, exemplifies remembrance via literary and spatial figures, which the There entries invite. In this example, There can be found on the same longitude as the Here entries, two of which this entry sits between. Such near-parallel would seem to reveal at once the foreignness of There and the propensity of other places to weave, be woven, into one's sense of home and of self. This entry aligns specific trees with specific poets: "Narayan singing the deodar, Derek Walcott the / caesurina, each invents ways to touch the sky."[26] New logs this entry as if the cultural artifacts encountered There, with their connection to trees, comprise an imaginative version of a neighbourhood of trees, a designation applied to Kitsilano in this entry and others. The Narayan and Walcott allusions function as landmarks in the speaker's psychic forest, each text a memory planted. The neighbourhood of trees is always "nearby," both imaginatively—via memory—and textually: as the speaker implores his addressee in this entry to remember this psychic proximity, he segues into a Here entry. Looking back (or north) to Kitsilano points to "the maple scrub" where "children again are building swings—."[27] The following Here entry reads in its entirety:

> horsefeather, goosefeathers, catkins, cream,
> the suddenness of silliness, basswood, bream,
>
> mulberry, nannyberry, elderberry, yew,
> swing upon a sycamore: I will too – [28]

The childlike singsong cadence reveals nostalgia as one way to connect There with Here.[29] Human geography echoes scientific geography—and neither is necessarily accurate to the point of perfection.

Entries could easily be selected at random to cultivate an argument regarding the orienting potential of There, each a seed pack biographically latent and lyrically ardent. I focus on a few There entries that, taken together, articulate New's underwood poetics by attending to tensions between natural forests and language, the act of moving through physical space and the work of writing:

> 49° 12'N 68° 10'W: you walk daily into the forest: trees
> enclose you:
>
> you do not notice,
> lost in birling metaphor[30]

Forest inhabitants include "eagles coursing overhead" (which go unnoticed), manitou ("the grey wolf howling / *soon*"), a minotaur (who's "never there"), and finches (eagles' prey). Experience and imagination animate There, informing memories and a sense of geography (physical and psychic). New's speaker reads (or hears) of manitou and minotaur, for example, and goes on to write himself:

> 51° 30'N 0° 5'W: the poet sits at the Underwood, tapping out
> two-fingered tomes in daily measures:
>
> ....
> he writes to find imaginary space,
>
> but *after* eludes him: always
> he tangles the keys, loses himself
>
> in roots and origins, *before*
> calling him back, vermiculate[31]

Underwood in this entry demands to be read as both a typewriter with which to write and an in-between space from which to write. Here aligns with *after*, a spatiotemporal correlative to the *before* that experiences There provide. The act of writing—whether at home or abroad—captures the overlapping qualities of spatial and temporal categories. Just as every moment occurs after another and anticipates the next in the sinuous flow of time, place recalls origins and looks to other destinations, which inevitably become part of a longer, larger process of travel. No small part of this process, Elsewhere beckons beyond and after There:

> 41° 53'N 87° 40'W: and yet why type except to write yourself
> elsewhere, *into* the underwood: take the
>
> beeline, the el track, the A-train,
> all will get you near,
>
> ....
>
> when will *there* become *here*, you-and-I,
> alongside the branch water,
>
> limbs tangling in trumpet flowers,
> spelling the world in ease and green desire:[32]

Writing, with its concomitant rules (grammar, punctuation, spelling), enacts travel between There and Here; its ontological work coalesces as geographical movement from Point A to Point B—as the public transit references suggest—and as relatively static movement in a liminal space. The "branch water" alongside which "you-and-I" tumble might just be a stream, but New's choice of words both continues the poem's arboreal vocabulary while suggesting a possible direction—an offshoot along a path—and nods to the "nip of scotch beside" the poet in the earlier entry, branch water being the source water whisky distilleries and drinkers use to cut their drink.[33] The temporal inflection—"when will *there* become *here*"—again implies that Here and There exist as much spatially as temporally, as much in physical as imaginative experience. One does not have to be a writer, however, to experience Elsewhere.

Elsewhere

> *For Nomads the notion of past and future is perhaps subservient to the experience of elsewhere. Something that has gone, or is awaited, is hidden elsewhere in another place.*
> —John Berger, "The Red Butterfly"[34]
>
> *Elsewhere: it is at once place, no place, a desire—or an imposed or chosen place or process or idea. It can also be configured as an alternative style or manner of shaping worlds into narrative and meaning.*
> —Marta Dvorak and W.H. New, "Introduction: Troping the Territory"[35]

Once Canadian literature is considered by a critic outside of Canada, Sugars suggests, "here" becomes "there," and when this happens, "we are back in the space of the cultural periphery; back in the oppositional divide of Smith's dichotomy: native or cosmopolitan; colonial or international. The potentially disruptive middle term, the 'postcolonial,' has disappeared."[36] Elsewhere performs a version of postcolonial disruption, even if it does so in ways that postcolonial theory might not recognize at first glance. The disruptions are spatially and temporally, not necessarily culturally, inflected; and yet they present a strategy for standing back from easy generalizations and dichotomies in order to question a dominant mode of being in the world. The project of knowing and mapping place recurs in postcolonial theory as a colonial project. Even if New's poem avoids blatant reference to colonialism and postcolonialism, it devotes much energy to undermining the idea that place can ever fully be known and mapped. It also reveals how efforts to know and map implicate personal experience and individual memory while celebrating the best of what globalization has to offer—namely, access to a world of literature from which we might learn. Reading widely emerges as a viable way to negotiate the uncertainties of Elsewhere.

One of the problems raised in *Underwood Log*, which resonates with a postcolonial interest in mapping, is how to name accurately. Perhaps *saying + apprehending = naming*. Even so, there remains an echo of apprehension, Elsewhere resonating in linguistic lack and self-doubt. Understanding and reluctance go hand in hand through the forest, dropping, you hope, some crumbs along the way. Names connect disparate places, so that one can be both Here and Elsewhere at the same time. If "you have a name for everything" Here, "names of elsewhere / take you traveling," according to the poem's opening lines (and the first Elsewhere entry).[37] If "hereness" is nothing "except the politics of status and region," as New claims in his essay "Writing Here," then elsewhereness enables movement *between* Here and There, place and memories of other places.[38] If There is a place where getting lost is a real threat, Elsewhere extends the threat to include the forest of the mind. New's Elsewhere entries seem to dramatize the speaker's attempts to bring an old friend back from the Elsewhere of memory loss and toward a (re)claiming of subjectivity. An early entry records one such attempt:

where is here?

Kitsilano: Kitsilano's edge

a place of trees?
a grid of cross-streets: larch, balsam,

arbutus, yew,
the avenues are numbered

> and this is where you live?
> in the intervals, along the margins
>
> you call it home?
> why do you ask?
>
> *say I*:³⁹

The gentle interrogatory tone of this entry invokes Here as home, at the edge of a neighbourhood whose tree-named streets intersect with numbered avenues to form a grid, the better to know where you are. This "place of trees" again provides an echo of Stanley Park—the novel and the place—as an urban forest. But this home has, in Canadian poet Don McKay's terms, "acquired a frayed edge," much like the meadow near his home in Lobo Township "where the permeable membrane between place and its otherwise first became apparent to [him]."⁴⁰ The slightly cryptic responses in New's entry unsettle the questioner's desire for clarity, culminating in the final, still gentle imperative to "say I." The place of trees, Here, evades the simple logic of grid systems and seeds coordinates of Elsewhere, which tempts the self out of and away from the self. The speaker wants his friend to "say I" because that common claim to subjectivity represents a self-knowledge that, he hopes, might defeat the loss of memory and accompanying loss of self, might hold a flame to the frayed edge of home.

This remembering of home/Here would counter what poet Maureen Scott Harris identifies as disorientation. Harris wants to correct her disorientation—not, it should be noted, a medically induced disorientation such as the one the figure in *Underwood Log* seems to suffer from—by surrounding herself with "names which *don't* insist on elsewhere, names native to *this* place, arising from the countryside itself, presenting it directly, unfiltered by nostalgia or desire for some place else."⁴¹ The arboreal street names would seem to act as anchors for New's increasingly lost friend in Harris's framework; but the pull of Elsewhere remains powerful, and *Underwood Log* records the ambivalent ways that Elsewhere frees the self even as self disorients and repositions. In "Writing Here," New acknowledges "empirical *fact* of a changing place (sea and land both in motion, the geologic plates tectonically unstable) ... the sun and the rain and the sockeye run and the changing people."⁴² In *Underwood Log*, he extends this notion both to lament and to embrace the state of being, or becoming, lost.

Don McKay poses an apposite question and scenario in the prose poem "Five Ways to Lose Your Way" (2005). "How can you be lost?" he writes; then, "At the first sick surge of lostness the interrogative gets to work dividing self from self, finds a you to interrogate among the glum bureaucracies of I ... *How Can You Be Lost?*"⁴³ Of course, McKay's speaker is speaking to himself upon realizing that he is, indeed, lost in the woods. Despite the myriad methods humans have developed since antiquity to prevent such a thing. Despite the maps with latitude and longitude coordinates; despite access to GPS; despite the copious flags of "neon tape" in the bush that read "Falling Boundary" and "Road Location," which seem "to imply" both border and direction.[44] Eventually, with the speaker finding a stump to sit on, the rhetorical question settles into a real question of ontological import. A question that resonates with the concerns illuminated in New's Elsewhere entries. McKay's speaker "reflect[s] on the paradox embedded in the common expression 'to get lost': you can't get lost any more than you can fall asleep. Lostness gets you; sleep gathers you inward to itself. But being lost—can a person dwell in such a space, somewhere outside the ubiquity of plans, the falling boundaries and logging roads projected into the future? 'Deep in the forest,' says Tomas Tranströmer from the depths of your backpack, 'there's an unexpected clearing which can be reached only by someone who has lost his way.'"[45] Lost in this formulation does not represent one side of a dichotomy: lost and found. Rather, being lost insinuates you into an Elsewhere at the manic edges of Here and There. To "dwell in such a space" is surely possible; but to dwell with careful and clear mind is another matter. What's *in* that clearing deep in Tranströmer's forest? What's in it for you? What's in it for me? Presumably, a clarity that comes with the efforts of having worked our way through the forest, with and without coordinates to direct us. Presumably, a catharsis that comes from having kept moving, kept travelling, kept remembering the names of Elsewhere. Presumably, a reward for having spoken "I" enough times while There to find our way back to Here along a trail of ontological crumbs.

But is finding the clearing an arrival or a return? New's first Elsewhere entry ends with the following lines:

> only Away:
>
> for only by going away, gambling your life
> into the halting underwood,
> can you or I,
> chilled and welcome, once
> again
> return – [46]

Return where? Here from There. There from Here. Perhaps Elsewhere, perhaps travelling, is a state of perpetual returning enabled by story, memory, language, imagination, knowledge. How else to record and recall a life if not by embracing the "halting underwood" as it embraces us, gambling our lives on the uncertainties that lie ahead and remain behind? How else to rethink our relations among Earth's inhabitants and ecologies if not by dwelling in an Elsewhere that is also Here and There? Ways of thinking, ways of knowing, ways of being. *Underwood Log* engages with all of these, creating layer upon layer of erudition and humility, both of which help make the committed environmental critic attuned to the "random ... seed[s]" of others' disciplinary work.

Notes

1. Tim Lilburn, "How to Be Here?" in *Poetry & Knowing: Speculative Essays and Interviews*, ed. Tim Lilburn (Kingston: Quarry, 1995), 167.
2. Alan Morantz, *Where Is Here? Canadian Maps and the Stories They Tell* (Toronto: Penguin, 2002), xv.
3. W.H. New, *Articulating West: Essays on Purpose and Form in Modern Canadian Literature* (Toronto: New Press, 1972), xi–xii.
4. W.H. New acknowledges this myth in *Articulating West*. Daniel Francis develops an approach to a history of Canadian culture in "The Ideology of the Canoe, The Myth of Wilderness," from his book *National Dreams: Myth, Memory, and Canadian History* (Vancouver: Arsenal Pulp Press, 1997).
5. Tim Lilburn, "How to Be Here?," 164.
6. Timothy Taylor, *Stanley Park* (Toronto: Random House, 2001), 335.
7. Frye's question is posed in his Conclusion to the *Literary History of Canada* (1965), though he is not, of course, proposing to conclude Canadian literature or literary scholarship.
8. Siwash Rock, a rocky outcrop situated in the shallows of Burrard Inlet, is one of Stanley Park's attractions.
9. While New has achieved acclaim and respect for his work over a long career as an influential critic, editor, and teacher, attention to his poetry, which he began to publish in 1996, has been slow to emerge. To my knowledge, the only other article that offers critical analysis of New's poetry is my own. See Travis V. Mason, "Literature and Geology: An Experiment in Interdisciplinary, Comparative Ecocriticism," in *Greening the Maple: Canadian Ecocriticism in Context*, ed. Ella Soper and Nicholas Bradley (Calgary: University of Calgary Press, 2013), 475–509. That *Underwood Log* is the first of his poetry books to be nominated for a major literary prize suggests that he has achieved something, if not essentially Canadian, then suitably complex and accessible in both subject and form.
10. Alison Calder, "What Happened to Regionalism?," *Canadian Literature* 204 (2010), 113.
11. Another Canadian long poem from the first decade of the twentieth century, Dionne Brand's *Inventory* (2001), can also be read as a series of entries. That Brand's entries explicitly log humanity's numerous violences against the earth and its inhabitants perhaps explains the vast amount of critical and popular attention it has received compared with New's book. According to Carrie Dawson, *Inventory* is worth considering for "its call for a situated or grounded affect, its cultivation of a way of thinking that is self-conscious (as perhaps Frye's was not) about the deployment of the non-human world as an imaginative resource that creates affect." Carrie Dawson, "How Does Our Garden Grow?" *Canadian Literature* 204 (2010), 111. The situatedness of this reading resonates with New's continual return to a specific place in *Underwood Log*. Claiming Brand's poetical "declaration that 'that ravaged world is here' (47) might be understood as a warning to forego the familiar 'Where is here?' in favour of a carefully grounded materialist practice wherein the appeal to generalized structures of feeling goes hand in hand with an examination of particular structures of being" (Dawson,

"How Does our Garden Grow?" 112), Dawson clarifies the value of layering place in active, imaginative, and affective responses.

12 An email exchange with New, on 13 April 2012, confirms Kitsilano as the more accurate location, and New admits to having "used the gazetteer at the back of an old atlas" when writing *Underwood Log* and necessarily "approximated a lot of the co-ordinates." So, while the impetus to map New's long poem using online resources makes sense, it also reveals a disjuncture between the lyric entries and the physical world. Another comment in our email exchange corroborates this point: "For me," New writes, "writing the book, the key thing was NOT to be specific, but to create a disparity between the apparent specificity of a numerical grid (in lat/long or in street pattern) and the resistance to specificity (and permanence or fixity) that emerges in the poems themselves." That I have been tempted to locate the entries in *Underwood Log*, despite my argument for the way they resist such attempts at location, says a lot about the importance of tracking New's lyric interventions and listening to the music resonating between places.

13 W.H. New, *Underwood Log* (Lantzville: Oolichan, 2004), 22.

14 Ibid.

15 Ibid.

16 New's overt use of mapping coordinates poses a challenge to some critical conceptions of the long poem in Canada. Referring to Smaro Kamboureli's *On the Edge of Genre: The Contemporary Canadian Long Poem* (1991), for example, Reinhold Kramer suggests that "the long poem is unlocalizable and indeterminate in its positioning of the self. Not simply parodying epic form, the long poem arrives at an unmappable form." Reinhold Kramer, "The Contemporary Canadian Long Poem as System: Friesen, Atwood Kroetsch, Arnason, McFadden," in *Bolder Flights: Essays on the Canadian Long Poem*, ed. Frank M. Tierney and Angela Robbeson (Ottawa: University of Ottawa Press, 1998), 102. Logging on to Google Earth, one can effectively "map" *Underwood Log*, though, in keeping with Siwash's distrust of maps, the resulting map would be inaccurate. The latitude and longitude titles are simply not detailed enough to locate the entries accurately in a consistent manner. My marginalia for the entry "23°20'N, 123°10[']W" (New, *Underwood Log*, 116–17) places this part of the poem in the middle of the "Pacific Ocean, near Mexico."

17 New teases a lot of puns from his pithy title. Underwood evokes the typewriter model made famous by Faulkner and Fitzgerald. Log also suggests a catalogue, a form that, despite the poem's list of locations, New seems to recognize is impossible to mimic, perhaps much the way space is impossible to mimic with maps; Graham Huggan applies Homi Bhabha's notion of mimicry to consider shortcomings with the "discursive system" of "the map, itself split between its appearance as a 'coherent,' controlling structure and its articulation as a series of differential analogies." Graham Huggan, *Interdisciplinary Measures: Literature and the Future of Postcolonial Studies* (Liverpool: Liverpool University Press, 2008), 22–23. The completeness implied by the term "catalogue," moreover, calls to mind Robert Kroetsch's ironically titled long poem collection *Completed Field Notes* (1989). According to D.M.R. Bentley, "the long poem is modestly catalogic in its comprehension of external reality and tends to use the catalogue and its more pictorial cognates (the panoramic survey and the picturesque *tableau*) either to order the subject environment (and thus to indicate its successful colonization) or to suggest its immense expanse (and thus to suggest its openness or, perhaps, resistance to colonization)." D.M.R. Bentley, "Colonial Colonizing: An Introductory Survey of the Canadian Long Poem," in *Bolder Flights: Essays on the Canadian Long Poem*, ed. Frank M. Tierney and Angela Robbeson (Ottawa: University of Ottawa Press, 1998), 9.

18 W.H. New, "Writing Here," *BC Studies* 147 (2005), 5–6.

19 Ibid., 13.

20 In "Regionalism, Postcolonialism and (Canadian) Writing" (1998), Herb Wyile offers a strategy for thinking about regionalism through a postcolonial lens: "Rather than giving in to the temptation to look at regionalism as a way out of unitary conceptions of a national and/or postcolonial literature," he writes, "I want to argue instead that past and present issues in postcolonialism [...] have their parallels in regionalism, and that regionalism, rather than providing a ready means of modifying those conceptions with a necessary sense of diversity and heterogeneity, instead presents a familiar and somewhat analogous set of problems." Herb Wyile, "Regionalism, Postcolonialism and (Canadian) Writing: A Comparative Approach for Postnational Times," *Essays on Canadian Writing* 63 (1998), 10.
21 Cynthia Sugars, "Can the Canadian Speak? Lost in Postcolonial Space," *ARIEL: A Review of International English Literature* 32, no. 3 (2001), 124.
22 Ibid., 120.
23 New, *Underwood Log*, 11.
24 "All the ululation in the world," writes New in another Here entry, "can't guarantee the gymnast won't / slip off the balance beam, // the planted sapling grow." New, *Underwood Log*, 69.
25 Henri Lefebvre, *The Production of Space*, translated by Donald Nicholson-Smith (London: Blackwell, 1991), 292.
26 Ibid., 116.
27 Ibid., 116.
28 Ibid., 117.
29 New has published three picture books (illustrated by Vivian Leigh) and one chapter book. Like Margaret Atwood and Dennis Lee, New explores the roots of lyric cadence by returning to children's verse.
30 New, *Underwood Log*, 33.
31 Ibid., 91.
32 Ibid., 95.
33 Ibid., 91.
34 John Berger, "The Red Butterfly," *Brick: A Literary Journal* 71 (2003), 10.
35 Marta Dvorak and W.H. New, "Introduction: Troping the Territory," in *Tropes and Territories: Short Fiction, Postcolonial Readings, Canadian Writing in Context*, ed. Marta Dvorak and W.H. New (Montreal and Kingston: McGill-Queen's University Press, 2003), 11.
36 Sugars, "Can the Canadian Speak?" 124.
37 New, *Underwood Log*, 9.
38 New, "Writing Here," 19.
39 New, *Underwood Log*, 17.
40 Don McKay, *Deactivated West 100* (Kentville, NS: Gaspereau, 2005), 27.
41 Maureen Scott Harris, "Being Homesick, Writing Home," in *Fresh Tracks: Writing the Western Landscape*, ed. Pamela Banting (Victoria: Polestar, 1998), 255; original emphases.
42 New, "Writing Here," 15; original emphasis.
43 McKay, *Deactivated West 100*, 87; original emphasis.
44 Ibid., 89.
45 Ibid., 92.
46 New, *Underwood Log*, 9.

CHAPTER 17

Re-envisioning Epic in Jon Whyte's Rocky Mountain Poem *The fells of brightness*

Harry Vandervlist

The fells of brightness is the overall title of Jon Whyte's ambitious poem (planned for five volumes) on the subject of the Rocky Mountains. Before his death in 1992, Whyte published the first two sections. *The fells of brightness, [first volume]: some fittes and starts* appeared with Edmonton's Longspoon Press in 1983, and *The fells of brightness, second volume: Wenkchemna* was published by the same press in 1985. Volume three, entitled *Minisniwapta* and focused on the Bow River, was composed but not published. The typescript remains among Whyte's papers at the Whyte Museum of the Canadian Rockies, however, and in 1988 CBC Radio broadcast the poem as a performance for four voices.[1] Jon Whyte took several years to decide on the generic name for this gargantuan project. He uses the term "epic" to describe an earlier work, when in the preface to his 1981 book-length poem *Homage, Henry Kelsey*, he declares, "The poem began to shape itself into epic."[2] In a 1983 CBC interview discussing the early stages of *The fells of brightness*, he first asserts "the poem is anatomical."[3] By 1986, he returns to the term "epic," when he writes, "My poems declare their epic intent."[4] Yet to what extent, and in what sense, can *The fells of brightness* be considered epic? In order to suggest what Whyte was able to do with epic as he re-envisioned it, it will be necessary first to explore how he finds a way of thinking about epic that suits his project, and then to place *Fells* in the context of some works that accommodate themselves to the genre's multifaceted history in similar ways.

How far, then, does the term "epic" genuinely suit Whyte's ambitious poem? And by extension, is "epic"—with its connotations of often imperialistic heroism and domination—an appropriate term for any poem that aims to incorporate the kind of vision that would now be described as ecological? (Such a vision was increasingly important to Whyte, and by the 1980s, a well-established part of his work as both a poet and an activist.) For Whyte, and for several poets with whom he shares affinities, one answer is to re-envision epic in ways that develop selected elements of the genre's long heritage. For example, a group of English poets, and in particular the Birmingham poet Roy Fisher, have been seen as sharing a mode provisionally labelled "pocket epic." This modest-sounding designation recognizes that they turn away, each in his or her own manner, from the more heroic modernist ambitions of Pound and Joyce. (At the same time, these poets share a problematic relationship to academic designations of postmodernism.) They also tend to be "regional" writers and to some degree autodidacts, or at least constructors of alternative canons. Whyte's poetry fits into this context less uneasily than most others, I will suggest.

It is also useful to inquire into the way Whyte pursues his own particular re-envisioning of epic, because of the way this helps to highlight the reflective and critical force of his Rockies poem. Whyte unquestionably valued the Rockies, and in particular the Bow Valley, from an ecological or environmentalist standpoint. Yet his poetry embodies a highly developed awareness of the power of multiple representations—artistic, scientific, touristic—both to define that which must be valued, and to suggest ways to go about doing so. Because of the way in which Whyte's poem juxtaposes cosmology and culture, and the way it meditates upon "landscape" and perception, it might best be approached in ways that answer the demand from ecocritics Dana Phillips and Timothy Morton: an approach to writing about places in an ecological way that avoids an anti-intellectual sense of "Reverence, awe, piety, and mystical oneness." Dana Phillips calls these—when isolated from critical reflectiveness—"antiseptic responses to nature."[5] Whyte took far too much pleasure in his own intellectual explorations to fall victim to such a dualism. In this way, he is not the kind of writer "attracted to prayerful ways of 'beholding nature'"[6] if this is seen as one half of a "disparity between the reality one savours and the reality one knows," a disparity decried by Phillips in much ecocriticism.[7] Whyte would reject such a distinction: for him the Rockies are at once an object of knowledge as, for example, in the talks he gave on "Art and Culture of the Canadian Rockies,"[8] and an immersive experience to be savoured, as so many of his *Crag and Canyon* columns demonstrate.[9] At the same time, Whyte's own spiritual attachment to the earth is perhaps summed up in the concrete poem he crafted entitled "Epitaph": it consists of an endless circle created by the word "Earth," containing as it does the words "heart," "hearth," and the phrase "hear the art," depending on where one begins to read the circle.[10]

Whyte's writing can perhaps be understood using terms proposed by Timothy Morton when he characterizes critical ecological writing as writing that turns the here and now into an open question that asks "What is here?" Whyte takes advantage of the freedom afforded by a renewed idea of epic—one that emphasizes the genre's encyclopedic and cosmological aspects, along with its verbal and stylistic freedom, while maintaining ironic focus on the subjective and placed perception of the poet. He does so in order to lead the reader through histories, etymologies, and narratives both personal and mythic, to confront precisely the questions "What is here, now?" and "What might here become?" Instead of setting up a permanent state of aporia, questioning in Whyte seems to me allied with the kind of excited, attentive lure toward onwardness he evokes in writing about mountain passes in *Wenkchemna*.[11]

From an early age, Whyte was impressed by the literary chutzpah of the first epic he encountered, E.J. Pratt's (1952) *Towards the Last Spike*: "When I was in high school I read a lot of E.J. Pratt ... and I thought that I would go out and compete with Pratt. In the fifties I guess he was considered the grand old man of Canadian poetry. So at that time I decided to write a poem on an epic subject, a disaster. I chose to write about the Frank Slide."[12] When Whyte returned to live in Banff after being educated at the University of Alberta and Stanford, he "began to more elaborately inquire about the means of approaching landscape literarily."[13] This led him to compose his long poem *Homage, Henry Kelsey*, which he later cast as a sort of prologue to the much larger, multifarious poem on the Rockies, which he spent the 1980s pursuing.

This longer poem would certainly contain strong ecological elements, if only because by the 1980s Whyte had become a passionate advocate for the mountain environment. He was a member of the Bow Valley Naturalists from the 1970s on, and became a trustee of the organization now known as the Canadian Parks and Wilderness Society.[14] He had also developed an informed respect for the history and contributions of the local First Nations peoples, the Nakoda (Stoney). This he learned from his uncle Peter and aunt Catherine Whyte, founders of the Whyte Museum of the Canadian Rockies, and is reflected in his choice of title for the three volumes of his Rockies poem. He explains his title for the first volume as follows: "The Fells of Brightness, which is to say 'Assine Watche,' directly translated from Cree, 'brilliant mountains,' the Rockies."[15] He titles his second volume *Wenkchemna*, the Nakoda word for "ten," as in the Valley of the Ten Peaks; and the third, "Minisniwapta," the Nakoda name for the Bow River.

The epic genre's traditions ought to raise several problems for a poet like Whyte, who so respected both the ecology and the original human inhabitants of the Rockies. For one thing, even in its most immediate manifestation, such as E.J. Pratt's Canadian epics, the genre seems tainted by a still-colonial mission to subdue, rather than understand, the territory Whyte saw as home: "In both [*Brebeuf and His Brethren* and *Towards the Last Spike*], faith and vision are pitted against chaos and inertia in heroic endeavours that attempt to give human shape to the physical and spiritual wilderness of Canada."[16] The terms used here attribute "chaos" and "inertia" to the very place whose local history and uniqueness Whyte aimed to valorize. What is more, Pratt's epic narratives have been described in terms of a mastering, even a dominating imagination, rather than the kind of receptive and inquiring imagination I hope to show Whyte's poem implies:

> Both in their own size and in the stories they tell, these poems glorify the imagination working on the grand scale. For breadth and audacity of vision they approach, within their own spheres, the achievements of the heroes that they celebrate. As Van Horne laid steel rails over some of the most intractable regions of mountains and muskeg in the world, so Pratt, on the level of the poetic imagination, surveyed the terrain, discovered the passes, and made accessible to the imagination vast expanses of Canadian history that must have seemed, at first glance, very uninviting to poetry and myth.[17]

As this language shows, Pratt's epic is delivered from a confidently central, even hegemonic cultural position—a notion affirmed by Northrop Frye when he calls Pratt a poet who "has so central a relation to his society, there is no break between him and his audience: he speaks for, as much as to, his audience, and his values are their values."[18] Jon Whyte, fully aware of the strong regionalist spirit in western Canadian literature in the 1970s and after, understood that such a position was not available to a poet speaking from a very particularized place, a lone mountain valley—albeit a well-publicized one—in the culturally marginalized Canadian West.

Pratt, the most obvious Canadian antecedent in the epic genre, already demonstrates these reasons for Whyte to shun a genre seemingly ill-adapted to giving a sympathetic voice to "intractable regions." Things only get worse when the larger generic history of epic is considered. From Vergil onward, the epic becomes increasingly entangled with imperialism, especially during the early modern period. Reed Way Dasenbrock argues that epic poets such as Camoes and Tasso construct their own Vergilian "vision of discovery and conquest"[19] with such a degree of success that "Vergilian epic can thus be seen as writing the script for European expansion in the Renaissance."[20] At this stage of its history, Dasenbrock argues, the epic genre experiences a "flattening out of moral complexity in an increasing subordination to ideology."[21] During imperial expansion, this often entailed a "contemptuous depiction of the indigenous other."[22] Nothing could have been less sympathetic, it would seem, to a poet like Whyte.

As for more recent versions of epic such as Pound's *Cantos*, these embody a modernist aesthetic that Whyte appears to have found partly attractive, and partly impossible. In his essay "Making Modernist Masterpieces," Michael André Bernstein enumerates attributes of such "masterpieces": their aspiration to universality, their totalizing ambitions, their difficulty, and their impersonality.

Whyte does perhaps echo a kind of modernist hope for universality when he states, "My Rockies are, I hope, an archetype of anywhere, a complex of folk tale and anecdote, personal experience and Earth, a geography of climate, passions, and place."[23] And although Whyte certainly had no difficulty being difficult, modernist impersonality (à la Flaubert or Joyce) is the last thing aimed for by this poet, who includes within his poem narratives of his own specific childhood revelations on landscape, shares his hiking experiences, and explores highly personal moments of perception. It is telling, for example, that Whyte's notes for the first volume of *Fells* begin with the phrase "the ripples begin where I begin; they wash up on that larger circle, the world."[24] Whyte continues: "The lot, the neighbourhood, the milieu, the family, the town, the community, the ring of mountains containing the Bow Valley first, then the greater ranges of the Rockies beyond, the national parks of the continental crest: these are the expanding—the receding—horizons of consciousness 'Some Fittes and Startes' involves."[25] The bulk of these notes involve genealogical and biographical notes on specific members of Whyte's family, and on local figures important to Banff history, who also happened to be neighbours and dinner guests.

The final quality of a modernist masterpiece, Bernstein writes, is that such a work "simultaneously seeks to escape and to incorporate both the historical-social world from which it springs and the literary tradition whose summa it intends to be. It must, that is, stand in an explicitly dialectical relationship to all previous texts, resolving their antitheses and partialities into the final synthesis of its absolute form."[26] Can poems such as Pound's *Cantos*, Williams' *Paterson*—or any poem—actually achieve such an absolute and self-contained quality? Not at all. In fact, Bernstein writes,

> The modernist masterpiece shuttles continuously between a totalizing project it intermittently recognizes as inaccessible, if not outright paranoid, and a contradictory attentiveness to the particular moment and the purely local, contingent realization. If the first impulse is encyclopedic and epic, then the second can be thought of as essayistic, flourishing precisely by refusing to subordinate its curiosity to any larger systems. Essayism's complete absorption in the immediate and local resists subordination to the lure of the whole, and the strain of shuttling between the epic and the essayistic fractures the modernist masterpiece as it is producing it.[27]

Whyte might well have been attracted by the potential geological imagery in the phrase "fractured modernist masterpiece." And perhaps, it is the importance of the "particular moment and the purely local, contingent" that makes the very aspiration to such an impossible totalization and resolution or closure foreign to Whyte's long poem. Yet at the same time as Whyte's long poem shares some of the attributes deemed modernist, he also shares some of the attitudes Bernstein attributes to postmodernism, which "dismisses the idea that art can redeem lost time or damaged lives, and counters assertions of the masterpiece's universality with an insistence that every work explicitly acknowledge its own historicity and partiality."[28] In 1986 Whyte called himself a "post-post-modernist."[29]

If Whyte mainly rejects modernism's artist-hero ambitions to create a new totalizing epic, and also cannot accept an earlier epic heroism associated with dominance and imperialism, then why lay claim to the tainted epithet in the first place? What remaining aspects of epic are sufficiently attractive to a poet like Whyte? The clearest of these is epic's encyclopedic scope. In a 1984 interview, Whyte describes the Rockies as "a subject big enough for a big poem that could encompass cosmology, geology, and I suppose, ideas of culture. I believe a cultured man should be aware of what James Hutton meant by uniformitarianism, and I think he should be aware of what J. Tuzo Wilson has brought to geomorphology, and I think a cultured man should be aware of the story of his tribe. So I created structures that encompass my idea of a cultured man being a person who is finally aware of as much as possible while he's at home."[30] Whyte's phrase "the story of his tribe" echoes Pound's famous use of the phrase to describe epic. Epic remains the genre most associated with the ambition to tell the story of the tribe—yet the story will need to be told in a new way.

Epic poets of the early modernist period began to expand the definition of epic in precisely this direction, at the very same time as the genre was also serving imperialist ends. Van Kelly argues that for Tasso, and for less known early modern epic poets such as Desmarets de Saint-Sorlin (author of *Clovis*), "the epic becomes an imitation of the creation."[31] Kelly writes, "Early modern theorists far exceed Aristotle's description of the epic as one of two high-mimetic genres: the epic is an epistemological touchstone, an encyclopaedia of all knowledges, observations, techniques and arts."[32] Such a movement toward cosmological, intellectual, and cultural inclusiveness continues through the Romantic period (Kelly offers Victor Hugo as an example) and informs the modernist poets as well.

In order to accommodate this strand of the genre's history, Kelly arrives at a redefinition of epic, which might have been highly attractive to Whyte:

> the epic is a creative habit, mentality and practice, expressed in verse and/or in prose, that (1) attempts to adequate a spatiotemporal vastness, and a

sense of the multiplicity of sociopolitical and cultural frames that transcend individual interests, with a continuous narrative (or several narratives linked into a plausible semblance of continuity ...); and (2) simultaneously attempts to merge that wide field of referents into a highly-dense symbolic, or mythic, structure that is exploited and "worked" continuously and intensively as if the context were lyrical, not narrative.[33]

Such an understanding of epic might seem too loose or flexible to distinguish the genre, so Kelly pins the genre's specificity on its use of an "overarching frame": "Frequently the epic makes use of the dialogue that typifies drama, of the highly individualistic perspective and verse rhythms that typify lyric enunciation, of the special ease, continuity, and flow that typify novelistic prose, of the sociopolitical and documentary actuality that history furnishes; yet the epic maintains these elements in an overarching, symbolic frame that distinguishes it from the treatise or encyclopedia."[34] These inclusive accounts of epic's encyclopedic tendency sort well with Whyte's 1986 description of epic as "a cosmological word that, in my wordhoard, means something akin to "'the narrative that defines the universe at a time.'"[35]

Whyte also understands that his universe-defining narrative gets defined, reciprocally, by what it aims to describe. As I have written elsewhere, "The autobiographical elements in *Fells* serve as a case study in the formation of perception and memory in the interplay with environment."[36] It is here that Whyte's epic procedures overlap with those of the small group of poets studied under the rubric "pocket epic" (in the spring 2000 issue of the *Yale Journal of Criticism* entitled "Pocket Epics: British Poetry After Modernism"). According to Nigel Alderman, this group includes "a diverse and still largely unexamined genre of minor or local epics, works positioned between modernism and postmodernism and in the political tension between the canonical ambitions of their predecessors and the particular demands of their own historical geographies."[37] The designation "pocket epic" is offered as a "heuristic term" that "implies a position secreted within the folds of the major canonical economy of the epic and perhaps hints at some form of obdurate resistance." As in Kelly's description of some early modern epics, these works display their own series of tensions:

> between the grandiose projects of high modernism and the diminished anti-modernism for which the Movement can stand as the marker; between the Poundian line as a basic formal unit and the demands of English syntax; between the lyric and longer narrative forms; between individual subjectivity and some larger social collective; between a hegemonic history and forgotten histories; between the emerging devolution of various regional and national sites and the residual power of an imperial center; between a desire for some new local center and a belief in the productivity of the periphery; between a global economy and a local one.[38]

John Kerrigan suggests that poets writing from such a position "within the folds"—or as he puts it, from "faded or overlaid centers"—might include (at certain points in their careers) Geoffrey Hill, John Montague, Ted Hughes, and Gillian Clarke. Their "long poems, or broken sequences" tend to "[cross] localized memoir with historical excavation": "Required by the scope of their material to transcend the free-standing lyric, these poets cannot give the thickly mediated events of the past the coherence of continuous verse narrative. Their curiously jigsawed texts have rather different trajectories, but all find multiplicity and irresolution in the past, rather than tales of national revival."[39] The phrase "curiously jigsawed" might have been invented to describe the way so many of Whyte's texts appear on the page. Kerrigan's description of the need to counter "tales of national revival" with texts less hegemonic and resolute also captures an element of Whyte's response to E.J. Pratt. Yet the English writers discussed under the "pocket epic" rubric display one key difference from Whyte. While Whyte, as a conservationist, fought various expressions of a globalizing economy that threatened the unique value of the Bow Valley—for example, a massive development plan for Lake Louise in the 1970s, funded by international capital—his position in Banff National Park did not confront him with the scale of industrialization seen in, for instance, Birmingham. In his own different fashion, however, Whyte does confront the way Banff and the Rockies faced the problem of a region that faces definition from outside, while aiming to continue defining itself from within, based on more local history and knowledge.

Whyte certainly sets up a specific sort of relationship between history and the present, which Kerrigan finds in the "pocket epic" poets, who write from the "center" of a world subjected to a new hegemony, now based much further afield: "Because these residual centers are the sites of a childhood geography that has been overlaid not just by individual adult experience but by rapid and widespread socio-economic modernization, history tends to telescope in these works and archaic forms of life merge with how things were in the 1930s and 1940s."[40] Whyte graphically illustrates how far Banff is, for him, the "site of a childhood geography" when he maps the Bow River shoreline and the site of Peter and Catherine Whyte's house in a manuscript document he titles "When the World Was Five Years Old."[41] Whyte never misses an opportunity to conflate personal and family history with regional history and literary heritage, as in the opening notes for the first volume of *Fells*, where he casts Mary T.S. Schäffer's 1911 *Old Indian Trails of the Canadian Rockies* as a "source "of both his own work and his life: "my mother visited Mary in 1931, met my father as a consequence of that visit, and ten years later I was a further consequence."[42]

The kind of "telescoping of history" and "crossing" of "localized memoir with historical excavation," which Kerrigan observes in pocket epic, appears in graphic form within the second volume of Whyte's poem, where he juxtaposes historic narratives with his own present-day retracing of the same walks. The plainest example of such juxtaposition comes in *Wenkchemna*, where Whyte crosses S.E.S. Allen's account of crossing three mountain passes in the 1896–97 issue of the *Alpine Journal* with a personal narrative of hiking the same territory. Allen's text runs down the left side of the page, with Whyte's own narrative on the right:

As I advanced	|	*The sun's heat has awakened*
up the snow troughs	|	*gravel slurries, water runnels*
by the side of the glacier's	|	*streams cascading into crevasses*
lateral moraine	|	*ogives' circular cracks.*
I kept a continual look-out	|	*until we reached the bergschrund*
for 'shooting' gullies	|	*(or randkluft? the big one.)*
in the cliffs	|	*Pamela's in running shoes,*
which covered the vicinity	|	*we have no rope, just hope,*
with debris.	|	*it's only two o'clock;*
At last I gained the glacier,	|	*once we reach the pass*
and it proved quite free	|	*it'll be swift descent*
from crevasses.	|	*to tea, the bus, and back.*[43]

Of the several poets designated as practitioners of pocket (or, for Kerrigan, local) epics, the Birmingham poet Roy Fisher strikes me as offering the most interesting parallels to Whyte. This is not the place for an extended exploration of Fisher's poetics, or even for a fully developed comparison between those of Fisher and Whyte. My purpose in outlining affinities between the two poets' procedures is meant instead to help contextualize the way Whyte's poem claims some of the cosmological and encyclopedic properties of the epic genre, but in a way that aims to take account of pre-existing representations and seeks to "juxtapose the contents with the frame."[44] Like Whyte, Fisher also acknowledges that place and perception have a mutual shaping influence, for him. As he puts it, "Birmingham's what I think with."[45] This formulation, it seems to me, might have been used by Jon Whyte with reference to the Bow Valley. When Peter Barry describes the way Fisher writes about the two rivers in Birmingham in his poem *A Furnace*, once again the terms lend themselves strongly to a description of Whyte's own procedures in writing about the Bow Valley: "Such writing seems the product of research in local history libraries and walks on the terrain itself, ambulatory, antiquarian, poetic, easy-paced, eclectic…. Indeed, Fisher's whole urban poetic project has this intellectual eclecticism, believing in the efficacy of an amateur kind of general reflective learnedness, rather than the professional's narrower 'single-issue' mode."[46] The author of five volumes of local history, and very much an eclectic intellectual as revealed in his weekly columns for the Banff *Crag and Canyon*, Whyte fits this description quite immaculately. Much as Whyte does with the Bow Valley, Fisher constructs a kind of collage or "composite-epic" poem around Birmingham, which "allows maximum independence to the component parts, and encourages intratextual seepage between them."[47]

Like Whyte, Fisher is "deeply influenced by visual art"[48] and makes the exploration of perception a prominent part of his poetic enterprise. As I would suggest Whyte's later poems do, Fisher's poem *City* incorporates an "'I' that perceives, creates, moves among his images and, finally, becomes an object in an overall design."[49] When Fisher describes himself as depicting "a landscape that 'didn't need to be rendered fictive because it was so,'"[50] he echoes Whyte's evocation, in *Wenkchemna*, of "the postcard Lake Louise" which cannot be separated from the "real thing."

Despite their differences, Fisher's and Whyte's poetics do appear to overlap to the extent that they both exemplify a certain kind of critical approach to representing an environment. Such a critical approach, one that recognizes the role played by previous representations among other factors, is called for by Peter Morton: "Ecomimesis is above all a practice of juxtaposition. Avant-garde art values juxtaposition as collage, montage, bricolage or rhizomatics. But it all very much depends upon what is being juxtaposed with what. If it is to be properly critical, montage must juxtapose the contents with the frame. Why? Simply to juxtapose contents without bringing form and subject position into the mix would leave things as they are."[51] Whyte's work continually refers to his sense that his perception of place is inseparable from both previous representations and present observation. If the Bow Valley is "what he thinks with," then his relationship with it is not a dualistic one, but a participatory one. This prevents him from experiencing anything "pure," or purely "natural," as when he is unable to see the actual Lake Louise without the intrusion of the "postcard Lake Louise."[52] Here he echoes Morton's point that sometimes "nature is already the quintessence of kitsch."[53] But the kitsch is also part of "what he thinks with," as he recalls his early musings upon the *mise-en-abyme* effect of both the mirrors in the Banff barbershop, and the cereal boxes of his childhood. For Whyte, as for Freud, "'Home' is the strangest place,"[54] but it is that way for a reason because of a history of layered perceptions and ideas about perceptions. Emphasizing the encyclopedic elements in a revised version of epic allows Whyte to write about home in an epically inclusive way. This approach allows his poem to show "its eyes getting wider"[55] as he pursues the sense that place has its own inbuilt questioning quality. In a punning and literal fashion, Whyte uses concrete poetry strategies to drive this point home with reference to the "beckoning" quality of Wenkchemna Pass:

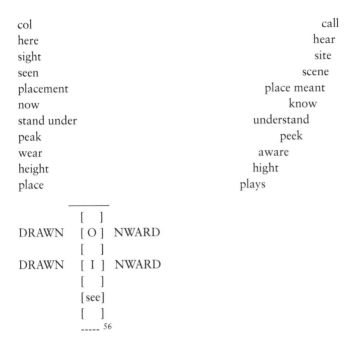

```
         col                              call
         here                             hear
         sight                            site
         seen                             scene
         placement                    place meant
         now                              know
         stand under                 understand
         peak                             peek
         wear                             aware
         height                           hight
         place                            plays

                    [   ]
         DRAWN    [ O ]   NWARD
                    [   ]
         DRAWN    [ I ]   NWARD
                    [   ]
                  [ see ]
                    [   ]
                    ----- 56
```

What Whyte's re-envisioning of epic allows him to do, then, is to pursue this kind of critical ecological representation that incorporates the viewer as an intellectual and cultured presence and aims not to reify what is observed. In undertaking a new presentation of his home, he reflects the pro-regionalist tenor of quite a bit of western Canadian writing in the 1970s and 1980s. In aiming to do so in a way that remains true to his own developing ecological sensibility, he is willing to try everything, in just the kind of intellectual eclecticism attributed to Roy Fisher, above. Epic inclusiveness provides him with a framing procedure (and a linguistic permissiveness he revelled in) while also licensing the essayistic (in Bernstein's sense) elements of his Rockies poem.

Notes

1 Jon Whyte, *Minisniwapta: Voices of the River*, S28/1 to 39, broadcast on "State of the Arts," Canadian Broadcasting Corporation (CBC), 1987, audio recording, Jon Whyte Fonds, Whyte Museum of the Canadian Rockies, Banff, Alberta (hereafter Whyte Fonds).
2 Jon Whyte, *Jon Whyte: Mind over Mountains, Selected and Collected Poems*, ed. Harry Vandervlist (Calgary: Red Deer Press, 2001), 108.
3 Jon Whyte, CBC Radio interview with Carol L. Foreman, S28/40, audio recording, Whyte Fonds.
4 Jon Whyte, "Cosmos: Order and Turning," *Trace: Prairie Writers on Writing*, ed. Birk Sproxton (Winnipeg: Turnstone Press, 1986), 269–74.
5 Dana Phillips, *The Truth of Ecology: Nature, Culture, and Literature in America* (Oxford: Oxford University Press, 2003), 209.

6 Ibid., 206.
7 Ibid., 247.
8 Jon Whyte, "Art and Culture of the Canadian Rockies," undated ts, M88 /E/disc 14, Whyte Fonds.
9 Jon Whyte, *Mountain Chronicles: A Collection of Columns on the Canadian Rockies from the Banff Crag and Canyon, 1975–1991*, ed. Brian Patton (Banff: Altitude, 1992).
10 This poem can be viewed online at http://www.whyte.org/jonwhyte/earthpoem.html.
11 Jon Whyte, *The fells of brightness, second volume: Wenkchemna* (Edmonton: Longspoon Press, 1985), 84–87.
12 Whyte, CBC Radio interview.
13 Ibid.
14 Mike McIvor, "Former CPAWS Trustee Dies in Banff," *Borealis* 3, no. 3 (1992), 51.
15 Whyte, *Mind over Mountains*, 111.
16 James F. Johnson, "'Brébeuf and His Brethren' and 'Towards the Last Spike': The Two Halves of Pratt's National Epic," *Essays on Canadian Writing* 29 (1984), 143.
17 Ibid., 149.
18 Northrop Frye, "Silence in the Sea," in *The Bush Garden: Essays on the Canadian Imagination* (Toronto: House of Anansi, 1971), 184, 186. Quoted in Johnson, "'Brébeuf and His Brethren,'" n1.
19 Reed Way Dasenbrock, "Constructing a Larger Iliad: Ezra Pound and the Vicissitudes of Epic," in *Epic and Epoch: Essays on the Interpretation and History of a Genre*, ed. Steven M. Oberhelman, Van Kelly, and Richard J. Golsan (Lubbock: Texas Tech University Press, 1994), 253.
20 Ibid., 251.
21 Ibid., 253.
22 Ibid.
23 Whyte, *Mind over Mountains*, 147.
24 Ibid.
25 Ibid.
26 Michael André Bernstein, "Making Modernist Masterpieces," *Modernism/Modernity* 5, no. 3 (1998), 10.
27 Ibid., 18.
28 Ibid., 12.
29 Whyte, "Cosmos: Order and Turning," 273.
30 Whyte, CBC Radio interview.
31 Van Kelly, "Introduction," in *Epic and Epoch: Essays on the Interpretation and History of a Genre*, ed. Steven M. Oberhelman, Van Kelly, and Richard J. Golsan (Lubbock: Texas Tech University Press, 1994), 12.
32 Ibid.
33 Ibid., 17.
34 Ibid., 19.
35 Whyte, "Cosmos: Order and Turning," 273.
36 Harry Vandervlist, "Jon Whyte," in *Dictionary of Literary Biography*, vol. 334, *Twenty-First-Century Canadian Writers*, ed. Christian Riegel (Detroit: Gale Research, 2006), 278.
37 Nigel Alderman, "Introduction," in "Pocket Epics: British Poetry after Modernism," ed. Nigel Alderman and C.D. Blanton, special issue, *Yale Journal of Criticism: Interpretation in the Humanities* 13, no. 1 (2000), 1.
38 Ibid., 2.
39 John Kerrigan, "Divided Kingdoms and the Local Epic: Mercian Hymns to the King of Britain's Daughter," in "Pocket Epics: British Poetry after Modernism," ed. Nigel

Alderman and C.D. Blanton, special issue, *Yale Journal of Criticism: Interpretation in the Humanities* 13, no. 1 (2000), 3–21.
40 Ibid., 4.
41 Whyte, *Mind over Mountains*, 3.
42 Ibid., 114.
43 Whyte, *Mind Over Mountains*, 160.
44 Timothy Morton, *Ecology Without Nature: Rethinking Environmental Aesthetics* (Cambridge, MA: Harvard University Press, 2007), 143.
45 Roy Fisher, quoted in Peter Barry, "'Birmingham's What I Think With': Roy Fisher's Composite-Epic," in "Pocket Epics: British Poetry after Modernism," ed. Nigel Alderman and C.D. Blanton, special issue, *Yale Journal of Criticism: Interpretation in the Humanities* 13, no. 1 (2000), 87.
46 Ibid., 97.
47 Ibid., 102.
48 John Mathias, "The Poetry of Roy Fisher," in *Contemporary British Poetry: Essays in Theory and Criticism*, ed. James Acheson and Romana Huk (Albany: SUNY Press, 1996), 38.
49 Ibid., 41.
50 Ibid., 42.
51 Morton, *Ecology without Nature*, 143.
52 Whyte, *Mind over Mountains*, 131–32.
53 Morton, *Ecology without Nature*, 173.
54 Sigmund Freud, quoted in Morton, *Ecology without Nature*, 175.
55 Morton, *Ecology without Nature*, 173.
56 Whyte, *The fells of brightness*, 87.

CHAPTER 18

Ware's Waldo
Hydroelectric Development and the Creation of the Other in British Columbia

Daniel Sims

In *Simulacra and Simulation*, French philosopher Jean Baudrillard, like a latter-day McLuhan-Nietzschean Zarathustra, proclaims the death of reality, sacrificed to image, representation, and simulation.[1] He describes a process by which the image supplants reality by first reflecting a basic reality, then masking and denaturing[2] that basic reality, followed by the masking of the absence of that basic reality, and finally bearing no relationship to basic reality at all, becoming a simulacrum.[3] This is not a new idea and, indeed, was proposed by nineteenth-century German philosopher Friedrich Nietzsche in *Twilight of the Idols*.[4] And while it is debatable whether or not all reality is truly dead, notable historians like H.V. Nelles in *The Art of Nation-Building* and Ian McKay in *The Quest of the Folk* have alluded to how images of reality can be decontextualized and separated from reality through time, space, the construction of identity, significance in the case of the former author, and creation of the *volk* in the latter.[5] In particular, both cite the problem of photographs as representations of the past and/or reality, and how their meanings change based on the context in which they are viewed, and the information already known, gathered, or in the case of captions, provided.[6] These problems hold true for motion pictures, particularly documentaries and news reports, which, as a rule, generally claim to accurately represent reality, to be factually correct, and to tell the truth.[7] This, of course, should be obvious to anyone trained in any form of source analysis. However, it is worth noting here, because people often assume that, even if photographs do not accurately reflect reality because of their spatially and temporally static nature, motion pictures can or do because of their ability to capture reality in motion and over a duration of time.

This chapter is an examination of two documentaries, produced in the 1970s, examining the effects of hydroelectric development on the Peace and Columbia rivers on the local inhabitants: the 1970 *CBC Hourglass* documentary on the Tall Grass Indians and Williston Lake[8] and the 1975 production, *The Reckoning*.[9] These documentaries reflect Baudrillard's procession of the simulacra: first, by representing the effects of hydroelectric development on the Columbia and Peace river systems; second, by masking and denaturing this reality by selectively choosing the sources used, people interviewed, images shown, music played, and narration/context told; third, by masking the absence of reality by the very nature of motion pictures themselves, which when viewed by an audience appear as a reality that at once seems to exist and yet does not exist; and finally, as the disconnect from reality became apparent, by representing a simulacrum completely separated from the reality of the people and environment of the Kootenays, Columbia, and Peace River country. I argue that this process is aimed toward the construction of the local inhabitants of these regions as the "other": in the case of the Kootenays and Columbia, as the *volk* of these regions (if not British Columbia and Canada as a whole), and in the case of the Peace, as the "Indian" other.

Background/Historiography[10]

In the 1960s, the Province of British Columbia began major hydroelectric developments on the Peace and Columbia rivers.[11] Called the "two river policy," the dams included in this policy were completed during a period starting in 1967 with the W.A.C. Bennett Dam on the Peace River and the Duncan Dam on the Duncan River (a tributary of the Kootenay River, which itself is a tributary of the Columbia River), and ending in 1973 with the Mica Dam on the Columbia River.[12] Much ink has been spilled over the wisdom and effects of these hydroelectric developments on the Peace and Columbia rivers,[13] the equity of the Columbia River Treaty,[14] and indeed, even over whether or not the "two river policy" was a priori or a posteriori.[15] This chapter is not another salvo in this dispute over water and dams in British Columbia, per se. Rather, as mentioned previously, it is an examination and comparison of two documentaries dealing with the effects of these hydroelectric developments: *CBC Hourglass* (1970) and *The Reckoning* (1975).

History of the Documentaries

The backstory of both documentaries is quite interesting. The *CBC Hourglass* documentary was broadcast on the CBC current events program *Hourglass*, which was divided into two parts, news and in-depth analysis, on 20 November 1970.[16] The documentary in question aired under the latter part. I originally learned of the documentary in my home community of Tsay Keh Dene (a Tse Keh Nay community on the north end of the Williston Lake Reservoir) while conducting research at the Treaty Office for my MA thesis on Tse Keh Nay history and ethnic identity.[17] The documentary had been provided to the community by BC Hydro through their Aboriginal Relations video series and is paired with a 2002 documentary titled *In the Words of Kwadacha* (another Tse Keh Nay community to the north of Tsay Keh Dene also known as Fort Ware or just Ware).[18] On the DVD, the film is titled *Tsay Keh Dene: CBC Hourglass Documentary*, but given the fact the documentary refers to the Tsay Keh Dene as the Tall Grass Indians, it is clear that this was not the original title.[19] Neither the BC Hydro website nor BC Hydro Aboriginal relations revealed anything about the original title, but my research did reveal a 1974 Simon Fraser University MA thesis on *Hourglass*, which notes the documentary under the title "Tall Grass/Williston."[20] Whether this is the official title or a working description is unclear, however. Besides the issue of the original title, I realized that the recording of the documentary not only is not from the original airing (as revealed in the transition monologue of the first half of the program), but also has been edited by BC Hydro to remove the end credits.[21] (The original broadcast aired on 15 November 1970 on the program *Weekend*, and credits Mike Halleran as the reporter and Mike Poole as the producer.)[22] Why BC Hydro chose not to use the original broadcast and then chose to edit the re-airing from *CBC Hourglass* is unclear, but it serves to decontextualize the documentary from those who produced it outside of the parent program, *Weekend*, and the reporter Mike Halleran, who is named in the previously mentioned transition monologue.[23] It is for this reason I refer to the documentary as *CBC Hourglass* and not *Weekend*; BC Hydro has transformed it.

The Reckoning has a better-known backstory. Originally aired on 16 February 1975, the copy I have is from a *CBC Sunday Best* rebroadcasting from 11 May of the same year.[24] Unlike the untitled *CBC Hourglass* documentary, to which academic citations remained elusive, historian Tina Loo cites *The Reckoning* in her 2004 article, "People in the Way," which focuses on the hearings surrounding hydroelectric development on the Columbia River, and the alternative McNaughton plan for said development that focused on reversing the Kootenay River into the Columbia by damming it near the U.S.–Canada border.[25] This use of the documentary is problematic as, in his memoirs, the co-chair of BC Hydro at the time, and the individual in charge of the Columbia, Hugh Keenleyside, calls into question the accuracy and objectivity of the documentary.[26] Indeed, the documentary was so controversial that the Wednesday after the original broadcasting on 16 February 1975, its factual basis, in particular concerning a claimed secret BC Hydro committee was debated in the BC Legislature.[27] Even when one considers that Keenleyside is defending his legacy by discounting *The Reckoning*—and that politicians will debate anything if it offers a political advantage—the fact that he goes into precise detail regarding why the documentary was wrong and biased raises questions regarding the use of this documentary in later historical research. Further complicating the use of this documentary as a source is authority: as the credits reveal, rather than being a reporter, as in *CBC Hourglass*, Mike Halleran is the narrator and scriptwriter, with Mike Poole again serving as the producer and director.[28] And while it remains possible that Mike Halleran and Mike Poole filled exactly the same roles for *CBC Hourglass*, it does not appear they are credited as such and this raises the question: what is the difference between Mike Halleran the reporter and Mike Halleran the scriptwriter/narrator or Mike Poole the producer and Mike Poole the producer/director? After all, both documentaries splice and edit footage, still images, and sounds to provide a narrative arc to build an argument using the creation of the "other." In this creative strategy, all three roles meld into each other, echo the collapsing of poles (reporter versus narrator/scriptwriter; non-fiction versus fiction), and lead to the procession of the simulacra:[29] Mike Halleran reports the reality, masks and denatures it with his narration and scriptwriting, and finally covers and disconnects it from reality through a combination of the first two actions, leading to a simulacrum where the audience cannot tell what is real, what is false, or what is from the past, what is from the present. The opening act of this simulacrum begins the process of constructing the "other," thereby supplanting reality.

Opening Act

In the first step of Baudrillard's procession of simulacra, the image represents/reflects reality. This is evident immediately in both documentaries. With *CBC Hourglass*, it appears in the opening scene where our reporter/narrator Mike Halleran is standing on the shore of the Williston Lake Reservoir, a shore choked with floating timber (see Figure 18.1).[30] Almost immediately, however, this reality is masked and denatured. Although it is clear Mr Halleran is on the edge of a lake (presumably Williston Reservoir), it is unclear exactly where he is on Williston Lake, and whether the shore he is on is typical or the exception, with regards to the amount of debris found in the reservoir.[31] And although he is presumably included in the right third to give perspective to the size and amount of debris along the shoreline, without knowing Mr Halleran's height or the angle of the shot it is hard to judge the depth of the debris. Further compounding the masking and denaturing of this representation of reality is the narration, which begins after this image as the camera zooms in on Mr Halleran's face, figuratively removing the environment/set from the scene, a literal denaturing: "This is Williston Lake. It is named after the Minister of Land, Forests and Water Resources for British Columbia, the Honourable Ray Williston. It was created by the Bennett Dam, named after William Andrew Cecil Bennett of British Columbia, the Premier and Minister of Finance. A good deal has been said and written about the prosperity that this dam and its power project have brought to British Columbia. But little or nothing has been said about the sad and tragic experience of a small group of Indian people, who lived at the junction of the Finlay and Parsnip rivers, a spot now covered by three hundred feet of water."[32] In *The Quest of the Folk*, Ian McKay points out not only how the meaning of an image changes with regards to context, but also how a caption is a key element, if provided.[33] What emerges and proves telling, here, based on McKay's premise, is that the narration for this image makes no mention whatsoever of the debris taking up the bottom two-thirds of the scene; instead, it opts to contextualize what is seen by talking about politicians, prosperity, and the local Aboriginal population. In this, nature becomes the set for Mr Halleran's opening monologue. Arguably, this is done because the debris, while a visual manifestation of the problems discussed in the documentary, is not the true focus of it, as the narration reveals. The problem, however, by showing and not explaining the debris in the opening scene, creates questions (if not a cognitive dissonance)

Figure 18.1 Mike Halleran on the shore of the Williston Lake Reservoir, from *CBC Hourglass*. Image courtesy of CBC Vancouver.

in the minds of the viewer, which draws the audience in, and calls for answers and resolution. Indeed, the manner in which the narration is drawn out by the use of full names, titles, and honorifics makes the absence of any mention of the debris even more noticeable.

This representation of reality, followed by its masking and denaturing occurs also in the opening scene of the *The Reckoning*. Instead of an image of Mike Halleran standing on the edge of the Duncan Lake Reservoir, the Arrow Lakes, the Kinbasket Lake Reservoir, or the Koucanusa Lake Reservoir, the viewer is confronted with the sight of Waldo, British Columbia, burning (see Figure 18.2).[34] There is no need to place Mr Halleran in the shot here for context because everyone is more or less familiar with the basic dimensions of wood-frame buildings. In other words, *The Reckoning* presents an environment more familiar to *Homo sapiens*: a town burning, whereas *CBC Hourglass* situates a human in a "natural" environment and refers to human individuals and a human social group to contextualize the scene.[35] Indeed, this divide continues in the narration for this *Reckoning* scene, in which Halleran places Waldo spatially and historically in British Columbia and international politics: "On the western slope of the Rocky Mountains in Canada, the historic town of Waldo, British Columbia, is put to the torch by the BC government to make way for floodwaters rising behind an American dam. We were required to perform this act to fulfill our obligations under the terms of the Columbia River Treaty."[36] And yet, even here the image remains masked and denatured. Where on the western slope of the Rocky Mountains is Waldo? After all, the Rockies run from the U.S.–Canada border to the Liard River. Why is Waldo historic? Indeed, are not all towns historic, and if not, what makes a town historic? Furthermore, is "historic" an important adjective in this narration, which based on the words and tone, is clearly supposed to rally Canadian nationalists against the Columbia River Treaty? Are we supposed to infer that had Waldo not been historic that its destruction would somehow matter less? I would argue the answer is yes, as the environment shown in *CBC Hourglass* is neither presented as historic nor asked to be mourned by the reporter/narrator.[37] But more important even is that in their opening scenes both documentaries have not only represented reality—a debris-filled Williston Lake Reservoir and the destruction of Waldo by fire—but also they have begun to mask and denature this reality by not mentioning key characteristics of the nature of the environment and its inhabitants (mentioned in *CBC Hourglass* and implied by the presence of Waldo in *The Reckoning*) and its historic nature. Instead, they have constructed the local inhabitants through their environment as the "other" in comparison with the rest of British Columbia and Canada: the "natural" Tall Grass Indians affected by individual politicians and hydroelectric development and the "historic" *volk* of Waldo and the Kootenays and Columbia affected by government, foreigners, treaties, and hydroelectric development.

Figure 18.2 Waldo, BC, burning, from *The Reckoning*. Image courtesy of CBC Vancouver.

Enter the Tragic Hero Stage Right

This construction of the other through visuals of the environment is seen when each documentary introduces the people affected, unable to prevent their tragedy due to their "other" nature. In *CBC Hourglass* the opening monologue transitions visually to an image of Finlay Forks from the air, thereby removing all "traces" of "unnatural" human change to the environment (see Figure 18.3). These traces, however, are quickly shown in the form of the Fort Grahame Tse Keh Nay village of Finlay Forks, one of three Fort Grahame Tse Keh Nay villages mentioned in the documentary: Finlay Forks, Fort Grahame, and Ingenika (the latter commonly referred to as Old Ingenika to distinguish it from the post-reservoir community of the same name).[38] It is a community, however, denuded of the local population; a picturesque collection of three cabins located in a forest. As previously mentioned, given the fact Finlay Forks was 300 feet underwater both images are representations of a past that was (a representation of the absence of the original reality), and yet in the case of the latter, by removing the human population it is a masked and denatured representation of that absent reality. Indeed, it is not entirely clear if this is Finlay Forks simply denuded of its population for the camera (a simulation) or Finlay Forks, the dead village, awaiting inundation (reality).[39] A picture from Bernard McKay's *Crooked River Rats* not only suggests the latter is the case,

Figure 18.3 Finlay Forks from the air, from *CBC Hourglass*. Image courtesy of CBC Vancouver.

Figure 18.4 Village of Finlay Forks, from *CBC Hourglass*. Image courtesy of CBC Vancouver.

Figure 18.5 Village of Finlay Forks, from McKay, *Crooked River Rats*, 172. Image courtesy of Hancock House Publishers, http://hancockhouse.com.

but also reveals that the supposed reality of Finlay Forks in the documentary is, in fact, a trick of the camera, done to suggest a community in the midst of a forest.[40] Rather, it is a community in a clearing surrounded by a forest, much like most communities of any size in northern British Columbia west of the Rocky Mountains (see Figure 18.4).[41]

It would appear then that the choice of these two shots (see Figures 18.4 and 18.5) for Finlay Forks in the documentary was meant to convey an image that suited a particular conception of the local inhabitants, namely, harmony with nature, and not reality per se. In other words, not only does it represent the destroyed Finlay Forks to the audience (a representation of the absence of reality), but it recreates the "real" Finlay Forks via the media to its own ends (the simulacrum) regardless of reality.

This image of human development in harmony with nature, the narrator directly states in the next scenes, explains who the Fort Grahame Tse Keh Nay or Tsay Keh Dene are. Transitioning to an image of a riverman moving toward the camera and the bottom third, and then two rivermen moving from the left third to the right third (see Figure 18.6), Halleran states: "Except for the few log cabins on the riverbank, it had not changed much since the time of Mackenzie and Simon Fraser. The Natives were people of the land, known as the Tall Grass People: rivermen, hunters, and wanderers for thousands of years. For them, the North was not some wild enemy to be tackled and tamed. It was home."[42] Here were a people untouched by time, change, and progress; people, unlike normal British Columbians and Canadians, who did not seek to tame the wild, but instead saw it as their natural habitat and were part of it: people out of touch with the modernity sweeping British Columbia during this era.[43] Furthermore, as Paige Raibmon argues in *Authentic Indians*, these traits were but a few of those that supposedly diametrically divided

312 Daniel Sims

Figure 18.6 "Tse Keh Nay" rivermen, from *CBC Hourglass*. Image courtesy of CBC Vancouver.

Euro-Canadians from the "Indian" other.[44] And yet even in using images of nature, rivers, and rivermen, a closer examination reveals that the documentary masks and denatures the reality it purports to show its audience. The entire riverboat scene of "Tall Grass Indians" unchanged since Mackenzie and Fraser, is drowned out by the sound of a petrol outboard motor.[45] Clearly, more has changed than just a "few log cabins on the riverbank," and yet the Tsay Keh Dene are portrayed as static, unchanging "Indian" others that are one with nature: *Homo naturalis*.

While the Tsay Keh Nay are portrayed as the Tall Grass "Indian" other in *CBC Hourglass*, the residents of the Kootenays, Columbia, and adjoining areas of British Columbia are presented as the *volk* in *The Reckoning*. As Ian McKay notes in *The Quest of the Folk*, central to the idea of the volk was that there was "within the population a subset of persons set apart ... characterized by their own distinctive culture and isolated from the modern society around them."[46] The symbol of this isolation from modern society, which the documentary directly cites as such, is the sternwheeler SS *Minto* (see Figure 18.7), an outdated piece of technology from an earlier energy regime and era.[47] Or, as Mike Halleran puts it: "The sternwheeler *Minto*, a symbol of the peaceful and leisurely way of life along the Arrow Lakes in British Columbia. In these mountain valleys great change would soon occur. International negotiations

Figure 18.7 SS *Minto*, from *The Reckoning*. Image courtesy of CBC Vancouver.

seemed remote for the realities of traditional lifestyle. The change was imposed from afar, by people of another world."[48] Where is this "other world?" Mars? Yuggoth?[49] No, the Lower Mainland and Vancouver Island, the seat of power of the Laurentian Empire,[50] and a cession from Henrietta Maria's province:[51] Vancouver/Victoria, Ottawa, and Washington, DC, respectively. The *Minto*, however, is not the only symbol of the inherent volk nature of the residents of the region. Like in *CBC Hourglass,* one of the communities of the volk is immediately shown after this opening scene, although unlike Finlay Forks it is never geographically located in the region in any way.[52] In doing so, it is presented as the universal volk village, which in its distilled state represents any, and all, volk communities in the Kootenay and Columbia regions.[53] In this, the reality of the village is represented by film, masked, and denatured by never being named or placed spatially or geographically, thereby representing a "typical" village that apparently existed prior to the hydroelectric development of the Columbia and subsequent flooding (a representation of the absence of reality); it is the volk village, completely universalized, separated from the basic reality of the original village filmed.

One of the first things evident in this scene (Figure 18.8) is the two individuals in the centre and middle thirds. Clearly, these are not natural, but man-made structures, or at least that needed to be portrayed to the audience. Separating them from "modern" British Columbia and Canada is the gravel

314 Daniel Sims

Figure 18.8 Houses in the Columbia River Valley, BC, from *The Reckoning*. Image courtesy of CBC Vancouver.

road, a rarity in a province that under the Social Credit government of British Columbia (the same government that started construction on the hydroelectric developments in question) had not only adopted a general policy of paving roads, but had done so to the point of the Minister of Highways Phil Gaglardi becoming a caricature, as illustrated in a *Vancouver Sun* editorial cartoon from 16 September 1964.[54] Also of note, unseen in the scene of Finlay Forks from *CBC Hourglass*, are the fences that clearly delineate nature from the site of human habitation. These fences echo a combination of themes found in *Authentic Indians* and *The Quest of the Folk* about Euro-Canadians and the volk, namely, that even if they are in nature, they are not of nature; in this case, the line is the fence boundary.[55] Indeed, even in the next shot, where the house is inundated with nature (see Figure 18.9), the fence line is still evident at the height of grass, and to the left side of the house.[56] And yet the angle of the shot suggests it seeks to mask and denature even this fact.

While the scenes separate the Kootenay and Columbia regions from "modern" British Columbia, the narration separates the volk from outside politicians and international agreements, namely, the Columbia River Treaty. As Mr Halleran states: "Seeing their fate discussed by international negotiators whom they did not know, and whom they could not reach, the valley residents of the Columbia River were helpless to plead their own case. With the whole

Figure 18.9 Houses in the Columbia River Valley, BC, from *The Reckoning*. Image courtesy of CBC Vancouver.

future at stake, they placed their trust in the politicians, that the system would protect their rights, that they would receive justice."[57] With this, we no longer have a village of British Columbians in the Kootenays or Columbia, but the victims of hydroelectric development on the Columbia River, disconnected from modern British Columbia and Canada technologically, developmentally, and politically, in but not of nature: the volk of British Columbia, a simulacrum of real people.

Dénouement

The concluding scenes of both documentaries firmly reinforce this divide between the Indian and the volk. In *CBC Hourglass*, gone is the debris visible in the first scene. In its place are muskeg and swamp, if not a beaver pond, the Peace in its "natural" and "real" state, the perfect homeland for the Tsay Keh Dene, or as Mike Halleran classed them, the Tall Grass Indians (see Figure 18.10).[58]

Following this is a scene of one of the cabins constructed on the banks of the reservoir as it rose, even more a part of nature than those at Finlay Forks in its utilitarian ad hoc form.[59] This reality is not mentioned, however, and the viewer is left wondering (if not believing) if this is a typical Tall Grass

Figure 18.10 Muskeg in the Peace River Country, BC, from *CBC Hourglass*. Image courtesy of CBC Vancouver.

Figure 18.11 Cabin on the banks of the Williston Lake Reservoir, from *CBC Hourglass*. Image courtesy of CBC Vancouver.

structure (see Figure 18.11). Again, reality is masked and denatured from its original form and context.

The next scene is a montage of some of the "Tall Grass Indians" met in the documentary: the first one, Keom Pierre, is followed by the unnamed Suzanne Pierre and Robert Tomah.[60] In the case of the latter, these two individuals are transformed from widow and son, to unnamed victims of the "flood," who, like the previously mentioned volk village in *The Reckoning*, have been universalized to the point of being detached from their own reality: simulacrum.[61] All the while Halleran calls into question the future of *Homo naturalis*, the Indian other: "When we left the Peace River country, the fall rains had begun. Snow will follow soon and the land will be locked in winter until April or May. Next spring Williston Lake will be raised again. The homes on the shoreline will have to be moved once more. The wilderness calls these people. And when spring comes maybe they can make their move, over the big new lake to Fort Ware or Ingenika. If they have help that is."[62] A move away from their new reserves described in the documentary—reserves designed to bring them within close proximity of Mackenzie, British Columbia, and modernity, as exemplified by the hydroelectric development on the Peace and the planned nature of Mackenzie.[63] But more importantly, it is a move back to "wilderness" (whatever that may be), the antithesis of environmental development and modernity as represented by the Williston Lake Reservoir and the W.A.C. Bennett Dam. They can escape this reality, according to the documentary, because they are the "Indian" other, *Homo naturalis*, already detached from their own personal reality and therefore having no ties to the wider reality of modern British Columbia.

And what options do the volk of British Columbia have? Not many. But, then again, as Ian McKay points out, the volk are more for the salvation of modern society from itself than for their (the volk's) own benefit.[64] Reflecting this, the scenes ending *The Reckoning* are not about an escape to a mythic wilderness seen in the opening of *CBC Hourglass*. Rather, the closing scene is that of destruction; the same kind of debris in the first scene of *CBC Hourglass*, except at low water (see Figure 18.12).[65] This is followed by a montage of the Buerges, a simple volk family, walking along the new waterline of their property, a waterline that had destroyed their farm and self-sufficiency.[66] Scenes of construction, more debris, and the Buerges walking again, all to a melancholy tune to evoke sympathy and sorrow follow this.[67] Three separate scenes and realities edited into one for the viewers' benefit, masked, and denatured from temporality and spatial geography, and with the aid of music moving beyond merely masking the absence of this edited unitary "reality" to that of the simulacrum. The escape from this "reality" (in the future) is the volk way. As Halleran states:

What did we learn from this treaty? Must we use primitive economies as the only criterion for measuring the gains and losses? Can we ever learn to assess the value of the so-called intangibles? What about protecting the rights of people? What about the creation of a sentiment among whole communities that the political process has betrayed them? We are clumsy at measuring these costs as well. They are very real. This experience clearly shows that if you limit the size of the commitment, you limit the extent of the risk. Perhaps we should proceed more slowly in future and not always go for the big scheme. Why not let the people in? Give them access to all information. Make public hearings a matter of law and hold them before the fact, not after. Why not consider the common values? They are held by common people, and it is they who must meet the final cost. There is a poem by Robert Service called "The Reckoning."[68]

Clearly, then, the solution here is not to run away, to escape "reality" as the Tall Grass can. For the volk, one cannot escape modernity, here in the form of the Columbia Treaty dams.[69] Rather, one can merely change it to make it more humane. This change can be accomplished by learning from the volk, or as Mike Halleran calls them in the above quote, the "common people." Then, again, he would know. You see, Mr Halleran grew up in Nelson (one

Figure 18.12 Debris at low water of a reservoir in the Columbia River Valley, BC, from *The Reckoning*. Image courtesy of CBC Vancouver.

of the communities affected by hydroelectric development in the Kootenays and Columbia) and died there in 2008; our reporter/narrator/scriptwriter was apparently one of the volk.[70]

Conclusion: Why the Other?

I never knew I was a Tall Grass Indian until I watched *CBC Hourglass*. Similarly, I never knew my step-grandparents were the volk until I watched *The Reckoning*. In the case of the latter, it is truly not a big deal. Some of their ancestors had previously been constructed as the Pilgrim Fathers in the Anglo-American world and *das Volk* in Germany by two Grimm brothers. In the case of learning I was a Tall Grass, it was surprising. When I first watched the documentary, my cousins sitting next to me, they were just as surprised as I was; we had always been told we were Tsay Keh Dene, Tsay Keh Nay, Tse Keh Nay, Sekani, or some variation of, but not Tall Grass.[71] Indeed, reflecting the static anachronistic portrayal of the Tsay Keh Nay as one with nature, the last Euro-Canadian to refer to any Tse Keh Nay group proper like that was Simon Fraser in his first journal from 1806.[72] He had called the "apparent" ancestors of the Tsay Keh Nay, the Meadow Indians.[73] Then again, Halleran has stated the Tsay Keh Nay and their environment had not changed much since the time of Fraser and Mackenzie. What better way to prove that than to use a variation of an old name? But more importantly, why were the people of the Kootenay and Columbia regions, along with the Tsay Keh Nay constructed as the "other" in both documentaries? There are a few possible reasons for this. In the case of the former documentary (*CBC Hourglass*), it could simply be a matter of discourse fundamental to colonization in Canada and the world, namely, presenting Indigenous peoples as having no history and being ahistorical.[74] A discourse that also meant, however, that nothing could be learned from them and their plight. Hence, the documentary's suggested solution: their return to the "wilderness" from whence they came, because as Mike Halleran notes, "they do not believe that there is any place for them in the Whiteman's world."[75] Whether the Tsay Keh Nay believe this or Halleran believed they believed this, however, is impossible to determine. Things get interesting in the case of the latter documentary (*The Reckoning*), however. Here, although constructed as the volk other, this "otherness" does not diametrically separate them from Euro-Canadians as the "Indian" other does.[76] Rather, they represent all the "good" things supposedly lost with modernization.[77] Because of this essence, *The Reckoning* is a morality tale. The volk are created to temper government policy regarding hydroelectric development in the province and/or on international rivers. Neither the volk nor the Indian are accurate representations of reality, however. Rather, through the process Baudrillard describes, they have first denatured and masked that reality, only to reach a point where they represent the absence of the original reality (a point aided by the passage of time from the actual events, the documentaries and the original airings of the documentaries), and eventually became disconnected from any relationship to their original reality at all. In "othering" to the volk and the "Indian" other, they are assigned traits that make them tragic heroes: the volk are too out of touch with modernity, the "Indians" are too much a part of nature to truly comprehend it.[78] Why is this done? Why the tragic hero? This leads to perhaps the most disturbing reason for "othering" the people of the Kootenays, Columbia, and Peace. As Baudrillard puts it, one reason for creating a simulacrum is that "forgetting extermination is part of extermination, because it is also the extermination of memory, of history, of the social, etc. This forgetting is essential as the event, in any case unlocatable by us, inaccessible to us in its truth. This forgetting is still too dangerous, it must be effaced by an artificial memory."[79] In this sense, the artificial memory is the volk, the Tall Grass Indians, and each group's untouched environment, while the tragedy is hydroelectric development and its effects on the artificial memory, recreated in the documentaries, too late to prevent the tragedy, and yet in a medium that ends the "shameful legacy of the not-said," thereby alleviating societal guilt.[80] Perhaps this is the reason why these two documentaries were made.

Notes

1. Jean Baudrillard saw his philosophy as the logical outcome of Marshall McLuhan's as he states on page 82ff. Jean Baudrillard, *Simulacra and Simulation: The Body, in Theory: Histories of Cultural Materialism*, translated by Sheila Faria Glaser (Ann Arbor: University of Michigan Press, 1994); Laurence Gane and Piero, *Introducing Nietzsche* (Thriplow: Icon Books, 2005), 169.
2. In this sense, the word means to alter a thing, to change its original nature or properties. It does not inherently refer to removing and/or changing nature or the environment, although theoretically this is an outcome.
3. Baudrillard, *Simulacra*, 6.
4. Gane and Piero, *Introducing Nietzsche*, 140–41, 168–70; Friedrich Nietzsche, "How the World Finally Became a Fable," in *The Twilight of the Idols; Or How to Philosophise with a Hammer* (London: T. Fisher Unwin, 1899), 124–25.
5. Ian McKay, *The Quest of the Folk: Antimodernism and Cultural Selection in Twentieth-Century Nova Scotia* (Montreal and Kingston: McGill-Queen's University Press, 1994); H.V. Nelles, *The Art of Nation-Building: Pageantry and Spectacle at Quebec's Tercentenary* (Toronto: University of Toronto Press, 1999).
6. McKay, *The Quest of the Folk*, xi–xvii, passim; Nelles, *Art of Nation-Building*, 3–17, passim.
7. For those in doubt about this claim think about the evolution of motion pictures, from early silent films with caption interludes to modern movies with narrator voice-overs and dialogue.
8. *Tsay Keh Dene: CBC Hourglass Documentary* (CBC Television, 1970); hereafter referred to as *CBC Hourglass*.
9. *The Reckoning* (CBC Television, 1975).
10. The works cited in this historiography are not exhaustive, and I have chosen not to include general histories such as Martin Robin's two-volume *The Company Province*

or Jean Barman's *The West Beyond the West.* Jean Barman, *The West Beyond the West: A History of British Columbia,* 3rd ed. (Toronto: University of Toronto Press, 2007); Martin Robin, *The Company Province: 1871–1933, 1934–1972,* 2 vols. (Toronto: McClelland & Stewart, 1972–73).

11 Like all historical events, the genealogy of hydroelectric development preceded actual construction on the Peace and Columbia rivers. From this perspective, hydroelectric development can be said to have begun on the Peace in 1956 with Axel Wenner-Gren and the Columbia in 1943 with the United States and 1944 with the International Joint Commission. Furthermore, in the case of the W.A.C. Bennett Dam construction ended in 1967, but the reservoir did not begin to fill until the following spring (of 1968). Karl Froschauer, *White Gold: Hydroelectric Power in Canada* (Vancouver: UBC Press, 1999), 175, 179, 180, passim; Hugh Keenleyside, *On the Bridge of Time,* vol. 2, *Memoirs of Hugh L. Keenleyside* (Toronto: McClelland & Stewart, 1982), chapter 15; Tina Loo, "Disturbing the Peace: Change and the Scales of Justice on a Northern River," *Environmental History* 12, no. 4 (2007), 899–900, passim; Tina Loo, "People in the Way: Modernity, Environment and Society on the Arrow Lakes," *BC Studies,* no. 142/143 (2004), 163, passim; Patrick McGeer, *Politics in Paradise* (Toronto: Peter Martin, 1972), chapter 3; David Mitchell, *W.A.C. Bennett and the Rise of British Columbia* (Vancouver: Douglas & McIntyre, 1983), 285–89, passim; Jeremy Mouat, *The Business of Power: Hydro-Electricity in Southeastern British Columbia, 1897–1997* (Victoria: Sono Press, 1997), 139, passim; Earl Pollon and Shirlee Smith Matheson, *This Was Our Valley* (Calgary: Detselig, 2003), 153–59, passim; Paddy Sherman, *Bennett* (Toronto: McClelland & Stewart, 1966), 211–21, passim; Gordon Shrum, *Gordon Shrum: An Autobiography with Peter Stursberg* (Vancouver: UBC Press, 1986), chapters 8, 9, 11, passim; Neil Swainson, *Conflict over the Columbia: The Canadian Background to an Historic Treaty* (Montreal and Kingston: McGill-Queen's University Press, 1979), 19, passim; John Wedley, "Infrastructure and Resources: Governments and Their Promotion of Northern Development in British Columbia, 1945–1975," PhD dissertation, University of Western Ontario (1986), chapter 7, passim; Eileen Williston and Betty Keller, *Forests, Power and Policy: The Legacy of Ray Williston* (Prince George, BC: Caitlin Press, 1997), 170–78.

12 This list includes only those dams implemented during the W.A.C. Bennett administration, 1952–1972, or included in the Columbia River Treaty ratified by said administration. Arguably, succeeding dams could be considered the "spiritual" successors to this "two river policy" though. Froschauer, *White Gold,* 179, 181, passim; Loo, "People in the Way," 161–64; McGeer, *Politics in Paradise,* chapter 3; Mitchell, *W.A.C. Bennett,* 289; Pollon and Matheson, *This Was Our Valley,* 159; Sherman, *Bennett,* 211–21; Shrum, *Gordon Shrum,* 81–82; Swainson, *Conflict,* 84, 95, 118, passim; Stephen Tomblin, "W.A.C. Bennett and Province-Building in British Columbia," *BC Studies* 85 (1990), 51, passim; Williston and Keller, *Forests, Power and Policy,* chapter 7.

13 Froschauer, *White Gold,* chapter 7; Keenleyside, *On the Bridge of Time,* vol. 2, chapter 15; Loo, "Disturbing the Peace"; McGeer, *Politics in Paradise,* chapter 3; Loo, "People in the Way;" Bernard McKay, *Crooked River Rats: The Adventures of Pioneer Rivermen* (Surrey: Hancock House, 2000), 7, 171–75; Mitchell, *W.A.C. Bennett,* chapters 9 and 10, passim; Mouat, *Business of Power,* 170–71; Pollon and Matheson, *This Was Our Valley;* Sherman, *Bennett,* chapters 9 and 10, passim; Shrum, *Gordon Shrum,* chapters 9–11, passim; Swainson, *Conflict;* Tomblin, "W.A.C. Bennett"; James Waldram, *As Long as the Rivers Run: Hydroelectric Development and Native Communities in Western Canada* (Winnipeg: University of Manitoba Press, 1988), 15–16; Wedley, "Infrastructure and Resources," chapter 7, passim; Williston and Keller, *Forests, Power and Policy,* chapter 7, passim.

14 Froschauer, *White Gold,* chapter 7; Keenleyside, *On the Bridge of Time,* vol. 2, 530–37, passim; Loo, "Disturbing the Peace," 900–101; Loo, "People in the Way"; McGeer,

Politics in Paradise, chapter 3; Mitchell, *W.A.C. Bennett*, chapters 9 and 10, passim; Mouat, *Business of Power*, 147–48; Sherman, *Bennett*, chapters 9 and 10; passim; Shrum, *Gordon Shrum*, 93, passim; Swainson, *Conflict*; Tomblin, "W.A.C. Bennett," 53–55; Williston and Keller, *Forests*, 220–22, passim.

15 Loo, "Disturbing the Peace," 899–900; Loo, "People in the Way," 161–64; Mitchell, *W.A.C. Bennett*, 289, 297, passim; Pollon and Matheson, *This Was Our Valley*, 127–28, 159, passim; Sherman, *Bennett*, 221; Shrum, *Gordon Shrum*, 81–82; Tomblin, "W.A.C. Bennett", 51; Wedley, "Infrastructure," 300–301, 309, 451–52, 454; Williston and Keller, *Forests*, 170–71, 175–78.

16 Linda Johnston, "The Feasibility of Broadcasting Material Produced by Citizens' Groups on the CBC TV Network with Its Current Structure and Policy," MA thesis, Simon Fraser University (1974), 61; Lisa Szabo-Jones, email messages to Ken Puley, 10–13 Apr. 2012.

17 Daniel Sims, "Tse Keh Nay-European Relations and Ethnicity, 1793–2009," MA thesis, University of Alberta (2010).

18 *In the Words of Kwadacha* (BC Hydro, Television, 2002); *CBC Hourglass*.

19 *CBC Hourglass*.

20 Part of the problem is the website itself, as well as the fact that the DVD is apparently from an older series of Aboriginal Relations videos than the most recently ones posted in 2005, which themselves are apparently now outdated and have become orphaned on the site due to changes in its structure. My inquiries with the BC Hydro Aboriginal relations coordinator in March 2011 revealed the DVD is no longer available. BC Hydro, "BC Hydro: For Generations," www.bchydro.com; BC Hydro, "Video Series: BC Hydro: For Generations," http://www.bchydro.com/community/aboriginal_relations/initiatives/video_series.html#An Untold BC History; Johnston, "Feasibility of Broadcasting Material," 191, 195.

21 *CBC Hourglass*.

22 L. Szabo-Jones, email to Daniel Sims, 13 Apr. 2012. This email includes email exchanges between Szabo-Jones and CBC Archive librarian Ken Puley from 10 to 13 April 2012.

23 Szabo-Jones, ibid.; *CBC Hourglass*.

24 The BC *Hansard* cites an original airing of 16 Feb. 1975. Tina Loo cites the broadcast date as 1977, presumably another rebroadcast. I originally learned of the documentary during a reading class with Dr Jeremy Mouat. British Columbia, Legislative Assembly, *Official Report of Debates of the Legislative Assembly* (19 Feb. 1975), 11 (L.A. Williams, MLA); *The Reckoning*; Tina Loo, "People in the Way," 167.

25 Loo, "People in the Way," 167, 182–83.

26 Keenleyside, *On the Bridge of Time*, vol. 2, 497–500, 531–37; Shrum, *Gordon Shrum*, 90–91.

27 British Columbia.

28 *The Reckoning*.

29 Baudrillard, *Simulacra*, 31, passim.

30 *CBC Hourglass*.

31 This is particularly relevant because, as noted by local residents, the debris was not uniform on the reservoir. McKay, *Crooked River Rats*, 174; Pollon and Matheson, *This Was Our Valley*, 190, chapter 22.

32 *CBC Hourglass*.

33 McKay, *The Quest of the Folk*, xi–xv.

34 *The Reckoning*.

35 *CBC Hourglass*.

36 *The Reckoning*.

37 *CBC Hourglass*.

38 Sims, "Tse Keh Nay-European Relations," chapter 3; *CBC Hourglass*.

39 McKay, *Crooked River Rats*, 172–75.
40 Ibid., 172.
41 Ibid.
42 Ibid.
43 Chris Dummitt, *The Manly Modern: Masculinity in Postwar Canada* (Vancouver: UBC Press, 2007), chapter 1.
44 Paige Raibmon, *Authentic Indians: Episodes of Encounter from the Late-Nineteenth-Century Northwest Coast* (Durham: Duke University Press, 2005), 7–8, passim.
45 *CBC Hourglass*.
46 McKay, *The Quest of the Folk*, 9.
47 *The Reckoning*.
48 Ibid.
49 This is a reference to the name for the dwarf planet Pluto found in H.P. Lovecraft, "The Whisperer in the Dark," which was written from February to September 1930, right around the time Pluto was discovered, and is fitting for this chapter as Yuggoth truly is a simulacrum of the actual dwarf planet. H.P. Lovecraft, "The Whisperer in the Dark," in *The Call of Cthulhu and Other Weird Stories*, ed. S.T. Joshi (Toronto: Penguin, 1999), 215, 401, 406n30, 200–67.
50 This is a reference to Donald Creighton, *The Commercial Empire of the St Lawrence, 1760–1850* (Toronto: Ryerson Press, 1937).
51 The Province of Maryland, which became the State of Maryland was named after Queen Henrietta Maria, the spouse of Charles I.
52 *The Reckoning*.
53 Indeed, I have to confess that even though I know it is not Fruitvale, it reminds me of my step-grandparents' farm at Fruitvale.
54 Note the difference between how David Mitchell describes the extensive paving of roads in British Columbia, compared with John Wedley. Dummitt, *Manly Modern*, 10, passim; Loo, "People in the Way," 162–63; Mitchell, *W.A.C. Bennett*, 208–9, 260–63; Martin Robin, *Pillars of Profit*, vol. 2, *The Company Province, 1934–1972* (Toronto: McClelland & Stewart, 1973), 194–95, passim; Wedley, "Infrastructure," 152–53.
55 McKay, *The Quest of the Folk*, 12; Raibmon, *Authentic Indians*, 7.
56 *The Reckoning*.
57 Ibid.
58 *CBC Hourglass*.
59 Ibid.
60 They are my paternal grandmother proper and uncle. Ibid.
61 *The Reckoning*.
62 *CBC Hourglass*.
63 Dummitt, *Manly Modern*, chapter 1; *CBC Hourglass*.
64 McKay, *The Quest of the Folk*, chapter 1.
65 *The Reckoning*; *CBC Hourglass*.
66 Ibid.
67 Ibid.
68 Ibid.
69 Dummitt, *Manly Modern*, chapter 1.
70 My step-grandmother went to school with him in Nelson. BC Wildlife Federation, "In Memory of Mike Halleran," www.bcwf.net/oldsite/memorials/Mike_Halleran.html (accessed 22 Jan. 2011; site now discontinued).
71 Interestingly, Baudrillard notes how attempts to recreate the real are too real. I think this is evident with regards to naming for the Tse Keh Nay. Baudrillard, *Simulacra*, 107.

72 While it is true that Diamond Jenness does refer to the T'Lotona or Long Grass Indians in his 1937 work, he describes them as a Tse Keh Nay-Gitxsan group, distinct from the Tsay Keh Nay. Simon Fraser, "First Journal of Simon Fraser from April 12th to July 18th 1806," in *The Letters and Journals of Simon Fraser 1806–1808*, ed. W. Kaye Lamb (Toronto: Dundurn, 2007), 182–246; Diamond Jenness, *The Sekani Indians of British Columbia* (Ottawa: J.O. Patenaude I.S.O., 1937), 13–15.
73 Fraser, ""First Journal of Simon Fraser," 182–246, passim; Raibmon, *Authentic Indians*; Sims, "Tse Keh Nay-European Relations," chapter 1, passim.
74 Raibmon, *Authentic Indians*.
75 *CBC Hourglass*.
76 Raibmon, *Authentic Indians*.
77 McKay, *The Quest of the Folk*, chapter 1.
78 *The Reckoning*; *CBC Hourglass*.
79 Baudrillard, *Simulacra*, 49.
80 Ibid., 49.

AFTERWORD

Humming Along with the Bees
A Few Words on Cross-Pollination

Pamela Banting

"People, years later, blamed everything on the bees; it was the bees, they said, seducing Vera Lang, that started everything."[1] This is the opening sentence of the canonical Canadian novel *What the Crow Said*, by our literary genius and genius of place, Robert Kroetsch. One spring, a farm girl, a young woman, who drowses asleep, nude, on the warm spring prairie amid the "blue-purple petals," "silken stems," and "pollen-yellow tongues" of prairie crocuses, yellow buffalo beans, violets, buttercups, and pink shooting stars awakes to a homing swarm of bees covering her body, buzzing and nuzzling her nakedness. As one might expect, her terrifying ecstatic and orgasmic experience, which results in her becoming pregnant, reverberates far beyond the personal to the whole community. This luscious scene represents one of the most remarkable interspecies encounters and instances of cross-pollination in all of North American literature.

Years later, people may say, the "Cross-Pollinations Workshop: Seeding New Ground for Environmental Thought and Activism across the Arts and Humanities," held at the University of Alberta in March 2011, and the collection of essays, images, and poems that have been extracted from it, had similarly reverberant effects on the community of academics, writers, artists, and activists in western Canada. People may blame it on the bees. Universities have been incorporating the words "interdisciplinary," "multidisciplinary," and "cross-disciplinary" into their mandates, mission statements, public relations initiatives, websites, and official documents for a long time. Nevertheless, the majority of the scholarly work carried out and valorized in the academy remains resolutely disciplinary. It is far more common for a new topic, issue, or object to work its way into recognition within a discipline than for disciplinary boundaries themselves to be eroded, transgressed, or ruptured. Perhaps it has to do with the word "discipline" itself for, despite its positive associations with professional training, order, organization, and depth, the word "discipline" also bears the unpleasant associations of mischief, misbehaviour, crime, and punishment. "Discipline" conjures memories of being taught proper grammar, decorum, and table manners through repetitive instruction and censure. It reminds us of getting caught skipping in the classroom and having to stay after school to write ad nauseum, "I will not skip in the classroom." Or even worse forms of corporal punishment than writing. "Discipline" evokes Foucauldian prisons, panopticons, and techniques of surveillance. In his article "Discipline and Discipleship," John Frow points to both the positive and negative senses of academic discipline:

> a discipline is not a body of knowledge alone, but is as well an organization of practices, a mode of institutional control, and a principle of limitation operating on discourse. This limitation, it must be stressed, is not the repression of a spontaneously developing knowledge but is precisely productive of knowledge; energy is an effect of structure, not its opposite.... Thus we could say that disciplines not only operate with a specific domain of objects, methods, techniques, and protocols for the recognition of true propositions, but they also act as mechanisms for the generation of new propositions within a particular conceptual and technical framework. These propositions may be either true or false; what defines them as valid propositions is that they are constructed in accordance with the rules for the formation of disciplinary objects and concepts. *But disciplines function as much to reproduce an existing structure of knowledge as to produce new knowledge.* For most disciplines, the complex interrelation between these two functions is bounded by the educational apparatus and established in the process of transmission of knowledge to "disciples" and in their accreditation as legitimate knowing subjects.[2]

Frow discusses the productive transmission of knowledge and the reproduction of disciplinary subjects in the training of psychoanalysts, religious disciples, and graduate students in English literature, and he examines the entanglements among desire, the desire-to-know, feelings of loneliness and self-alienation, and the role of transference love inherent in these processes of production and reproduction. Perhaps it is not surprising then that, at a subconscious level, many of those who have been disciplined into a given field of study and undergone the trials of the subject-in-process feel that inter-, multi-, and cross-disciplinary work is tantamount to skipping in the classroom.

The title of the workshop out of which flowed the essays, meditations, artwork, prose poems, and poetry included here, however, "Cross-Pollination," seems to me to be an evocative and fertile alternative metaphor for work that moves simultaneously within and across disciplinary and, even more problematically, university departmental borders to create new knowledge, new forms of knowledge, and new objects, practices, and methodologies. The term "cross-pollination," which refers to "the transfer of pollen from the anthers of one flower to the stigma of another flower by the action of wind, insects, etc.,"[3] is not entirely unlike what transpired between Vera Lang and the bees and has the potential to take us outside disciplinary silos, even on occasion outside the ivory tower and into collaboration with artists, writers, community activists, local communities, and even other species. Cross-pollination plus collaboration leads to community just as Vera Lang's ecstasy flowed outward beyond the personal into the community beyond, and affected even the weather.

Vera's announcement to her mother of her pregnancy coincides with or instigates a snowfall in late spring. So, too, our actions (and inactions) are altering the climate of the earth. Given that virtually all scientists worldwide not in the employ of harmful industries agree that climate change has begun, and that we are at the tipping point after which ongoing climate disruption will be irreversible, this workshop on environmental thought and activism was timely. With such a dire prospect hanging over our heads and those of the generations to follow, the separation of the disciplines is no longer an option.

Moreover, we cannot afford to relegate the future to economists, corporations, and politicians. Artists and academics in arts faculties have an important role to play in offering alternative visions of ecologically and socially sustainable lifeways. As the contents of this anthology demonstrate, work in the arts and culture can offer increased clarity of thought and expression of ideas, the incisive and necessary critique by which we can discern both how to move forward and how to learn to dwell in the present and in place, an integrated and more holistic vision of problems and solutions, broader and deeper images of the "big picture," and inspiration and fellow feeling, thus creating the grounds on which true and viable communities—not just groupings based on shared demographics—can coalesce.

Knowledge, which is multifarious and flourishes everywhere, far too often is corralled and categorized in terms of boundaries, borders, and property, as the intellectual capital of a given group, social class, or profession, for instance. After we demonstrate a function, a skill, or a new way of thinking, we may ask our interlocutor, "Did you grasp that?" Although "grasping" knowledge may suggest proprietorial control, alternatively it may suggest something transferred from hand to hand: an object, a tool, a definition, a concept. Or a limb of a tree. It is not impossible that our sense of knowledge goes back all the way to the time when our own species lived in trees, and "knowledge" may have been a firm clasp of the next sturdy limb or branch as we made our rapid transit through the forest canopy. Maybe knowledge is the ability to avoid a painful or deadly fall, be it through the branches or into environmental catastrophe. Just as for the *bricoleur*, the handyman or handywoman, knowledge itself is not segmented into disciplinary territories or, if it is, the bricoleur is the figure with the courage to blur and transgress those boundaries, so, too, we in the academy need to supplement the notions of mastery and expertise with the attributes and practices of adaptability, aesthetics, creativity, design, equality, ethics, flow, function, and process.

In the essays, propositions, meditations, homages, inquiries, poetry, poetic prose, recollections, probes, and narrative scholarship collected here, the figure of the bricoleur/bricoleuse recurs in various contexts and matrices. He appears, for instance, in the guise of Warren Cariou's uncle Eli, recently retired manager of the Ituna Landfill in central Saskatchewan, who has a lot in common with Edward Burtynsky and the Cree trickster Wisakaychak, both of them aficionados of the nuisance grounds and possessors of knowledge about consumption, excess, surplus, waste, and recycling. Likewise in Christine Stewart's "Propositions from Under Mill Creek Bridge," we encounter some of the denizens of the "underbridge," the men who live in Edmonton's ravine pathway system amid various material discards and remainders. They make do as best they can: some of this detritus becomes arrangement, comfort, bundle, or expressive scatter. Through her walks in the ravine, Stewart experiments with how to attend theriomorphically, phytomorphically, and geomorphically to place.

Stewart's meditation on horizontal and spatial reading practices under the "underbridge" chimes with Rita Wong's psychogeographical essay on how to make sense of the "understory" of our world—particularly the stories of water, which underwrites our very existence and yet is so often taken for granted, neglected, or deliberately contaminated. She reminds us of the streams and creeks that have been paved over in the urbanization of Vancouver and ponders what a renewed, embodied, and active sense of "respect" might feel like. In Wong, poet of the politics of H_2O, we encounter the figure of the bricoleuse, the fixer-upper whose mission is to recover, restore, re-story, and reuse.

Embodiment and handwork also come through in Trevor Herriot's essay, in which he implores readers to "walk the talk" by becoming stakeholders and speculators "in the distant future by choosing the right way to act today." He wants us not just to invest our money in other money or pretend-money such as compound derivatives, but in "grass futures." Herriot suggests that we learn not only to hold onto hope in our minds but to *embody* it by planting trees in deforested lands and restoring natural grass prairie and to "trust that, with the right care and attention, it will come to good." Along a similar vein, Nancy Holmes shows how the Woodhaven Eco Art Project in Kelowna, BC, an inter-arts, community-based, activist collaboration, transformed artists and other local citizens into stewards of the Woodhaven Nature Conservancy. She writes that eco-art is less specular and more participatory than traditional art practices: "My experience over the many months of creating in Woodhaven is that the artist or knower needs to be 'booted up' by physical contact with the place, through feet and hands and eyes and ears and scents.... In Woodhaven, many of the artists felt they were literally creating the work for the place itself to absorb and to enjoy. We have numerous photographs of animals watching us." The hands-on, multi-, and cross-disciplinary projects of Woodhaven felt like co-creation with place and even seemed at times to offer the tantalizing prospect of cross-species collaboration. For many animals, to acquire information, knowledge, and understanding is to make "scents" of things, to read the air, to watch and pay attention, to practice a sensitivity to one's surroundings that we human animals need to relearn.

Collaboration with the natural world can be seen in the work of visual artist Lyndal Osborne, whose work represented, to my senses, the most enticing example of bricolage of the entire workshop. Claude Lévi-Strauss's description of the bricoleur as the one who collects and retains objects on the principle that someday they may come in handy finds perfect expression in Osborne's astonishing and joyous projects. A list of the components of just one of her mixed-media installations reads like poetry and illustrates the surprising extent of her activity of collection. Her work *ab ovo*, for example, consists of

> glass, foam, wood, lights, clove-studded oranges, lime grass, gourds, lemons, sponge cakes, bananas, poinciana pods, DNA models, chestnuts, Port Jackson shark eggs, used grape skins, avocado skins, sunflowers, grapefruit skins, coconuts, fungus, ginseng, corks, corncobs, cycad seeds, eucalyptus seeds, seafoam, Tasmanian gumnuts, crab shells, bones, kelp, redwood pine cones, durian, beets, persimmon, moth flower, and laboratory glove compartment.

Like many other participants in the workshop, Osborne is a walker whose peripatetic wanderings are part of her research. Like Cariou's uncle, Osborne attempts to "interrupt the cycle of consumption and waste" through recycling and reinterpreting found and saved materials. Like Herriot, she finds that "discovering the energy that binds global issues to local interpretation is found in making with my hands" and in the work's labour-intensiveness. Like Stewart, Osborne wants to slow down the process of interpretation.

One of the most important acts of interdisciplinary derring-do in this text can be found in Jon Gordon's essay. One of the earliest researchers in the newly emergent field of petrocultures, Gordon wades into the murky and sometimes treacherous zones of bitumen extraction, tailings ponds, oil spills, toxicity, and geopolitics to raise this central ethical question that remains unsatisfactorily addressed in Canada and North America today: "What Should We Sacrifice for Bitumen?" Trafficking in poetry, non-fiction, cultural studies, the rhetoric of oil and gas executives, company reports, oil industry economics, and ethics, Gordon's project illustrates the rich potential of ranging across disciplines and genres. Literature, he writes, paraphrasing Robert Kroetsch, requires us to dwell in the present and has the potential to disrupt oil capital's (e)utopian imaginary.

Taking chances, getting outside, paying attention, walking the talk, grounding knowledge, thinking with other-than-human animals, creating wonder, recycling, making do, staying put, setting down roots, working with one's hands, even getting one's hands dirty, foraging for and forging new knowledges, collaborating, learning true respect, and inspiring community: these are just some of the new modes of observation, reading, data collection, thinking, and representational practices that open up in tandem with disciplinary border crossings. To return one last time to the image of Vera asleep on the blooming prairie, the scene in literature in which someone falls asleep outside is a signature of ease and integration with the natural world. Unlike Snow White, who is awakened by the kiss of a handsome prince and who, the story suggests, is rewarded for her virtue with the expectation of social advancement, prairie girl Vera Lang is awakened by the kisses of hundreds of industrious and amorous bees. In this age of extinctions, being receptive to the bees seems to me to offer one of the best opportunities for planetary survival, and cross-pollination is a superb model for working and collaborating across disciplines.

Notes

1 Robert Kroetsch, *What the Crow Said* (Don Mills, ON: General Publishing, 1978), 7.
2 John Frow, "Discipline and Discipleship," *Textual Practice* 2, no. 3 (1988), 307 (emphasis added).
3 "Cross-pollination," Online Free Dictionary, http://www.thefreedictionary.com/.

BIBLIOGRAPHY

Abram, David. *The Spell of the Sensuous: Perception and Language in a More-Than-Human World*. New York: Vintage, 1997.
Agamben, Giorgio. *The Open: Man and Animal*. Stanford, CA: Stanford University Press, 2004.
——. *Remnants of Auschwitz: The Witness and the Archive*. New York: Zone, 2002.
Aitken, Sally N., Sam Yeaman, Jason A. Holliday, Tongli Wang, and Sierra Curtis-McLane. "Adaptation, Migration or Extirpation: Climate Change Outcomes for Tree Populations." *Evolutionary Applications* 1, no. 1 (2008): 95–111.
Alaimo, Stacy. *Bodily Natures: Science, Environment, and the Material Self*. Bloomington: Indiana University Press, 2010.
Alberta Federation of Labour. "Poster Project." *Project 2012*. http://www.project2012.ca/default.asp?mode=posters&id=11.
Alderman, Nigel, and C.D. Blanton, eds. "Pocket Epics: British Poetry after Modernism." Special issue. *Yale Journal of Criticism: Interpretation in the Humanities* 13, no. 1 (2000).
Alighieri, Dante. *The Divine Comedy*. Digital Dante, 1997. http://dante.ilt.columbia.edu/comedy/.
Allison, Leanne. *Being Caribou/Vivre comme les caribous*. DVD. Montreal: National Film Board of Canada, 2004.
Anderson, Benedict. *Imagined Communities: Reflections on the Origin and Spread of Nationalism*. New York: Verso, 2000 [1991].
Andrews, Megan. "On Collaboration." In *Across Oceans: Writings on Collaboration*, ed. Maxine Heppner, 35–50. Toronto: Across Oceans, 2008.
Apetagon, Byron, ed. *Norway House Anthology: Stories of the Elders*, vol. 2. Winnipeg: Frontier School Division #48, 1992.
Armstrong, Jeannette. "Place, Story, and Aboriginal Identity." Plenary address, Association for the Study of Literature and the Environment eighth biennial conference, Victoria, British Columbia, 3 June 2009.
Assisted Migration Adaptation Trial (AMAT). June 2011. http://www.for.gov.bc.ca/hre/forgen/interior/AMAT.htm.
Auden, W.H. *Prose*, vol. 3, *1949–1955*. Edited by Edward Mendelson. Princeton, NJ: Princeton University Press, 2008.
——. *Collected Poems*. Edited by Edward Mendelson. New York: Vintage, 1991.
——. *Forewords and Afterwords*. Selected by Edward Mendelson. New York: Random House, 1973.

———. "Freedom and Necessity in Poetry." In *The Place of Value in a World of Facts: Proceedings of the Fourteenth Nobel Symposium, Stockholm, September 15–20, 1969*, ed. Arne Tiselius and Sam Nilsson, 135–42. New York: Wiley, 1970.

Auld, Jerry. *Hooker and Brown: A Novel*. Victoria: Brindle and Glass, 2009.

Babiuk, Colin. *Oil Sands and the Earth: Framing the Environmental Message in the Print News Media*. Charleston, NC: VDM Verlag Dr Muller, 2008.

Baker, Nadia, Patrick Lilley, Toshiko Sasaki, and Heather Williamson. "Investigation of Options for the Restoration of Camosun Bog, Pacific Spirit Regional Park." Environmental Studies 400 thesis, University of British Columbia, 2000. http://www.ensc.ubc.ca/about/pdfs/theses/baker_et_al.pdf.

Barman, Jean. *The West Beyond the West: A History of British Columbia*. 3rd ed. Toronto: University of Toronto Press, 2007.

Basheer, Chanbasha, Hian Kee Lee, and Koh Siang Tan. "Endocrine Disrupting Alkylphenols and Bisphenol A in Coastal Waters and Supermarket Seafood from Singapore." *Baseline/Marine Pollution Bulletin* 48 (2004): 1145–67.

Bass, Rick. *Caribou Rising*. Seattle: Mountaineers, 2004.

Bataille, Georges. *The Accursed Share*, vols. 2 and 3, *The History of Eroticism and Sovereignty*. Translated by Robert Hurley. New York: Zone, 1993.

———. *The Accursed Share*, vol. 1, *Consumption*. Translated by Robert Hurley. New York: Zone, 1991.

Baudrillard, Jean. *Simulacra and Simulation: The Body, in Theory: Histories of Cultural Materialism*. Translated by Sheila Faria Glaser. Ann Arbor: University of Michigan Press, 1994.

BC Hydro. "BC Hydro: For Generations." www.bchydro.com.

———. "Video Series: BC Hydro: For Generations." www.bchydro.com/community/aboriginal_relations/initiatives/video_series.html#An Untold BC History.

BC Wildlife Federation. "In Memory of Mike Halleran." www.bcwf.net/oldsite/memorials/Mike_Halleran.html (accessed 22 Jan. 2011; site now discontinued).

Beers, Don. *Jasper-Robson: A Taste of Heaven, Scenes, Tales, Trails*. Calgary: Highline, 1996.

Bell, Vikki. "Performativity and Belonging: An Introduction." In *Performativity and Belonging*, ed. Vikki Bell, 1–10. Thousand Oaks, CA: Sage, 1999.

Bentley, D.M.R. "Colonial Colonizing: An Introductory Survey of the Canadian Long Poem." In *Bolder Flights: Essays on the Canadian Long Poem*, ed. Frank M. Tierney and Angela Robbeson, 7–29. Ottawa: University of Ottawa Press, 1998.

Berger, John. "The Red Butterfly." *Brick: A Literary Journal* 71 (2003): 10–17.

Bernstein, Michael André. "Making Modernist Masterpieces." *Modernism/Modernity* 5, no. 3 (1998): 1–17.

Biespiel, David, ed. *Long Journey: Contemporary Northwest Poets*. Corvallis: Oregon State University Press, 2006.

Binnema, Theodore (Ted), and Melanie Niemi. "'Let the Line Be Drawn Now': Wilderness, Conservation, and the Exclusion of Aboriginal People from Banff National Park in Canada." *Environmental History* 11, no. 4 (2006): 724–50.

Bird, Louis. *The Spirit Lives in the Mind: Omushkego Stories, Lives, and Dreams*. Edited by Susan Elaine Gray. Montreal and Kingston: McGill-Queen's University Press, 2007.

———. *Telling Our Stories: Omushkego Legends and Histories from Hudson Bay*. Toronto: University of Toronto Press, 2005.

Blair, Danny. "The Climate of Manitoba." In *The Geography of Manitoba: Its Land and Its People*, ed. John Welsted, John Everitt, and Christoph Stadel, 31–42. Winnipeg: University of Manitoba Press, 1997.

Blanchot, Maurice. *The Unavowable Community.* Translated by Pierre Joris. New York: Station Hill Press, 1988.
Blaser, Robin. *The Holy Forest: Collected Poems of Robin Blaser.* Edited by Miriam Nichols. Berkeley: University of California Press, 2006.
Bocking, Stephen. "The Nature of Cities: Perspectives in Canadian Urban Environmental History." *Urban History Review* 34, no. 1 (2005): 3–8.
Bogue, Ronald. "The Landscape of Sensation." In *Gilles Deleuze: Image and Text,* ed. Eugene W. Holland, Daniel W. Smith, and Charles J. Stivale, 9–26. London: Continuum, 2009.
Bonta, Mark, and John Protevi. *Deleuze and Geophilosophy.* Edinburgh: Edinburgh University Press, 2004.
Botkin, Daniel. *Discordant Harmonies: A New Ecology for the Twenty-First Century.* Oxford: Oxford University Press, 1990.
Bott, Robert. *Canada's Oil Sands.* 2nd ed. Edited by David M. Carson and Tami Hutchinson. Calgary: Canadian Centre for Energy Information, 2007. Also available online http://www.centreforenergy.com.
Bracke, Astrid. "Redrawing the Boundaries of Ecocritical Practice." *ISLE* 17, no. 4 (2010): 765–68.
Bradford, Tolly. "A Useful Institution: William Twin, 'Indianness,' and Banff National Park, c. 1860–1940." *Native Studies Review* 16, no. 2 (2005): 77–98.
Brennan, Andrew. *Thinking about Nature.* Athens: University of Georgia Press, 1988.
Bringhurst, Robert. *Everywhere Being Is Dancing: Twenty Pieces of Thinking.* Kentville, NS: Gaspereau, 2007.
British Columbia. Legislative Assembly. *Official Report of Debates of the Legislative Assembly,* 19 Feb. 1975.
Brook, Isis. "Can Merleau-Ponty's Notion of Flesh Inform or Even Transform Environmental Thinking?" *Environmental Values* 14, no. 3 (2005): 353–62.
Brooymans, Hanneke. "Oilsands Boosts Toxic Metals in Athabasca Watershed: Study." *Edmonton Journal,* 31 Aug. 2010, A3.
———. "Fort Chipewyan Wants Answers about Cancer Rates." *Edmonton Journal,* 5 Sept. 2010, A1.
Brown, Laurence. Interview by Lisa Szabo-Jones and David Brownstein. Transcribed by Sandi Kingston. Camosun Bog, Vancouver, BC, 2 Sept. 2011.
Browning, Elizabeth Barrett. *The Works of Elizabeth Barrett Browning.* Edited by Sandra Donaldson. Vol. 2, *Poems 4th Edn (1856), Continued,* ed. Marjorie Stone and Beverley Taylor. London: Pickering and Chatto, 2010.
Brownstein, David. "Sunday Walks and Seed Traps: The Many Natural Histories of British Columbia Forest Conservation, 1890–1925." PhD dissertation, Institute for Resources, Environment and Sustainability, University of British Columbia, 2006.
Bruneau, Calvin. Presentation, University of Alberta, Dr Keavy Martin's class, 11 Mar. 2011.
Buchanan, Brett. *Onto-Ethologies: The Animal Environments of Uexküll, Heidegger, Merleau-Ponty, and Deleuze.* New York: SUNY Press, 2008.
Buell, Lawrence. "Foreword." In *Environmental Criticism for the Twenty-First Century,* ed. Stephanie LeMenager, Teresa Shewry, and Ken Hiltner, xiii–xvii. New York: Routledge, 2011.
———. *The Future of Environmental Criticism: Environmental Crisis and Literary Imagination.* Malden, MA: Blackwell, 2005.
———. *Writing for an Endangered World: Literature, Culture, and Environment in the U.S. and Beyond.* Cambridge, MA: Belknap Press of Harvard University Press, 2001.

———. *The Environmental Imagination: Thoreau, Nature Writing, and the Formation of American Culture*. Cambridge, MA: Belknap Press of Harvard University Press, 1995.
Burke, Edmund. *A Philosophical Enquiry into the Origin of Our Ideas of the Sublime and Beautiful*. Oxford: Oxford University Press, 1990.
Burroughs, John. "Real and Sham Natural History." *Atlantic Monthly*, Feb. 1903, 298–309.
Cairney, Richard. "Confronted." *SEE Magazine*, Feb. 1999.
Calder, Alison. "What Happened to Regionalism?" *Canadian Literature* 204 (2010): 113–14.
———. "Why Shoot the Gopher? Reading the Politics of a Prairie Icon." *American Review of Canadian Studies* 33, no. 3 (2003): 391–414.
Calvo-Merino, Beatriz, and Patrick Haggard. "Neuroaesthetics of Performing Arts." In *Art and the Senses*, ed. Francesca Bacci and David Melcher, 529–41. Oxford: Oxford University Press, 2011.
Campbell, Elizabeth M., S.C. Saunders, K.D. Coates, D.V. Meidinger, A. MacKinnon, G.A. O'Neill, D.J. MacKillop, S.C. DeLong, and D.G. Morgan. *Ecological Resilience and Complexity: A Theoretical Framework for Understanding and Managing British Columbia's Forest Ecosystems in a Changing Climate*. Victoria: BC Ministry of Forests and Range, 2009. http://www.for.gov.bc.ca/hfd/pubs/Docs/Tr/Tr055.htm.
Campbell, Maria, et al. *Give Back: First Nations Perspectives on Cultural Practice*. North Vancouver: Gallerie, 1992.
Canadian Climate Normals—Temperature and Precipitation, 1951–1980. Ottawa: Environment Canada, 1981.
Canadian Live-Stock and Farm Journal 5, no. 3 (1881): 85.
Canadian Stock Raisers' Journal 1, no. 6 (1884): 110.
Carlyle, W.J. "The Management of Environmental Problems on the Manitoba Escarpment." *Canadian Geographer* 24, no. 3 (1980): 255–69.
Carter, Sarah, Alvin Finkel, and Peter Fortna, ed. *The West and Beyond: New Perspectives on an Imagined "Region."* Edmonton: University of Alberta Press, 2010.
Carruthers, Beth. "Call and Response: Deep Aesthetics and the Heart of the World." Paper presented at Aesth/Ethics in Environmental Change workshop, Biological Station of Hiddensee, University of Greifswald, Germany, May 24–28, 2010.
———. "Through the Eye of the Heart: In Search of a Deep Aesthetics." Paper presented at Thinking Through Nature: Philosophy for an Endangered World summit, University of Oregon, Eugene, OR, June 19–22, 2008.
———. "PRAXIS: Acting as if Everything Matters." MA thesis, Lancaster University, 2006.
———. "Mapping the Terrain of Contemporary Ecoart Practice and Collaboration." Report commissioned by the Canadian Commission for UNESCO. Presented at "Art in Ecology: A Think Tank of Arts and Sustainability," Vancouver, BC, 27 Apr. 2006.
———. "Hybrid Forms." Video essay presented at "Between Nature: Explorations in Ecology and Performance," Lancaster, UK, 27–30 July 2000.
Cautley, R.W., and A.O. Wheeler. *Report of the Commission Appointed to Delimit the Boundary between the Provinces of British Columbia and Alberta*, part II, *1917 to 1921, From Kickinghorse Pass to Yellowhead Pass*. Ottawa: Office of the Surveyor General, 1924.
Cembalest, Robin. "Turning Up the Heat." *ARTnews* 107, no. 6 (2008): 102.

Census of the Canadas 1851–52.
Census of the Canadas 1860–61.
Census of Canada 1870–71.
Chandler, Katherine R., and Melissa Goldthwaite, eds. *Surveying the Literary Landscape of Terry Tempest Williams: New Critical Essays.* Salt Lake City: University of Utah Press, 2003.
Chisholm, Dianne. "The Art of Ecological Thinking: Literary Ecology." *ISLE* 18, no. 3 (2011): 569–93.
Christian, Carol. "Killed Workers Remembered." *Fort McMurray Today.* http://www.fortmcmurraytoday.com/ArticleDisplay.aspx?archive=true&e=1546066.
Clark, Karl. Papers. University of Alberta Archives, Edmonton, Alberta.
Clark Sheppard, Mary. *Oil Sands Scientist: The Letters of Karl A. Clark 1920–1949.* Edmonton: University of Alberta Press, 1990.
Coleman, A.P. *The Canadian Rockies, New and Old Trails.* Toronto: Henry Frowde, 1911.
———. "Mount Brown and the Sources of the Athabasca." *Geographical Journal* 5, no. 1 (1895): 53–61.
———. "Cree Words" (Morley, June 25, 1892) in *Notebook 12* (1892), 23–25, http://library2.vicu.utoronto.ca/apcoleman/rockies/first_nations.htm.
Collie, J. Norman. "The Canadian Rocky Mountains a Quarter Century Ago." *Canadian Alpine Journal* 14 (1924): 82–89.
———. "Exploration in the Canadian Rockies: A Search for Mount Hooker and Mount Brown." *Geographical Journal* 13, no. 4 (1899): 337–56.
Compton, Wayde. *Blueprint: Black British Columbian Literature and Orature.* Vancouver: Arsenal Pulp Press, 2003.
Con Create USL LP. "Whyte Avenue (Over Mill Creek) Bridge Rehabilitation Contract No. 0834 Edmonton, AB." 2008, http://www.uslltd.com/251.t1/images/usl.portfolio/Portfolio_48print.pdf.
Cook, Eleanor. *Against Coercion: Games Poets Play.* Stanford, CA: Stanford University Press, 1998.
Crate, Susan A., and Mark Nuttall, eds. *Anthropology and Climate Change: From Encounters to Actions.* Walnut Creek, CA: Left Coast Press, 2009.
Crawford, Robert, ed. *Contemporary Poetry and Contemporary Science.* Oxford: Oxford University Press, 2006.
Creighton, Donald. *The Commercial Empire of the St Lawrence, 1760–1850.* Toronto: Ryerson Press, 1937.
Cruikshank, Julie. "Glaciers and Climate Change: Perspectives from Oral Tradition." *Arctic* 54, no. 4 (2001): 377–93.
Davidson, Tonya K., Ondine Park, and Rob Shields, eds. *Ecologies of Affect: Placing Nostalgia, Desire, and Hope.* Waterloo, ON: Wilfrid Laurier University Press, 2011.
Davis, Paul. *Museums and the Natural Environment: The Role of Natural History Museums in Biological Conservation.* London: Leicester University Press, 1996.
Dawson, Carrie. "How Does Our Garden Grow?" *Canadian Literature* 204 (2010): 110–13.
Deleuze, Gilles. *Essays Critical and Clinical.* Translated by Daniel W. Smith and Michael A. Greco. Minneapolis: University of Minnesota Press, 1997.
———. *Spinoza: Practical Philosophy.* Translated by Robert Hurley. San Francisco: City Lights, 1988.
Deleuze, Gilles, and Félix Guattari. *What Is Philosophy?* Translated by Hugh Tomlinson and Graham Burchell. Minneapolis: University of Minnesota Press, 1994.

———. *A Thousand Plateaus: Capitalism and Schizophrenia*. Translated and foreword by Brian Massumi. Minneapolis: University of Minnesota Press, 1987.

Derrida, Jacques. "The Animal That Therefore I Am (More to Follow)." Translated by David Wills. *Critical Inquiry* 28, no. 2 (2002): 369–418.

———. *The Gift of Death*. Translated by David Wills. Chicago: University Chicago Press, 1995.

Dillingham, William B. *Rudyard Kipling: Hell and Heroism*. Gordonsville, VA: Palgrave Macmillan, 2005.

Donald, Dwayne. "Edmonton Pentimento: Re-reading History in the Case of the Papaschase Cree." *Journal of the Canadian Association for Curriculum Studies* 2, no. 1 (2004): 21–54. http://pi.library.yorku.ca/ojs/index.php/jcacs/article/viewFile/16868/15674.

Douglas, David. *Journal kept by David Douglas during his travels in North America, 1823–1827*. New York: Antiquarian Press, 1959 [1914].

Drainage Committee Report. GR 174, G 8373, file 13. Archives of Manitoba, Winnipeg, Manitoba.

Drummond, Thomas. "Sketch of a Journey in the Rocky Mountains and to the Columbia River in North America." In *Botanical Miscellany*, ed. William Hooker. London: John Murray, 1830.

Duke University. "Eastern U.S. Forests Not Keeping Pace with Climate Change, Large Study Finds." *ScienceDaily*, 31 Oct. 2011. http://www.sciencedaily.com/releases/2011/10/111031154132.htm.

Dummitt, Chris. *The Manly Modern: Masculinity in Postwar Canada*. Vancouver: UBC Press, 2007.

Duthie, Newby, Weber, and Associates. *Increased Load Limit Program B082 Mill Creek Bridge*. Edmonton: Author, 1987.

Dutton, Denis. "A Darwinian Theory of Beauty." TED talk given at TED 2010. http://www.ted.com/talks/denis_dutton_a_darwinian_theory_of_beauty.html.

Dvorak, Marta, and W.H. New. "Introduction: Troping the Territory." In *Tropes and Territories: Short Fiction, Postcolonial Readings, Canadian Writing in Context*, ed. Marta Dvorak and W.H. New, 3–13. Montreal and Kingston: McGill-Queen's University Press, 2003.

Edmonton (Alberta). Parks and Recreation Dept. *Mill Creek Ravine. Appendices: Prepared for Edmonton Parks and Recreation*. Edmonton: Butler Krebes, 1974.

"Edmonton Closes Tent City for the Homeless." *CBC News*, 15 Sept. 2007. http://www.cbc.ca/canada/edmonton/story/2007/09/15/tent-city.html.

Elliot, Robert. *Faking Nature: The Ethics of Environmental Restoration*. London: Routledge, 1997.

Ells, S.C. *Recollections of the Development of the Athabasca Oil Sands*. Ottawa: Department of Mines and Technical Surveys, 1962.

Emig, Rainer. "Auden and Ecology." In *The Cambridge Companion to W.H. Auden*, ed. Stan Smith, 212–25. Cambridge: Cambridge University Press, 2004.

Epp, Roger, and Dave Whitson, eds. *Writing Off the Rural West: Globalization, Governments, and the Transformation of Rural Community*. Edmonton: University of Alberta Press, 2001.

Essig, Laurie. "Your Inner 'Hysteric' May Just Be Right." *Chronicle of Higher Education*, 17 Mar. 2011. http://chronicle.com/blogs/brainstorm/interpreting-nuclear-disaster/33297.

Evanoff, Richard. *Bioregionalism and Global Ethics: A Transactional Approach to Achieving Ecological Sustainability, Social Justice, and Human Well-Being*. New York: Routledge, 2011.

Fairley, Bruce, ed. *The Canadian Mountaineering Anthology: 100 Years of Stories from the Edge*. Edmonton: Lone Pine, 1994.

Felstiner, John. *Can Poetry Save the Earth? A Field Guide to Nature Poems*. New Haven, CT: Yale University Press, 2009.

Fiege, Mark. *Irrigated Eden: The Making of an Agricultural Landscape in the American West*. Seattle: University of Washington Press, 1999.

Findlay, Len. "Is Canada a Postcolonial Country?" In *Is Canada Postcolonial? Unsettling Canadian Literature*, ed. Laura F.E. Moss, 297–99. Waterloo, ON: Wilfrid Laurier University Press, 2003.

Forests Forever Ecosystems Initiative (FFEI). N.d. http://www.for.gov.bc.ca/HFP/future_forests/.

Forest Practices Association of Canada (FPAC). "Carbon Neutral Pledge." *FPAC.ca: The Voice of Canada's Wood, Pulp and Paper Producers*. n.d. http://www.fpac.ca/index.php/en/carbon-neutral-pledge.

"Forum." *PMLA* 111, no. 2 (1996): 271–311.

Francis, Daniel. *National Dreams: Myth, Memory, and Canadian History*. Vancouver: Arsenal Pulp Press, 1997.

"Frank Oliver and the Michel Band, Excerpt from the Report of the Royal Commission on Aboriginal Peoples (RCAP)." *Michel First Nation*. http://www.michelfirstnation.net/frank-oliver--the-michel-band.html.

Fraser, Simon. "First Journal of Simon Fraser from April 12th to July 18th 1806." In *The Letters and Journals of Simon Fraser 1806–1808*, ed. W. Kaye Lamb. Toronto: Dundurn, 2007.

Friesen, Gerald. "The Evolving Meanings of Region in Canada." *Canadian Historical Review* 82, no. 3 (2001): 530–45.

Froschauer, Karl. *White Gold: Hydroelectric Power in Canada*. Vancouver: UBC Press, 1999.

Frow, John. "Discipline and Discipleship." *Textual Practice* 2, no. 3 (1988): 307–23.

Frye, Northrop. "Haunted by Lack of Ghosts: Some Patterns in the Imagery of Canadian Poetry." In *The Canadian Imagination: Dimensions of a Literary Culture*, ed. David Staines, 22–45. Cambridge, MA: Harvard University Press, 1977.

———. "Conclusion." In *The Literary History of Canada: Canadian Literature in English*, ed. Carl F. Klinck, 333–61. 2nd ed. Vol. 2. Toronto: University of Toronto Press, 1976.

Gablik, Suzi. "Alternative Aesthetics." *Landviews: Online Journal of LAND: Landscape, Art and Design* (2003). http://www.landviews.org/la2003/alternative-sg.html.

Gadd, Ben. *Handbook of the Canadian Rockies*. Jasper: Corax, 1992 [1986].

Gane, Laurence, and Piero. *Introducing Nietzsche: A Graphic Guide*. Thriplow: Icon Books, 2005.

Garrard, Greg. "Literary Theory 101." *ISLE* 17, no. 4 (2010): 780–83.

———. *Ecocriticism*. Abingdon, UK: Routledge, 2004.

Gayton, Don. *Impacts of Climate Change on British Columbia's Biodiversity: A Literature Review*. Kamloops, BC: Forrex, 2008.

Glissant, Édouard. *Poetics of Relation*. Translated by Betsy Wing. Ann Arbor: University of Michigan Press, 1997.

Glotfelty, Cheryll. "Flooding the Boundaries of Form: Terry Tempest Williams's *Unnatural History*." In *Change in the American West: Exploring the Human Dimension*, ed. Stephen Tchudi, 158–67. Reno: University of Nevada Press, 1996.

Gobster, Paul H. "Urban Park Restoration and the 'Museumification' of Nature." *Nature and Culture* 2, no. 2 (2007): 95–114.

Goudie, Andrew. *Encyclopedia of Geomorphology*, vol. 1, *A to I*. London: Taylor & Francis, 2003.

Gough, Barry M. "The Character of the British Columbia Frontier." *BC Studies* 32 (1976/77): 28–40.

Gouglas, Sean. "A Currant Affair: E.D. Smith and Agricultural Change in Nineteenth-Century Saltfleet Township, Ontario." *Agricultural History* 75, no. 4 (2001): 438–66.

Goyette, Linda. *Edmonton in Our Own Words*. Edmonton: University of Alberta Press, 2004.

Grainger, M. Allerdale. *Woodsmen of the West*. Toronto: McClelland & Stewart, 1964 [1908].

Grant, George. *Technology and Justice*. Toronto: Anansi, 1986.

———. *Technology and Empire*. Toronto: Anansi, 1969.

———. *Lament for a Nation: The Defeat of Canadian Nationalism*. Montreal and Kingston: McGill-Queen's University Press, 2005 [1965].

Gutnick, David. "Hiroshima." *The Current*, CBC Radio, 21 Mar. 2011, http://www.cbc.ca/thecurrent/episode/2011/03/21/hiroshima-david-gutnick/.

Haig-Brown, Roderick. *Timber*. Toronto: Wm. Collins, 1946 [1942].

Hak, Gordon. *Capital and Labour in the British Columbia Forest Industry, 1934–1974*. Vancouver: UBC Press, 2007.

———. *Turning Trees into Dollars: The British Columbia Coastal Lumber Industry, 1858–1913*. Toronto: University of Toronto Press, 2000.

Hanson, David T. *Waste Land: Meditations on a Ravaged Landscape*. New York: Aperture, 1997.

Harris, Maureen Scott. "Being Homesick, Writing Home." In *Fresh Tracks: Writing the Western Landscape*, ed. Pamela Banting, 253–64. Victoria: Polestar, 1998.

Harvey, Athelstan George. *Douglas of the Fir: A Biography of David Douglas, Botanist*. Cambridge, MA: Harvard University Press, 1948.

Harvey, Miles. *The Island of Lost Maps: A True Story of Cartographic Crime*. New York: Broadway Books, 2000.

Hejinian, Lyn. *Language of Inquiry*. Berkeley: University of California Press, 2000.

Henning, Michelle. *Museums, Media and Cultural Theory*. New York: Open University Press, 2006.

Henton, Darcy. "If Syncrude Convicted, Oilsands 'Doomed': Defence." *Edmonton Journal*, 29 Apr. 2010, A1.

Herriot, Trevor. *Grass, Sky, Song: Promise and Peril in the World of Grassland Birds*. Toronto: HarperCollins, 2009.

Hermansen, Sally, and Graeme Wynn. "Reflections on the Nature of an Urban Bog." *Urban History Review* 34, no. 1 (2005): 9–27.

Hertzog, Lawrence. "It's Our Heritage." *Real Estate Weekly* 24, no. 22 (2006).

Hessing, Melody, Rebecca Raglon, and Catriona Sandilands, ed. *This Elusive Land: Women and the Canadian Environment*. Vancouver: UBC Press, 2004.

Heuer, Karsten. *Being Caribou: Five Months on Foot with an Arctic Herd*. Seattle: Mountaineers, 2005.

Higgs, Eric S. *Nature by Design: People, Natural Process, and Ecological Design*. Cambridge, MA: MIT Press, 2003.

Hillman, James. "The Practice of Beauty." In *Uncontrollable Beauty: Toward a New Aesthetics*, ed. Bill Beckly and David Shapiro, 261–74. New York: Allworth, 1998.

Ho, Solarina. "Canada Tracks BPA Exposure, Finds in Most People." 16 Aug. 2010. http://www.reuters.com/article/2010/08/16/us-bisphenol-idUSTRE67F4NW20100816.

Hodgins, Jack. *The Resurrection of Joseph Bourne, or A Word or Two on Those Port Annie Miracles*. Toronto: McClelland & Stewart, 1998 [1979].

Hogan, Linda. "A Different Yield." In *Reclaiming Indigenous Voice and Vision*, ed. Marie Battiste, 115–23. Vancouver: UBC Press, 2000.

Holway, E.W.D. "New Light on Mounts Brown and Hooker." *Canadian Alpine Journal* 9 (1918): 45–48.

Homeless in Canada—Resources. http://intraspec.ca/homelessCanada.php#Edmonton.

Hooker, William, ed. *Companion to the Botanical Magazine* 2. London: Samuel Curtis, 1836.

———. *Flora Boreali-Americana; or, The Botany of the Northern Parts of British America*, vol. 1. London: H.G. Bohn, 1829.

Howarth, William. "Imagined Territory: The Writing of Wetlands." *New Literary History* 30, no. 3 (1999): 509–39.

Huggan, Graham. *Interdisciplinary Measures: Literature and the Future of Postcolonial Studies*. Liverpool: Liverpool University Press, 2008.

If a Tree Falls: A Story of the Earth Liberation Front. DVD. Directed by Marshall Curry and Sam Cullman. New York: Oscilloscope Pictures, 2010.

Ingold, Tim. *The Perception of the Environment*. London: Routledge, 2000.

Ingram, Annie Merrill, Ian Marshall, Daniel J. Philippon, and Adam W. Sweeting, eds. *Coming into Contact: Explorations in Ecocritical Theory and Practice*. Athens: University of Georgia Press, 2007.

———. "Introduction: Thinking of Our Life in Nature." In *Coming into Contact: Explorations in Ecocritical Theory and Practice*, ed. Annie Merrill Ingram, Ian Marshall, Daniel J. Philippon, and Adam W. Sweeting, 1–14. Athens: University of Georgia Press, 2007.

Iovino, Serenella. "Ecocriticism, Ecology of Mind, and Narrative Ethics: A Theoretical Ground for Ecocriticism as Educational Practice." *ISLE* 17, no. 4 (2010): 759–62.

Jeffers, Robinson. *The Wild God of the World: An Anthology of Robinson Jeffers*. Edited by Albert Gelpi. Stanford, CA: Stanford University Press, 2003.

Jenness, Diamond. *The Sekani Indians of British Columbia*. Ottawa: J.O. Patenaude I.S.O., 1937.

Jensen, Derrick. "Loaded Words: Writing as a Combat Discipline." *Orion Magazine* (Mar./Apr. 2012). http://www.orionmagazine.org/index.php/articles/article/6698/.

Jensen, Doreen. "Metamorphosis." In *Topographies: Aspects of Recent B.C. Art*. Vancouver Art Gallery. 29 Sept. 1996–5 Jan. 1997. http://www.ccca.ca/c/writing/j/jensen/jen001t.html.

Johnson, James F. "'Brébeuf and His Brethren' and 'Towards the Last Spike': The Two Halves of Pratt's National Epic." *Essays on Canadian Writing* 29 (1984): 142–51.

Johnston, Linda. "The Feasibility of Broadcasting Material Produced by Citizens' Groups on the CBC TV Network with Its Current Structure and Policy." MA thesis, Simon Fraser University, 1974.

Jones, Chris. *Climbing in North America*. Seattle: Mountaineers, 1997 [1976].

Jordan, Chris. *Midway*. http://www.midwayjourney.com/.

Jordan III, William R. *The Sunflower Forest: Ecological Restoration and the New Communion with Nature*. Berkeley: University of California Press, 2003.

Joyce, James. *A Portrait of the Artist as a Young Man*. Harmondsworth, UK: Penguin, 1992.

Judd, Richard W. "George Perkins Marsh: The Times and Their Man." *Environment and History* 10, no. 2 (2004): 169–90.
Justice, Daniel Heath. "'Go Away, Water!': Kinship Criticism and the Decolonization Imperative." In *Reasoning Together: The Native Critics Collective*, ed. Craig Womack, Daniel Heath Justice, and Christopher B. Teuton, 147–68. Norman: University of Oklahoma Press, 2008.
Kalman, Harold, et al. *The Rossdale Historical Land Use Study*. Edmonton: Commonwealth Historic Resource Management, 2004.
Kamps, Toby, and Ralph Rugoff. *Small World: Dioramas in Contemporary Art*. San Diego, CA: Museum of Contemporary Art, 2000.
Kant, Immanuel. *Critique of Judgment*. Translated by Werner S. Pluhar. Indianapolis: Hackett, 1987.
Katinas, Tom. "An Apology from Syncrude—and a Promise to Do Better." 2 May 2008. http://www.syncrude.ca/users/news_view.asp?FolderID=5690&NewsID=121.
Katz, Eric. "Another Look at Restoration: Technology and Artificial Nature." In *Restoring Nature: Perspectives from the Social Sciences and Humanities*, ed. Paul H. Gobster and R. Bruce Hull, 37–48. Washington, DC: Island Press, 2000.
Keenleyside, Hugh. *On the Bridge of Time*, vol. 2, *Memoirs of Hugh L. Keenleyside*. Toronto: McClelland & Stewart, 1982.
Kelly, Erin N., Jeffrey W. Short, David W. Schindler, Peter V. Hodson, Mingsheng Ma, Alvin K. Kwan, and Barbra L. Fortin. "Oil Sands Development Contributes Polycyclic Aromatic Compounds to the Athabasca River and Its Tributaries." *Proceedings of the National Academy of Sciences USA* 106, no. 52 (2009): 22346–51.
Kim, Myung Mi. "Generosity as Method: An Interview with Myung Mi Kim." By Yedda Morrison. *Tripwire: A Journal of Poetics* 1 (Spring 1998): 75–85.
Kipling, Rudyard. "If." http://www.swarthmore.edu/~apreset1/docs/if.html (accessed June 16, 2009).
Klingle, M.W. "Spaces of Consumption in Environmental History." *History and Theory* 42 (2003): 94–110.
Kramer, Reinhold. "The Contemporary Canadian Long Poem as System: Friesen, Atwood Kroetsch, Arnason, McFadden." In *Bolder Flights: Essays on the Canadian Long Poem*, ed. Frank M. Tierney and Angela Robbeson, 101–14. Ottawa: University of Ottawa Press, 1998.
Kroetsch, Robert. "Is This a Real Story or Did You Make It Up?" *Eighteen Bridges: Stories that Connect* 1 (Fall 2010): 18–19.
———. *What the Crow Said*. Don Mills, ON: General Publishing, 1978.
Kunuk, Zacharias, and Ian Mauro. *Inuit Knowledge and Climate Change*. 2010. http://www.isuma.tv/hi/en/inuit-knowledge-and-climate-change.
Kyba, Daniel. "Chasing the Giants." *Alberta History* 59, no. 1 (2011): 18–25.
Langer, Monika. "Merleau-Ponty and Deep Ecology." In *Ontology and Alterity in Merleau-Ponty*, ed. Galen A. Johnson and Michael B. Smith, 115–29. Evanston: Northwestern University Press, 1990.
Leach, Andrew. "Don't Let EU Define Our Emissions Policy." *Edmonton Journal*, 2 Mar. 2011, A15.
Lee, Abram. Diary. Erland Lee Museum, Stoney Creek, Ontario.
Lefebvre, Henri. *The Production of Space*. Translated by Donald Nicholson-Smith. London: Blackwell, 1991.
LeMenager, Stephanie, Teresa Shewry, and Ken Hiltner. "Introduction." In *Environmental Criticism for the Twenty-First Century*, ed. Stephanie LeMenager, Teresa Shewry, and Ken Hiltner, 1–15. New York: Routledge, 2011.

Leopold, Aldo. *Round River: From the Journals of Aldo Leopold*. Edited by Luna B. Leopold. Toronto: Oxford University Press, 1993.

Levant, Ezra. *Ethical Oil: The Case for Canada's Oil Sands*. Toronto: McClelland & Stewart, 2010.

———. "Shed No Tears for Ft. McMurray's Ducks." *National Post*, 5 May 2008, www.nationalpost.com/ (accessed 8 June 2008).

Levant, Ezra, and Andrew Nikiforuk. "Is Oil-Sands Oil the Most Ethical Oil on Earth?" *Q*, CBC Radio, 15 Sept. 2010, http://www.cbc.ca/q/blog/2010/09/15/is-oil-sands-oil-the-most-ethical-oil-on-earth/.

Lilburn, Tim. "How to Be Here?" In *Poetry and Knowing: Speculative Essays and Interviews*, ed. Tim Lilburn, 161–76. Kingston: Quarry, 1995.

Lioi, Anthony. "Part I: An Alliance of the Elements." *ISLE* 17, no. 4 (2010): 754–57.

Lippard, Lucy R. *The Lure of the Local: Sense of Place in a Multicentered Society*. New York: New Press, 1997.

Loo, Tina. "Disturbing the Peace: Change and the Scales of Justice on a Northern River." *Environmental History* 12, no. 4 (2007): 895–919.

———. "People in the Way: Modernity, Environment and Society on the Arrow Lakes." *BC Studies* 142/143 (2004): 161–96.

Love, Glen A. "Ecocriticism, Theory, and Darwin." *ISLE* 17, no. 4 (2010): 773–75.

———. *Practical Ecocriticism: Literature, Biology, and the Environment*. Charlottesville: University of Virginia Press, 2003.

Lovecraft, H.P. "The Whisperer in the Dark." In *The Call of Cthulhu and Other Weird Stories*. Edited by S.T. Joshi. Toronto: Penguin, 1999.

Lutts, Ralph H. *The Nature Fakers: Wildlife, Science, and Sentiment*. Charlottesville: University of Virginia Press, 1990.

Lutwack, Leonard. *Birds in Literature*. Gainesville: University of Florida Press, 1994.

Manning, Richard. *Grassland: The History, Biology, Politics, and the Promise of the American Prairie*. New York: Penguin, 1995.

Markham, Clements R. "The Present Standpoint of Geography." *Geographical Journal* 2 (1893): 481–504.

Mason, Travis V. "Literature and Geology: An Experiment in Interdisciplinary, Comparative Ecocriticism." In *Greening the Maple: Canadian Ecocriticism in Context*, ed. Ella Soper and Nicholas Bradley, 475–509. Calgary: University of Calgary Press, 2013.

Mathias, John. "The Poetry of Roy Fisher." In *Contemporary British Poetry: Essays in Theory and Criticism*, ed. James Acheson and Romana Huk, 34–63. Albany: SUNY Press, 1996.

McGeer, Patrick. *Politics in Paradise*. Toronto: Peter Martin, 1972.

McIvor, Mike. "Former CPAWS Trustee Dies in Banff." *Borealis* 3, no. 3 (1992): 51.

McKay, Bernard. *Crooked River Rats: The Adventures of Pioneer Rivermen*. Surrey: Hancock House, 2000.

McKay, Don. *Strike/Slip*. Toronto: McClelland & Stewart, 2006.

———. *Deactivated West 100*. Kentville, NS: Gaspereau, 2005.

———. *Vis à Vis: Field Notes on Poetry and Wilderness*. Wolfville, NS: Gaspereau, 2001.

McKay, Ian. *The Quest of the Folk: Antimodernism and Cultural Selection in Twentieth-Century Nova Scotia*. Montreal and Kingston: McGill-Queen's University Press, 1994.

McLuhan, Eric, and Frank Zingrone, eds. *Essential McLuhan*. Toronto: House of Anansi, 1995.

Merk, Frederick, ed. *Fur Trade and Empire: George Simpson's Journals*. Cambridge, MA: Harvard University Press, 1931.
Mitchell, Alanna. *Sea Sick: The Global Ocean in Crisis*. Toronto: McClelland & Stewart, 2009.
Mitchell, David. *W.A.C. Bennett and the Rise of British Columbia*. Vancouver: Douglas & McIntyre, 1983.
Mitman, Gregg, Michelle Murphy, and Christopher Sellers. "Introduction: A Cloud over History." *Osiris* 19 (2004): 1–17.
Monto, Tom. *Old Strathcona before the Depression*. Edmonton: Crang Press, 2008.
Morantz, Alan. *Where Is Here? Canadian Maps and the Stories They Tell*. Toronto: Penguin, 2002.
Morton, Timothy. "Guest column: Queer Ecology." *PMLA* 125, no. 2 (2010): 273–82.
———. *Ecology without Nature: Rethinking Environmental Aesthetics*. Cambridge, MA: Harvard University Press, 2007.
Morwood, William. *Traveler in a Vanished Landscape: The Life & Times of David Douglas, Botanical Explorer*. New York: Clarkson N. Potter, 1973.
Mouat, Jeremy. *The Business of Power: Hydro-Electricity in Southeastern British Columbia, 1897–1997*. Victoria: Sono Press, 1997.
Murphy, Peter J., with Robert W. Udell, Robert E. Stevenson, and Thomas W. Peterson. *A Hard Road to Travel: Land, Forests and People in the Upper Athabasca Region*. Durham, NC: Forest History Society, 2007.
Musqueam Ecosystem Conservation Society. http:// www.mecsweb.org/.
Nancy, Jean-Luc. *Being Singular Plural*. Translated by Robert D. Richardson and Anne E. O'Byrne. Stanford, CA: Stanford University Press, 2000.
Nash, Linda. "The Changing Experience of Nature: Historical Encounters with a Northwest River." *Journal of American History* 86, no. 4 (2000): 1600–29.
Nash, Roderick. *Wilderness and the American Mind*. New Haven, CT: Yale University Press, 2001 [1967].
Neilson, Ronald P., et al. "Forecasting Regional to Global Plant Migration in Response to Climate Change." *BioScience* 55 (2005): 749–59.
Nelles, H.V. *The Art of Nation-Building: Pageantry and Spectacle at Quebec's Tercentenary*. Toronto: University of Toronto Press, 1999.
New, W.H. "Writing Here." *BC Studies* 147 (2005): 3–25.
———. *Underwood Log*. Lantzville, BC: Oolichan, 2004.
———. *Articulating West: Essays on Purpose and Form in Modern Canadian Literature*. Toronto: New Press, 1972.
Nietzsche, Friedrich. *The Twilight of the Idols; Or How to Philosophise with a Hammer*. London: T. Fisher Unwin, 1899.
Nikiforuk, Andrew. *Empire of the Beetle*. Vancouver: Greystone, 2011.
Nyhart, Lynn. "Science, Art, and Authenticity in Natural History Displays." In *Models: The Third Dimension of Science*, ed. Soraya De Chadarevian and Nick Hopwood, 307–35. Stanford, CA: Stanford University Press, 2004.
Oakes, Jill, and Rick Riewe, eds. *Climate Change: Linking Traditional and Scientific Knowledge*. Winnipeg: Aboriginal Issues Press, 2006.
Oberhelman, Steven, Van Kelly, and Richard J. Golsan, eds. *Epic and Epoch: Essays on the Interpretation and History of a Genre*. Lubbock: Texas Tech University Press, 1994.
Oppermann, Serpil. "Ecocriticism's Theoretical Discontents." *Mosaic* 44, no. 2 (2011): 153–69.
———. "Ecocriticism's Phobic Relations with Theory." *ISLE* 17, no. 4 (2010): 768–70.

Order-in-Council # 30724, Report of a Committee of the Executive Council, 17 Jan. 1919, GR 1530, Archives of Manitoba, Winnipeg, Manitoba.

"Papachase Land Claim Resolved." Centre for Constitutional Studies. 2012. http://www.law.ualberta.ca/centres/ccs/rulings/papchaselandclaim.php.

Parfitt, Ben. *Managing BC's Forests for a Cooler Planet: Carbon Storage, Sustainable Jobs and Conservation*. Vancouver: CCPA-BC Office; BC Government and Service Employees' Union; Communications, Energy & Paperworkers Union; David Suzuki Foundation; Pulp, Paper and Woodworkers of Canada; Sierra Club BC; United Steelworkers District 3—Western Canada; and Western Canada Wilderness Committee, 2010.

Parr, Joy. *Sensing Changes: Technologies, Environments, and the Everyday, 1953–2003*. Vancouver: UBC Press, 2010.

Philippon, Daniel J. "Sustainability and the Humanities: An Extensive Pleasure." *American Literary History* 24, no. 1 (2012): 163–79.

Phillips, Dana. *The Truth of Ecology: Nature, Culture, and Literature in America*. New York: Oxford University Press, 2003.

Piper, Liza. *The Industrial Transformation of Subarctic Canada*. Vancouver: UBC Press, 2009.

———. "Subterranean Bodies: Mining the Large Lakes of North-west Canada, 1921–1960." *Environment and History* 13, no. 2 (2007): 155–86.

———. "Nature, History, and Marx." *Left History* 11, no. 1 (2006): 41–46.

Pollon, Earl, and Shirlee Smith Matheson. *This Was Our Valley*. Calgary: Detselig, 2003.

Potter, Will. *Green Is the New Red: An Insider's Account of a Social Movement under Siege*. San Francisco: City Lights, 2011.

Pritchard, Allan. "The Shapes of History in British Columbia Writing." *BC Studies* 93 (1991): 48–69.

"Province Won't Tear Down Edmonton's Tent City." *CBC News*, 18 July 2007. http://www.cbc.ca/canada/edmonton/story/2007/07/18/tent-city.html.

Raibmon, Paige. *Authentic Indians: Episodes of Encounter from the Late-Nineteenth-Century Northwest Coast*. Durham, NC: Duke University Press, 2005.

Rajala, Richard A. "Clearcutting the British Columbia Coast: Work, Environment and the State, 1880–1930." In *Making Western Canada: Essays on European Colonization and Settlement*, ed. Catharine Cavanaugh and Jeremy Mouat, 104–32. Toronto: Garamond, 1996.

———. "A Dandy Bunch of Wobblies: Pacific Northwest Loggers and the Industrial Workers of the World, 1900–1930." *Labor History* 37, no. 2 (1996): 205–34.

———. "The Forest as Factory: Technological Change and Worker Control in the West Coast Logging Industry, 1880–1930." *Labour / Le Travail* 32 (Fall 1993): 73–104.

Raulff, Ullrich. "Interview with Giorgio Agamben—Life, a Work of Art without an Author: The State of Exception, the Administration of Disorder and Private Life." *German Law Journal* 116 (2004), http://www.germanlawjournal.com/article.php?id=437.

The Reckoning. CBC Television, 1975.

Reed, Maureen G. "Uneven Environmental Management: A Canadian Comparative Political Ecology." *Environment and Planning A* 39 (2007): 320–38.

Rhor, Christian. "Man and Nature in the Middle Ages." Paper given at Novosibirsk State University, 29 Oct.–1 Nov. 2002. http://sharepdf.net/view/14095/man-and-nature-in-the-middle-ages.

Richards, I.A. *Science and Poetry*. London: Kegan Paul, Trench, Trubner, 1926.

Richardson, Jack. "Interventionist Art Education: Contingent Communities, Social Dialogue, and Public Collaboration." *Studies in Art Education* 52, no. 1 (2010): 18–33.

Ricou, Laurie. "Two Nations Own These Islands: Border and Region in Pacific-Northwest Writing." In *Context North America: Canadian/U.S. Literary Relations*, ed. Camille R. La Bossière, 49–62. Ottawa: University of Ottawa Press, 1994.

Riffaterre, Michael. *Semiotics of Poetry*. Bloomington: Indiana University Press, 1978.

River Valley Alliance. *Alberta Capital Region River Valley Park*. http://www.rivervalleyab.ca/media/uploads/rva-geological-history.pdf.

Robbins, David. "Sport, Hegemony and the Middle Class: The Victorian Mountaineers." *Theory, Culture, and Society* 4, no. 3 (1987): 579–601.

Robertson, Lisa. "Lastingness." *Open Letter* 13, no. 3 (2007): 47–61.

Robin, Martin. *The Company Province: 1871–1933, 1934–1972*. 2 vols. Toronto: McClelland & Stewart, 1972–73.

Robinson, Zac. "Storming the Heights: Canadian Frontier Nationalism and the Making of Manhood in the Conquest for Mount Robson, 1906–1913." *International Journal for the History of Sport* 22, no. 3 (2005): 415–33.

Robisch, S.K. "The Woodshed: A Response to 'Ecocriticism and Ecophobia.'" *ISLE* 16, no. 4 (2009): 697–708.

Rowley, Mari-Lou. "In the Tar Sands, Going Down." In *Regreen: New Canadian Ecological Poetry*, ed. Madhur Anand and Adam Dickinson, 63–66. Sudbury, ON: Your Scrivener Press, 2009.

Rueckert, William. "Literature and Ecology: An Experiment in Ecocriticism." *Iowa Review* 9, no. 1 (1978): 71–86.

"Ryerson University Students Design Innovative Wastewater Treatment for Removing Pharmaceuticals." 31 Mar. 2010. http://www.ryerson.ca/news/media/General_Public/20100331_rn_wastewat.html.

Sabin, Paul. *The Bet: Paul Ehrlich, Julian Simon, and Our Gamble over Earth's Future*. New Haven, CT: Yale University Press, 2013.

Saner, Reg. "Technically Sweet." In *The Four-Cornered Falcon: Essays on the Interior West and the Natural Scene*, 73–103. Baltimore: Johns Hopkins University Press, 1993.

Schama, Simon. *Landscape and Memory*. Toronto: Random House, 1995.

Schieder, Rupert. Introduction to *Woodsmen of the West*, by M. Allerdale Grainger, vii–xii. Toronto: McClelland & Stewart, 1964.

Scott, Chic. *Pushing the Limits: The Story of Canadian Mountaineering*. Calgary: Rocky Mountain Books, 2000.

Sedgwick, Eve Kosofsky. "Paranoid Reading and Reparative Reading, or, You're So Paranoid, You Probably Think This Essay Is about You." In *Touching, Feeling: Affect, Pedagogy, Performativity*, 123–152. Durham, NC: Duke University Press, 2003.

Selters, Andy. *Ways to the Sky: A Historical Guide to North American Mountaineering*. Golden, CO: American Alpine Club, 2004.

Sherman, Paddy. *Bennett*. Toronto: McClelland & Stewart, 1966.

Shiva, Vandana. *Water Wars: Privatization, Pollution, and Profit*. Toronto: Between the Lines, 2002.

Shklovsky, Victor. "Art as Technique." In *Russian Formalist Criticism: Four Essays*. Translated by Lee T. Lemon and Marion J. Reis, 3–24. Lincoln: University of Nebraska Press, 1965.

Shrum, Gordon. *Gordon Shrum: An Autobiography with Peter Stursberg.* Vancouver: UBC Press, 1986.
Shukin, Nicole. "Animal Capital: The Politics of Rendering." PhD dissertation, University of Alberta, 2005.
Silverstone, Peter. *World's Greenest Oil: Turning the Oil Sands from Black to Green.* Edmonton: PHS Holdings, 2010.
Sims, Daniel. "Tse Keh Nay-European Relations and Ethnicity, 1793–2009." MA thesis, University of Alberta, 2010.
Sinclair, Upton. *Oil!* 1927. New York: Penguin, 2007 [1927].
Siporin, Ona. "Terry Tempest Williams and Ona Siporin: A Conversation." *Western American Literature* 31, no. 2 (1996): 99–113.
Slemon, Stephen. "The Brotherhood of the Rope: Commodification and Contradiction in the 'Mountaineering Community.'" In *Renegotiating Community: Interdisciplinary Perspectives, Global Contexts*, ed. Diana Brydon and William D. Coleman, 234–45. Vancouver: UBC Press, 2008.
Slovic, Scott. "Part II: Elements of This New Alliance." *ISLE* 17, no. 4 (2010): 757–59.
Smith, Damaris. "Pioneer Wife." E.D. Smith Company Archives, Stoney Creek, Ontario.
Smith, E.D. Diaries, E.D. Smith Company Archives, Stoney Creek, Ontario.
Snyder, Gary. *Back on the Fire: Essays.* Berkeley, CA: Counterpoint, 2007.
———. *Danger on Peaks: Poems.* Washington, DC: Shoemaker & Hoard, 2004.
———. *The Practice of the Wild.* Berkeley, CA: Counterpoint, 1990.
Solnit, Rebecca. *A Paradise Built in Hell: The Extraordinary Communities That Arise in Disasters.* New York: Viking, 2009.
"Special Forum on Ecocriticism and Theory." *ISLE* 17, no. 4 (2010): 754–99.
Spittlehouse, David. *Climate Change, Impacts and Adaptation Scenarios: Climate Change and Forest and Range Management in British Columbia.* Victoria: BC Ministry of Forests and Range, 2008.
Stauffer, Robert C. "Haeckel, Darwin, and Ecology." *Quarterly Review of Biology* 32, no. 3 (1957): 138–44.
Stoll, Mark. "Rachel Carson's *Silent Spring*: A Book That Changed the World." Environment & Society Portal. http://www.environmentandsociety.org/exhibitions/silent-spring/personal-attacks-rachel-carson.
Stunden Bower, Shannon. *Wet Prairie: People, Land, and Water in Agricultural Manitoba.* Vancouver: UBC Press, 2011.
Stutchbury, Bridget. *Silence of the Songbirds: How We Are Losing the World's Songbirds and What We Can Do to Save Them.* Toronto: HarperCollins, 2007.
Stutfield, Hugh E.M., and J. Norman Collie. *Climbs and Explorations in the Canadian Rockies.* London: Longmans, Green, 1903.
Sugars, Cynthia. "Can the Canadian Speak? Lost in Postcolonial Space." *ARIEL: A Review of International English Literature* 32, no. 3 (2001): 115–52.
Swainson, Neil. *Conflict over the Columbia: The Canadian Background to an Historic Treaty.* Montreal and Kingston: McGill-Queen's University Press, 1979.
Syncrude Canada Ltd. "2007 Sustainability Report." http://sustainability.syncrude.ca/sustainability2007/.
Szabo, Lisa. "Wildwood Notes: Nature Writing, Music, and Newspapers." MA thesis, University of British Columbia, 2007.
Szeman, Imre. "System Failure." *South Atlantic Quarterly* 106, no. 4 (2007): 805–23.
Szerszynski, Bronislaw, Wallace Heim, and Claire Waterton. Introduction. In *Nature Performed: Environment, Culture, and Performance*, ed. Bronislaw Szerszynksi, Wallace Heim, and Claire Waterton, 1–14. Malden, MA: Blackwell, 2003.

Tallmadge, John. "Beyond the Excursion: Initiatory Themes in Annie Dillard and Terry Tempest Williams." In *Reading the Earth: New Directions in the Study of Literature and Environment*, ed. Michael P. Branch, Rochelle Johnson, Daniel Patterson, and Scott Slovic, 197–207. Moscow: University of Idaho Press, 2008.

Taylor, Alan. "'Wasty Ways': Stories of American Settlement." *Environmental History* 3, no. 3 (1998): 291–310.

Taylor, George Rogers, ed. *The Turner Thesis: Concerning the Role of the Frontier in American History*. Lexington, MA: D.C. Heath, 1949.

Taylor, John. Review of *Walt Whitman Bathing: Poems*, by David Wagoner. *Poetry* 171, no. 3 (1998): 229–32.

Taylor, Timothy. "It's Not about Ecology—It's about Gardening." *Globe and Mail*, 27 Aug. 2007. http://www.theglobeandmail.com/life/its-not-about-ecology---its-about-gardening/article4107735/.

———. *Stanley Park*. Toronto: Random House, 2001.

Terpak, Frances. "Diorama." In *Devices of Wonder: From the World in a Box to Images on a Screen*, ed. Barbara Maria Stafford and Frances Terpak, 325–29. Los Angeles: Getty Research Institute, 2001.

Thompson, David. *David Thompson's Narrative*. Edited by J.B. Tyrell. Toronto: Champlain Society, 1916.

Thorington, James Monroe. "Mounts Brown and Hooker: A Reply." *Canadian Alpine Journal* 17 (1929): 69–70.

———. "The Centenary of David Douglas' Ascent of Mount Brown." *Canadian Alpine Journal* 16 (1928): 185–97.

———. *The Glittering Mountains of Canada*. Philadelphia, PA: John W. Lea, 1925.

Tomblin, Stephen. "W.A.C. Bennett and Province-Building in British Columbia." *BC Studies* 85 (1990): 45–61.

Tsay Keh Dene: CBC Hourglass Documentary. CBC Television, 1970.

Tuan, Yi-Fu. *Topophilia: A Study of Environmental Perception, Attitudes, and Values*. New York: Columbia University Press, 1974.

Turner, Chris. *The War on Science: Muzzled Scientists and Wilful Blindness in Stephen Harper's Canada*. Vancouver: Greystone, 2013.

Turner, Frederick Jackson. *The Significance of the Frontier in American History*. London: Penguin, 2009.

Valenčius, Conevery Bolton. *The Health of the Country: How Americans Understood Themselves and Their Land*. New York: Basic Books, 2002.

Vandervlist, Harry. "Jon Whyte." In *Dictionary of Literary Biography*, vol. 334, *Twenty-First-Century Canadian Writers*, ed. Christian Riegel, 274–79. Detroit: Gale Research, 2006.

Venne, Sharon. "Treaties Made in Good Faith." In *Natives & Settlers, Now & Then: Historical Issues and Current Perspectives on Treaties and Land Claims in Canada*, ed. Paul W. DePasquale, 1–16. Edmonton: University of Alberta Press, 2007.

Vermeulen, Pieter. "Community and Literary Experience in (Between) Benedict Anderson and Jean-Luc Nancy." *Mosaic* 42, no. 4 (2009): 95–111.

Wagoner, David. *Traveling Light: Collected and New Poems*. Urbana: University of Illinois Press, 1999.

———. *Walt Whitman Bathing: Poems*. Urbana: University of Illinois Press, 1996.

———. "On a Mountainside." *Poetry* 163, no. 1 (1993): 1–3.

Waldram, James. *As Long as the Rivers Run: Hydroelectric Development and Native Communities in Western Canada*. Winnipeg: University of Manitoba Press, 1988.

Wang, T., A. Hamann, A. Yanchuk, G.A. O'Neill, and S.N. Aitken. "Use of Response Functions in Selecting Lodgepole Pine Populations for Future Climates." *Global Change Biology* 12, no. 12 (2006): 2404–16.

Ward, Roxane. *Tackle Climate Change—Use Wood*. Vancouver: BC Forestry Climate Change Working Group, 2009.

Waring, Richard H., Nicholas C. Coops, and Steven W. Running. "Predicting Satellite-Derived Patterns of Large-Scale Disturbances in Forests of the Pacific Northwest Region in Response to Recent Climatic Variation." *Remote Sensing of Environment* (2011). Online 3 Nov. 2011. doi: 10.1016/j.rse.2011.08.017.

Warren, Jim. "Placing Ecocriticism." *ISLE* 17, no. 4 (2010): 770–72.

Warshall, Peter. *Septic Tank Practices*. Garden City, NY: Anchor, 1979.

Washington, Bradford. *The Dishonorable Doctor Cook: Debunking the Notorious McKinley Hoax*. Seattle: Mountaineers, 2001.

Wedley, John. "Infrastructure and Resources: Governments and Their Promotion of Northern Development in British Columbia, 1945–1975." PhD dissertation, University of Western Ontario, 1986.

Weisman, Alan. *The World Without Us*. Toronto: HarperCollins, 2007.

Wheeler, Arthur O. "Mounts Brown and Hooker." *Canadian Alpine Journal* 17 (1929): 66–68.

———. "Expedition to Mount Robson." *Canadian Alpine Journal* 1, no. 2 (1908): 314–17.

White, Howard. *The Men There Were Then*. Vancouver: Arsenal Pulp Press, 1983.

White, James. *Place-Names in the Rocky Mountains Between the 49th Parallel and the Athabaska River: Transactions of the Royal Society of Canada, Section II*. Ottawa: Royal Society of Canada, 1916.

White, Richard. "'Are You an Environmentalist or Do You Work for a Living?': Work and Nature." In *Uncommon Ground: Rethinking the Human Place in Nature*, ed. William Cronon, 171–85. New York: W.W. Norton, 1996.

———. *The Organic Machine: The Remaking of the Columbia River*. New York: Hill & Wang, 1995.

———. *"It's Your Misfortune and None of My Own": A History of the American West*. Norman: University of Oklahoma Press, 1991.

White, Richard, and Patricia J. Limerick. *The Frontier in American Culture: An Exhibition at the Newberry Library, August 26, 1994–January 7, 1995*. Edited by James R. Grossman. Berkeley: University of California Press, 1994.

Whyte, Jon. Fonds. Whyte Museum of the Canadian Rockies, Banff, Alberta.

———. *Jon Whyte: Mind over Mountains, Selected and Collected Poems*. Edited by Harry Vandervlist. Calgary: Red Deer Press, 2001.

———. *Mountain Chronicles: A Collection of Columns on the Canadian Rockies from the Banff Crag and Canyon, 1975–1991*. Edited by Brian Patton. Banff, AB: Altitude, 1992.

———. "Cosmos: Order and Turning." In *Trace: Prairie Writers on Writing*, ed. Birk Sproxton, 269–74. Winnipeg: Turnstone, 1986.

———. *The fells of brightness, second volume: Wenkchemna*. Edmonton: Longspoon Press, 1985.

Williams, Terry Tempest. "Testimony." In *An Unspoken Hunger: Stories from the Field*, 125–31. New York: Vintage, 1995.

———. *Refuge: An Unnatural History of Family and Place*. New York: Vintage, 1992 [1991].

Williston, Eileen, and Betty Keller. *Forests, Power and Policy: The Legacy of Ray Williston*. Prince George, BC: Caitlin Press, 1997.

Wilson, Sara J., and Richard J. Hebda. *Mitigating and Adapting to Climate Change through the Conservation of Nature*. Vancouver: Land Trust Alliance of BC, 2008.

Wingrove, Josh. "Syncrude to Pay $3M for Duck Deaths." *Globe and Mail*, 22 Oct. 2010. http://www.theglobeandmail.com/report-on-business/industry-news/energy-and-resources/syncrude-to-pay-3m-for-duck-deaths/article1769027/.

Winterson, Jeanette. *Art Objects: Essays on Ecstasy and Effrontery*. Toronto: Alfred A. Knopf, 1995.

Wonders, Karen. *Habitat Dioramas: Illusions of Wilderness in Museums of Natural History*. Uppsala: Acta Universitas Upsaliensis, 1993

———. "Habitat Dioramas as Ecological Theatre." *European Review* 1, no. 3 (1993): 285–300.

Wong, Rita. *Poetic Statements KSW Positions Colloquium*. Vancouver, BC: KSW, VIVO, 2008.

———. *forage*. Gibsons Landing [Gibsons], BC: Nightwood, 2007.

Wood, Gillen D'Arcy. "What Is Sustainability Studies?" *American Literary History* 24, no. 1 (2012): 1–15.

Woodhaven Eco Art Project. http://www.woodhaven.ok.ubc.ca/.

Woodland, Malcolm. "Poetry." *University of Toronto Quarterly* 77, no. 1 (2008): 29–78.

Wright, Stephen. "The Delicate Essence of Artistic Collaboration." *Third Text* 18, no. 6 (2004): 533–45.

Wyile, Herb. "Regionalism, Postcolonialism and (Canadian) Writing: A Comparative Approach for Postnational Times." *Essays on Canadian Writing* 63 (1998): 139–62.

Yanchuk, A.D., J.C. Murphy, and K.F. Wallin. "Evaluation of Genetic Variation of Attack and Resistance in Lodgepole Pine in the Early Stages of a Mountain Pine Beetle Outbreak." *Tree Genetics and Genomes* 4 (2008): 171–80.

Young, G. Winthrop. "John Norman Collie." *Alpine Journal* 54 (May 1943): 62.

Zhu, Kai, Christopher W. Woodall, and James S. Clark. "Failure to Migrate: Lack of Tree Range Expansion in Response to Climate Change." *Global Change Biology* (2011): 1–11. doi: 10.1111/j.1365-2486.2011.02571.x.

Zwicker, Heather. "Dead Indians, Power Conglomerates and the Upper Middle Class: Commemorating Colonial Conflict in Edmonton's Rossdale." Presentation, Canadian Association of Cultural Studies Conference, Hamilton, Ontario, 2004. http://www.culturalstudies.ca/proceedings04/proceedings.html.

CONTRIBUTORS

Pamela Banting (English, University of Calgary) researches and teaches in the areas of ecocriticism and ecotheory, literature and the Anthropocene, animals and animality, wildness, bioregionalism, and psychogeography. Her recent articles include "Magic Is Afoot: Hoof Marks, Paw Prints and the Problem of Writing Wildly," published in *Animal Encounters*, and "The Ontology and Epistemology of Walking: Animality in Karsten Heuer's *Being Caribou*," in *Greening the Maple: Canadian Ecocriticism in Context*. She guest-edited the Canadian Literary Ecologies issue of the journal *Studies in Canadian Literature* (2014).

Nicholas Bradley is an associate professor in the Department of English at the University of Victoria and the coordinator of the department's graduate program in Literatures of the West Coast. His areas of research include Canadian literature and American literature. Among his recent publications is *We Go Far Back in Time: The Letters of Earle Birney and Al Purdy, 1947–1987* (Harbour Publishing, 2014).

David Brownstein is the principal of Klahanie Research Ltd, a Vancouver firm specializing in applied historical and geographical research and analysis. At present, he is co-editing a forthcoming series on Canada's forest history. He is also a sessional instructor in the Geography Department at the University of British Columbia, where he teaches classes in both historical and environmental geography.

Warren Cariou was born in Meadow Lake, Saskatchewan, Canada into a family of mixed Métis and European heritage. He has published works of fiction, criticism, and memoir about Aboriginal cultures in western Canada and has co-directed two films about Aboriginal communities in the oil sands region. He teaches at the University of Manitoba, where he holds a Canada Research Chair in Narrative, Community, and Indigenous Cultures, and he also directs the Centre for Creative Writing and Oral Culture.

Beth Carruthers is a transdisciplinary scholar, artist, curator, and consultant with more than two decades of experience in multiple aspects of arts, culture, and environmental change. Her writing is widely studied and taught across disciplines concerned with cultural practices and human–world relations in the Anthropocene, while her seminal 2006 report for the Canadian Commission for UNESCO helped define a necessary role for the arts in sustainability. Based on the wild west coast of Canada, she works internationally.

Dianne Chisholm is professor emeritus (University of Alberta). She researches and publishes in the fields of literary modernism, queer studies, and environmental humanities, and between the fields of ecology, cultural studies, and philosophy, usually within a feminist framework. She also conducts peripatetic fieldwork and wild analyses in territories outside the academy, including and especially the Canadian North and West. Her current book projects include "Becoming Ecologies: Art and Philosophy for a New Earth and People" (a critical monograph) and "Home on the DeRanged: Living, Walking and Perceiving the Cultural Stratigraphy of Alberta's Front Ranges" (creative/investigatory documentary).

Jon Gordon teaches Writing Studies at the University of Alberta. He has published on hog production, mountaineering literature, and bitumen. His current work, *Unsustainable Rhetoric: Facts, Counter-Facts, and Literature in the Debate over Alberta's Bituminous Sands*, builds on his chapter in this collection and examines the literature and rhetoric of bitumen extraction in Alberta; the book is forthcoming from the University of Alberta Press.

Sean Gouglas (PhD, McMaster University) is director of the Office of Interdisciplinary Studies and an associate professor in the Humanities Computing program at the University of Alberta. He is the director of the VITA Research Studio (CFI), which provides research infrastructure for projects in Computer Game Studies. Prior to working on game studies, his research concerned the environmental history of southern Ontario and the application of statistical and Geographic Information Systems technologies to colonial settlement histories.

Trevor Herriot is a naturalist, writer, and grassland advocate from Regina, Saskatchewan. His fourth book, *The Road Is How: A Prairie Pilgrimage through Nature, Desire, and Soul* was published in early 2014 by HarperCollins.

Nancy Holmes is an associate professor of Creative Writing at the University of British Columbia Okanagan. She has published five collections of poetry, most recently, *The Flicker Tree: Okanagan Poems* (2012). She is also the editor of *Open Wide a Wilderness: Canadian Nature Poems* (Wilfrid Laurier University Press, 2009). Currently, with Denise Kenney, she is working on several initiatives of a large-scale research/creation project called the Eco Art Incubator.

Travis V. Mason has taught ecocriticism, poetry, and poetics, and postcolonial and Canadian literatures. He received both a Mellon and a Killam postdoctoral fellowship. He has published several articles and reviews in Canadian and international journals and books; written *Ornithologies of Desire: Ecocritical Essays, Avian Poetics, and Don McKay* (Wilfrid Laurier University Press, 2013); and co-edited *Public Poetics: Critical Issues in Canadian Poetry and Poetics* with Bart Vautour, Erin Wunker, and Christl Verduyn.

Lyndal Osborne works in a wide range of media including print, drawing, and installation. The work has been exhibited extensively across Canada and internationally. Her installations utilize found and recycled materials and the work speaks poetically of the forces of transformation within nature, as well as commenting upon pressing issues relating to environmentalism. A survey exhibition *Bowerbird: Life as Art* was shown at the Art Gallery of Alberta, Edmonton, in 2014.

Richard Pickard is a professor in the Department of English at the University of Victoria, where he specializes in composition and the environmental humanities. Though his doctorate was in eighteenth-century British studies, he more frequently teaches Canadian and western North American literature. In 2009 he was the site host for the first biennial conference of the Association for the Study of Literature and Environment (ASLE) outside the United States, at the University of Victoria, and in 2010 he was president of the Association for Literature, Environment, and Culture in Canada (ALECC) during its first conference, at Cape Breton University in Nova Scotia.

Liza Piper is an associate professor at the University of Alberta where she teaches environmental history and the history of northern and western Canada. Her research and writing has focused on the histories of natural resource exploitation, climate, and health across Canada with particular attention to the role of science and the experiences of Indigenous people in the North. Her first book was *The Industrial Transformation of Subarctic Canada* (UBC Press, 2009).

Harold Rhenisch writes the environmental blog okanaganokanogan.com, and has authored twenty-seven books of poetry, translation, fiction, cultural criticism, memoir and environmental writing, including *Tom Thomson's Shack*, a view of contemporary Canada from its hinterlands; *Motherstone*, a portrait of the volcanic regimes of central British Columbia; and *The Wolves at Evelyn*, winner of the George Ryga Prize for Social Responsibility in British Columbia Literature. He lives in Vernon, British Columbia.

Zac Robinson is a historian and an assistant professor in the Faculty of Physical Education and Recreation at the University of Alberta. He has recently published *Conrad Kain: Letters from a Wandering Mountain Guide, 1906–1933* (University of Alberta Press, 2014), and he currently serves as the vice-president of Mountain Culture for the Alpine Club of Canada.

Daniel Sims is a Tsay Keh Dene historian from Prince George, British Columbia. Currently he is finishing his dissertation on the impact of the W.A.C. Bennett Dam, Williston Lake Reservoir, and Hart Highway on the larger Tse Keh Nay nation. This subject draws from a number of historical fields, including, but not limited to, Aboriginal history, British Columbian history, and environmental history, as well as the history of energy.

Stephen Slemon teaches postcolonial literatures and theory in the Department of English & Film Studies at the University of Alberta, with specific interest in comparative anglophone literatures, the literature of imperial management, and the literature of mountaineering from colonial beginnings to the global present. He is past director of the University of Alberta's Canadian Literature Centre/Centre de littérature canadienne, past chair of the Aid to Scholarly Publications Program with the Canadian Federation for the Humanities and Social Sciences, and past president of the Association of Canadian College and University Teachers of English.

Christine Stewart studies experimental poetics, Indigenous poetics, and creative research in the Department of English & Film Studies at the University of Alberta. She is also a founding member of the Writing Revolution in Place Research Collective. Among her publications are *from Taxonomy* (West House Press), *Pessoa's July: or the months of astonishments* (Nomados Press), *The Trees of Periphery* (above/ground press), and *Virtualis: Topologies of the Unreal* (BookThug).

Shannon Stunden Bower is an environmental historian primarily concerned with interactions between human and non-human nature on the Canadian Prairies. She is the author of *Wet Prairie: People, Land, and Water in Agricultural Manitoba* (UBC Press, 2011), which won numerous awards, including the Canadian Historical Association's Clio Prize in the Prairie Provinces. She has also published articles in edited collections and journals, including the *Journal of Historical Geography* and *Environmental History*.

Lisa Szabo-Jones (PhD, University of Alberta) is a Trudeau Foundation Doctoral Scholar, co-founder and co-editor of the online journal *The Goose*, and teaches in English and Film Studies at the University of Alberta. She has publications in *Canadian Literature* and *Greening the Maple*, and is a guest co-editor of a special issue of *ARIEL* on postcolonial ecocriticism.

Harry Vandervlist is an associate professor at the University of Calgary. He writes on Samuel Beckett's early work, and on Canadian literature, especially the Banff poet Jon Whyte, as in his recent essay "The Challenge of Writing Bioregionally: Performing the Bow River in Jon Whyte's 'Minisniwapta: Voices of the River,'" in *The Bioregional Imagination*, edited by Cheryll Glotfelty, Karla Armbruster, and Thomas Lynch (University of Georgia Press, 2012).

Angela Waldie teaches at Mount Royal University in Calgary. She recently completed her PhD at the University of Calgary, where her research focused on species extinction in Canadian and American literature. She is currently writing her first poetry collection, entitled "A Single Syllable of Wild," which explores wildlife conservation practices in the Canadian Rocky Mountain Parks.

Maria Whiteman has recently moved to Houston, Texas, to become a full-time artist after living in Canada for sixteen years. She was an assistant professor of Drawing and Intermedia in Fine Arts at the University of Alberta. Her current art practice explores themes such as art and science; relationships between industry, community, and nature; and the place of animals in our cultural and social imaginary. In addition to her studio work, she conducts research in contemporary art theory and visual culture.

Rita Wong is the author of three books of poetry: *sybil unrest* (co-written with Larissa Lai), *forage* (winner of Canada Reads Poetry 2011), and *monkeypuzzle*. Wong has received the Asian Canadian Writers Workshop Emerging Writer Award and the Dorothy Livesay Poetry Prize. Investigating the relationships between contemporary poetry, social justice, ecology, and decolonization, she works with the poetics of water. Wong serves as an associate professor at Emily Carr University of Art & Design.

INDEX

Page references followed by fig *indicate a figure.*

Aboriginal peoples: archaeological evidence of life of, 252; in British Columbia, 319; cabins of, 316*fig*; colonizers and, 4, 8; depiction on murals, 258n32; documentaries about, 311–12, 319, 320; family relationships among, 70; future of, 317–18; homelessness among, 250; impact of hydroelectric development on, 307; land claims by, 254–55; otherness of, 319–20; in poetry, 234; recognition of rights of, 68; relationships to nature, 4, 10; responsibility to preserve culture of, 225; smallpox epidemic and, 253–54; trading places of, 252, 253. *See also* Cree; Gwich'in; Métis culture; Musqueam; Papaschase First Nation; Stoney Nakoda; Tse Keh Nay
ab ovo (Osborne), 110, 112*fig*, 113*fig*
Abram, David, 38, 39
activism: as academic topic, 8, 12n16; in ecocritical debate, 119, 122; subtle, 65, 75
aesthetic: definition of, 71; of natural habitat, 51, 52; in photography, 26
aesthetic engagement, 65, 71
Agamben, Giorgio, 242, 244, 252, 268
agriculture: escarpments and, 159, 163–64; impact on environment, 17; land pollution and, 18; in Manitoba, 159–60; policy in North America, 204; research and practices, 18; use of pesticides in, 202
Alaska, 3, 8, 87, 104

Alaskan National Wildlife Refuge (ANWR), 87, 91, 97, 98, 101
Alberta, 16, 109, 245, 253–54. *See also* Treaty Eight; Treaty Six
Alberta Oil Sands (Burtynsky), 26
Alderman, Nigel, 295
Allen, S. E. S., 297
Allison, Leanne: depression of, 101; as environmental activist, 88, 92, 102, 107n37; as filmmaker, 87; thrumming experience of, 97; in "zone of proximity" to caribou, 93
Alpine clubs, 145, 147, 149, 150, 352
American Prairie Foundation, 19
Anderson, Benedict, 211, 212–13, 215, 216–17, 220, 228
Andrews, Howard L., 205
animals: damaged habitat of, 260; on display, 80, 81–82; humans and, 244; images of gazes of, 83–84; knowledge of, 84; in museums, 82; photographs of, 79*fig*, 81, 81*fig*, 83*fig*, 84*fig*; regulation of population of, 244; visualization of, 80
Anthropocene, 5, 11n9
Arbez, Reeve, 169
Archipelago (Osborne), 109, 111*fig*
Aristotle, 81
Armstrong, George, 170
Armstrong, Jeannette, 35, 40, 264
art: aim of, 106n11; consideration of place in, 111; interdisciplinary studies and, 327; practices, 65, 80; repetition of form in, 112; representation of

animals in, 80; role in post-humanist discourse, 80; *vs.* science, 95; use of scale in, 110
Articulating West (New), 273
artistic collaboration, 37–38
artistic way of thinking, 95
artists' role in cultural change, 72, 73
Art Objects (Winterson), 71
Assisted Migration Adaptation Trial (AMAT), 176
Association for the Study of Literature and Environment (ASLE), 134n4
"Astonished" (McKay), 118
astonishment, literary associations of, 127–29
Athabasca Pass, 141, 144, 155n15, 157n37
Athene cunicularia. See burrowing owl
atlases. *See* maps
atomic testing, 205
Auden, W. H., 124–25, 136n50, 137n65
Auld, Jerry, 149

Bailey, Kirk, 216, 218, 220
Bancroft, Hubert Howe, 149
Banff Springs Reserve, 146
Banks, Sir Joseph, 152n2
Barry, Peter, 297
Bass, Rick, 92, 107n32
Bataille, Georges, 23
Bate, Jonathan, 121, 134n4
Baudrillard, Jean, 303, 304, 319, 320n1
BC Hydro, 305, 306, 322n20
Bear River Migratory Bird Refuge, 194, 196, 197, 198, 206
Beaver Creek Wood Bison Ranch, 223, 224, *224*
becoming-animal, idea of, 90–93
becoming caribou: art of, 94–97; ethology of, 97–102; experience of, 92–93, 100; writing and, 94
becoming-minor, 90, 91, 102, 103, 104
Bédat, Claire, 74, 75
Beechey, Frederick, 155n10
bees, 325
Begg, Alexander, 156n24
Being, idea of, 89–90
Being Caribou (film), 87, 88, 93, 97, 102–3, 104–5

Being Caribou: Five Months on Foot with an Arctic Herd (Heuer), 87, 88, 91
Bell, Vikki, 60
Bennett, William Andrew Cecil, 307
Bentley, D. M. R., 286n17
Berger, John, 281
Bernstein, Michael André, 293
Biespiel, David, 129, 134n3
biology, 242, 253
biosphere, 267–68
Bird, Louis, 28, 29, 30
Bird, William, 248, 250
birds: annual migration, 199; benefits of study of, 230n39; burrowing owl, 196, 205–6; eagle, 233, 280; endangered, 201; farming and protection of, 206; grassland, 16–17, 200–201; humans and, 193, 195; in literature and mythology, 193, 194, 195–96, 199; loss of habitat of, 202–3; marsh wren, 74; plastic debris in dead, 266; poetic representation of, 233, 235; shorebirds, 99, 197; threats to, 208n50, 208n53; toxicity of environment and, 203–4; world without, 193–94. *See also* ducks
bison, 223–24, 224*fig*, 225, 253–54
bitumen industry: in Alberta, 222; bison as mascots of, 225; death of ducks and, 219; environmental impact of, 212–13, 214, 223; ethics of, 329–30; government regulations and, 213–14; health issues and, 218; justification of, 213, 221, 227; in literature, 225–28; oil extraction technology, 211; origin of, 221–22, 225; problem of sustainability, 218; public relations, 213; stories and entertainment, 228. *See also* oil exploration; petroculture
Blair Report (1950), 221
Blanchot, Maurice, 220
Blaser, Robin, 267, 268
"bodies politic," idea of, 104, 107n35
body: harmony of human and animal, 103–4; latitude and longitude of, 100; without organs, 93, 96, 104, 106n7
Botkin, Daniel, 188
Bott, Robert, 218
Bouton, Charles M., 50

Brand, Dionne, 285n11
Brathwaite, Kamau, 269
Brennan, Andrew, 36
Bringhurst, Robert, 134n3
Brink, Bert, 47
British Columbia: deforestation in, 177; eco-art projects in, 36; forest industry in, 175, 177–78; hydroelectric development in, 304, 308, 314; ideology of colonial governance, 180–81; logging culture in, 180–81; policy toward Aboriginal peoples, 68; transfer of land in, 77n8; tree species in, 188; unionization in, 187. *See also* Treaty Eight (1899)
Brook, Isis, 77
Brown, Laurence, 47, 52, 55, 57, 59
Browning, Elizabeth Barrett, 129
Buell, Lawrence: on ecocriticism, 118, 121, 137n80; on environmental crisis, 1; on toxic discourse, 202; works of, 126, 134n4
Buffalo Pound Provincial Park, 199
Burbridge, Jim, 34, 36, 40, 41
Burbridge, Joan, 34, 35, 36, 40, 41n1
Burke, Edmund, 82, 84
Burroughs, John, 49
burrowing owl, 196, 205–6
Burtynsky, Edward, 26, 328
Bush, George W., 88, 107n37

Calder, Alison, 275
Camosun Bog: aesthetic of, 52–53; age of, 46; artifices of, 57; characteristics of, 46, 48; description of, 43–44, 53; as ecological theatre, 51–52; educational panels in, 45, 46*fig*; history of, 45–47; as natural history diorama, 56; ritual activities and, 51, 62n31
Camosun Bog Restoration Group (CBRG), 43, 47, 51, 56–57, 59–60. *See also* Crazy Boggers
Camosun restoration project: accusations of fakery, 50, 54; benefits and outcome of, 60; characteristic of, 55; educational network and, 59; overview, 44–45, 47; participants of, 58; purpose of, 53, 54
Canada: colonial culture of, 70, 75–76; environmental racism in, 265; as nation, 221; perceptions of, 68; problem of self-identification in, 68, 69. *See also* Canadian literature
Canada v. *Lameman*, 254
Canadian Alpine Journal, 150
Canadian Alps, The (Sandford), 152
Canadian literature, 277, 287n20, 291–92, 296
Canadian Mountaineering Anthology, 150
Canadian Pacific Railway (CPR), 145
Canadian Stock Raisers' Journal, 168
Canadian Wildlife Service, 16
cancer, 202, 204, 205
capitalism, 267, 270
carbon, 178, 188
caribou, 87–88, 89, 92, 104
Caribou Rising (Bass), 92
Cariou, Warren, 6, 7, 328
Carruthers, Beth, 4, 8, 36
Carson, Rachel, 9
Carter (character), 181–83, 189
Cautley, Richard W., 157n52
CBC Hourglass, 304–7, 308, 314, 315, 316*fig*, 319
certainty, 161–62, 171, 233. *See also* uncertainty
Chan, Ficus, 259, 265
Chisholm, Dianne, 7, 134
Christian, Carol, 217
Christian, Dorothy, 262
circuitry, 270
Clark, James S., 176
Clark, Karl, 211, 222, 223
climate change: computer models of effect of, 176; cultural values and, 24; forest ecosystems and, 188; forest industry and, 7, 175–76; freedom and, 189; government reports on, 176, 177, 178; literary metaphor of, 327; research on, 176–77
climate data, 163, 173n11
Climbing in North America (Jones), 151
Coleman, Arthur P., 144–45, 146–47
Coleman, Lucius, 145
Collapse (Diamond), 269
collective imagining, 215
Collie, J. Norman, 147, 148–49
colonization, 8, 69, 262–63
Columbia Icefield, 148

Columbia River Treaty, 308, 314, 321n12
Columbia River Valley, 315*fig*, 318*fig*
Coming into Contact, 121
communities, 34–35, 211–12, 215, 218–19, 228
community-based projects, 37–38
Companion to the Botanical Magazine (Hooker), 149
Compton, Wayde, 269
Cook, Eleanor, 123, 125–26
Cook, Fredrick, 152
Cox, Ross, 157n39
coyotes, 243
Crawford, Alec (character), 183, 185–86, 187, 189, 192n36
Crazy Boggers, 47, 48, 51, 59, 60. See also Camosun Bog Restoration Group (CBRG)
Cree, 252–53, 255
Cross-Pollination workshop, 2–3, 8, 325, 327
cultural politics, 7
culture, in relation to nature, 52–53, 55

Daedalus, Stephen (character), 71
Daguerre, Louis-Jacques-Mandé, 50
Danger on Peaks (Snyder), 132
Dante Alighieri, 66, 71
Darwin, Charles, 143
Dasenbrock, Reed Way, 292
Davidson, Tonya K., 5
Davies, Charlie (character), 186
Davis, Peter, 58
Davis, Stephen, 201
Dawn Chorus Celebration, 73
Dawson, Carrie, 285n11
Deactivated West 100 (McKay), 127
death: characteristics of, 219–20, 221; communal perceptions of, 219; of ducks, 215, 230n29; immortality and, 217; justification of, 220; logic and meaning of, 212–13; in oil and coal industry, 222
debris, 306, 307–8, 318*fig*
Deleuze, Gilles: on animals, 107n23; on body, 100; on Europeanization, 107n38; on fabulation, 94; on haecceity, 106n5; on joy and sadness, 101, 107n29; philosophical ideas of, 89–90, 91; on writing and writers, 94, 98, 107n22; on "zone of proximity," 93, 106n14
Derrida, Jacques, 83–84, 220
Diamond, Jared, 269
Diefenbaker, John, 221
"different yield," concept of, 261–62
Dillingham, William, 225
dioramas, 49–50, 51–52, 62n55. See also habitat dioramas
discipline, 326
Dōgen, 38
Donald, Dwayne, 252
Donnan, John, 243
Douglas, David: Aemilius Simpson and, 156n37; as botanist, 141, 153; as cartographer, 143; death of, 144; description of Rocky Mountains by, 141–42, 150; fieldnotes of, 149, 150, 157n37; life and career of, 139–40, 141, 154n9; manuscript of, 142–43, 152–53; portrait of, 140*fig*; reputation of, 151–52; scholars and writers on, 150–52; scientific accomplishments of, 142, 152; as specimen collector, 143–44; travelogue of, 141–42, 144, 149
Douglas, James, 77n8
Douglas fir, 153, 177
Drummond, Thomas, 154n10, 156n37, 157n39
ducks, 212, 219, 230n29
Dutton, Denis, 75
Dvorak, Marta, 281

earth, fragility of, 234, 235
eco-art projects, 36, 329
eco-crisis, nature of, 65, 66
ecocriticism: characteristics of, 118–19, 120, 121, 122, 126; criticism of, 135n13; definition of, 118, 134n4; interdisciplinary nature of, 118; notion of place in, 137n80; observation as method of, 126; scholarly debates on, 121–23, 135n13; studies of, 134n4; theory of, 121–22
Ecocriticism (Garrard), 120
ecological modelling, 36
ecological restoration, 9, 54, 58
Ecologies of Affect, 5
ecology, 121, 195, 207n9, 207n10
Ecology without Nature (Morton), 120

ecosystems, 207n23
Edmonton: absence of cultural records in, 258n34; archaeology of, 252; bicycle and walking paths, 250; condo development in, 244; construction works in, 248; fur trade posts in, 252; homelessness in, 250, 251; murals in, 253, 258n32; natural disasters in, 248; population size, 248; social and economic dynamics, 249; water quality in, 250
Edmonton Yukon & Pacific Railway, 250
Eliot, T. S., 124
Elliot, Robert, 44
Ells, Sidney, 225, 226, 227
Elsewhere: definition of, 281, 282; disorientation of self and, 283; experience of, 281; lostness in, 284; overview, 276; as postcolonial disruption, 282. *See also* Here; There
Elverum, Duane, 266
Embers and the Stars, The (Kozah), 67
endangered species, 198, 201
Endless Forms Most Beautiful (Osborne), 110, 112
environment: agricultural impact on, 17; in documentaries, depiction of, 309; efforts to protect, 59, 263–64; human activity and, 5–6; human health and toxicity of, 204; impact of bitumen industry on, 212–13, 214, 223; importance of knowledge of, 161, 328; poetic representation of, 118, 123, 126–31, 133–34, 227, 236–37; pollution of, 235, 245; reading, 242; studies in literature and, 118–19, 123; urban, 244
environmental activism, 9, 12n16, 33–34, 88
environmental change, 68, 261
environmental education, 50–51
Environmental Imagination, The (Buell), 126
environmental policy, 8–9, 12n12, 12n14
environmental studies, 1–3, 4–5, 6
epic genre: in Canadian literature, 291–92; characteristics of, 292, 294–95; examples of, 292–93; inclusiveness of, 298–99; "pocket epic," 295–96; as poetic form, 289–90

escarpments: agriculture and, 159, 163–64; definition of, 159; impact on climate, 165–66; on maps, 160*fig*; in Ontario and Manitoba, comparison of, 161–62
Essig, Laurie, 222, 228
Ethical Oil (Levant), 213
ethology, 90, 98–102
Evanoff, Richard, 59
Evernden, Lorne Leslie Neil, 67

fabulation, concept of, 91, 94
Fairley, Bruce, 150, 151
farming: effective practices, 167; environmental knowledge and, 168, 170; in Manitoba, 159; in Ontario, 159, 166–67; profitability of, 168; protection of birds and, 206
fawns, photographs of, 81*fig*, 84*fig*
fells of brightness, The (Whyte), 289
Field Guide to Western Birds (Peterson), 195
figuration, 126
Findlay, Len, 212
Finlay Forks village, 309, 310*fig*, 311, 313, 314*fig*
fish, poetic representation of, 234–35, 236
Fisher, Roy, 290, 297, 298, 299
Fleming, Sanford, 202
Fletcher, Mary, 28
flooding, 165–66, 194, 199
Flora Boreali-Americana (Hooker), 143, 144, 154n10, 156n37
food, culture of consumption of, 17–18, 30–31
forage (Wong), 131, 132
forest: composition, changes in, 177–78; ecosystems and climate change, 188; experience of hiking in, 259–60; migration of, 176; in poetry, 276, 280; species in, 177
forest dark, metaphor of, 66
forest industry: climate change and, 175–76, 188–89; death accidents in, 179, 191n19; decision-making in, 178; discourse of independence and, 179–81, 187–88, 189; in fiction, 181–82, 183–88; historiography of, 177–78; operational risk, 179; origin

of, 179–81; research on, 178; technological change, 184; unionization in, 183, 184–85, 187
Fort Chipewyan, 218
Fort McMurray Today, 216
Francis, Daniel, 285n4
Franklin, John, 143, 155n10
Fraser, Simon, 234
"Freedom and Necessity in Poetry" (Auden), 124
Freeman, Mark, 111*fig*, 112*fig*, 113*fig*
"Friends of Science," 12n10
frontier, 180, 185, 226
Frow, John, 326
fruits, cultivation of, 167–68
Frye, Northrop, 69, 273, 285n7, 292
Fukushima nuclear disaster, 216
Future Forests Ecosystems Initiative (FFEI), 176
Future of Environmental Criticism, The (Buell), 1, 121

Gablik, Suzi, 36
Gaglardi, Phil, 314
garbage dumps, 27–28. *See also* waste
Gardens of Babylon Habitat Challenge, 73
Garrard, Greg, 120, 121, 135n13
Gayton, Don, 189
genetically modified organisms (GMOs), 110
geological sublime, 127–28, 129
Gift of Death, The (Derrida), 220
Gladu, Shirley, 255
Glissant, Édouard, 266, 268–69
Glittering Mountains of Canada, The, 150
Glotfelty, Cheryll, 195–96
Gobster, Paul, 51, 52, 53, 60, 61n29
Goodstoney, Jonas, 146, 147
Gordon, Jon, 9, 329
Gough, Barry, 180, 181
governance: community cooperation and, 20; European *vs.* Aboriginal, 255; models of, 19; values and better, 19–20; ways to improve, 20–21
Grainger, Martin Allerdale, 181, 189, 190
Grant, George, 69, 220, 221, 225, 227

Grass, Sky, Song (Herriot), 194, 201–3, 205
grassland: birds and, 200–201, 208n50; cultivation of, 17–18, 200, 235; ecoregions of, 16; future of, 21; incentives for protection of, 18, 19; investment in, 328; in literature, 200–201. *See also* landscapes; prairies
Grassland (Manning), 200
Grasslands National Park, 200, 208n39
Gray, Nelson, 72
Great Canadian Oil Sands (GCOS), 223
Great Pacific Garbage Patch, 266
Great Salt Lake, 194, 196, 197, 199
Guattari, Félix: on animals, 107n23; on Europeanization, 107n38; on haecceity, 106n5; philosophical ideas of, 89–90, 91; on writers, 107n22; on "zone of proximity," 93
Gwich'in, 104

habitat dioramas: aesthetic of, 52–53, 59; authenticity of, 50; concept of, 44–45; as ecological theatre, 51; interpretive pedagogy of, 53; in natural history museums, 50. *See also* dioramas
haecceity, 91, 106n5
Haeckel, Ernst, 207n10
Haig-Brown, Roderick, 183, 184, 185–86, 187, 189, 190
Hak, Gordon, 179, 181, 187, 190n
Halleran, Mike, 305–7, 308, 311, 315, 317–18, 319
Hanson, David T., 26
Hardy, Bob, 223
Harper, Stephen, 8
Harris, Howell, 184
Harris, Maureen Scott, 283
Harvey, Miles, 155n14
Heidegger, Martin, 67
Heim, Wallace, 55
Here: geographical location of, 279; as home, 278–79, 283; identity of, 277; overview, 276; transformation to There, 282; travel between There and, 281; uncertainty of, 281–83. *See also* Elsewhere; There
Herriot, Trevor, 8, 194, 199–200, 202–6, 328

Hertzog, Lawrence, 248
Hessing, Melody, 5
Heuer, Karsten: on becoming caribou, 89, 100–102, 107n24; on becoming-minor of Man, 102–3; on calving, 99–100, 101, 103; depiction of caribou by, 92, 97–98, 107n32; ecological activism of, 7, 87, 88; as environmental lobbyist, 102, 107n37; feelings of, 96, 101; on Gwich'in, 104; on nature of body, 99; nomadism of, 104–5; on thrumming, 96, 97, 106n21; views of, 95; writing methods of, 91, 94, 98, 105; in "zone of proximity" to caribou, 93
Higgins, Iain M., 134
Higgs, Eric, 55, 59, 62n55
Hillman, James, 71, 76
History of British Columbia (Bancroft), 149
Hobbes, Thomas, 66, 211
Hodgins, Jack, 132
Hogan, Linda, 261
Holmes, Nancy, 8, 329
Holt, Johnny (character), 183, 185–86, 186–87
Holy Forest (Blaser), 267
home, 8, 283, 298
homelessness, 250, 251, 274
Homo naturalis, 317
Hooker, William Jackson, 140–41, 143, 144, 154n9, 155n12
Hooker and Brown (Auld), 149
Hopkins, Gerard Manley, 132
Horticultural Society of London, 139, 140, 153n2, 154n9
Houston, Mary, 202
Houston, Stuart, 202
Howarth, William L., 198
Hryniuk, Margaret, 15
Hudson's Bay Company, 141, 146, 149
Huey-Heck, Lois, 37
Huggan, Graham, 286n17
humans: animals and, 244; danger of extinction of, 260–61; in European anthropology, 255
Hupfield, Maria, 264
Husserl, Edmund, 128
Hutton, James, 294
hydroelectric developments, 304, 314–15, 321n11

incompleteness, 41
independence: forest industry and idea of, 182, 189; frontier and myth of, 180–81; loggers', 184–85, 187–88
Indigenous nationhood *vs.* nation states, 267
Indigenous peoples. *See* Aboriginal peoples
Ingold, Tim, 67, 68, 72–73, 75
installations: consideration of place in, 111; media, 109–10, 111*fig*, 112*fig*, 113*fig*; repetition of form in, 112; use of scale in, 110
interdisciplinary studies, 1–2, 11n3, 325–26, 327, 330
Interdisciplinary Studies in Literature and Environment (ISLE), 118
Intergovernmental Panel on Climate Change (IPCC) report, 175–76
International Day of Mourning, 218
Inventory (Brand), 285n11
Island of Lost Maps, The (Harvey), 155n14
Ituna (Sask.) Landfill, 27

Jackson, Wes, 18, 40
Jamieson, Karen, 73
Jeffers, Robinson, 130
Jensen, Doreen, 264
Jones, Chris, 151
Jordan, Chris, 266
Jordan, William R., III, 57, 58
Joyce, James, 71
Juan de Fuca Provincial Park, 133–34
Justice, Daniel Heath, 267

Kafka, Franz, 90
Kain, Conrad, 149
Kaktovik village, 101, 105
Kalman, Harold, 253
Kamboureli, Smaro, 286n16
Kant, Immanuel, 82, 84
Katinas, Tom, 213
Katz, Eric, 44
Kelly, Erin, 213
Kelly, Van, 294–95
Kelowna (B.C.), 33, 37
Kerrigan, John, 296
Keyano College, 214, 230n39
Kilshaw, Denise, 37

Kim, Myung Mi, 269
Kipling, Rudyard, 225–26
Kitsilano, 275, 279, 282, 286n12
knowledge: accuracy of, 7; ancient view on, 39; certain and uncertain, 169–70, 172; characteristics of, 2, 327–28; as destination, 75; environmental, 171–72, 328; farming and construction of, 168, 170; forms of, 40, 80; information and, 72–73; limits of, 228; nature and, 7, 39, 161; perception of local, 168–71; production and transmission of, 326
Kohak, Erazim, 67, 75
Kramer, Reinhold, 286
Kroeber, Karl, 134n4
Kroetsch, Robert, 228, 325, 330
Kulchyski, Peter, 27
Kyba, Daniel, 155n15

Lament for a Nation (Grant), 220, 221
land: fragility of, 236; impact of technology on, 17; legal disputes over, 15; obligation to care for, 204; oil exploration and reclaim of, 227; possession and breaking, 15; *vs.* territory, 269. See also prairies
landscapes: cultivation of, 200, 202; definition of, 7; humans' connection to, 194, 195; long-term impacts of toxicity of, 204; perspectives of, 132–33; poetic representation of, 118, 126–32, 131; as refuge, 197; relatively stable, 163–66; restoration of, 57; sounds of, 95–96; underbridge, 245, 246. See also escarpments; grassland
Land Trust Alliance of British Columbia (LTA), 176
Lang, Vera (character), 325, 327
Langer, Monika, 66, 67
Lawrence, D. H., 90
LaxHösinsxw (honouring and respecting others), 264
Leach, Andrew, 213
Lee, Abram, 163
Lefebvre, Henri, 279
Legacy of Stone (Hryniuk), 15
Lehman, Otto, 50
Leopold, Aldo, 48
Levant, Ezra, 213, 219, 220, 227, 228
Lévi-Strauss, Claude, 329

lifeworld, concept of, 67
Lilburn, Tim, 273–74
Limerick, Patricia, 180
Linnaeus, Carolus, 81
Lippard, Lucy, 195
literary criticism, 119–20
Living City Forum, 73
logging industry. *See* forest industry
Log Poem, 37
Long Journey (Biespiel), 129
long poems, concept of, 286n16, 296
Loo, Tina, 305, 322n24
lostness, 283–84
Love, Glen A., 119, 120, 135n4
Lovecraft, H. P., 323n49
Lure of the Local, The (Lippard), 195
Lutwack, Leonard, 193

Macdonnell, Dan (character), 181–82, 189
Macoun, John, 202, 208n48, 208n49
Mairs, Lori, 35, 36, 40, 41
Malthusian theory, 5
Mandamin, Josephine, 262
Manitoba, 16, 159–60, 165, 168–69
Manitoba Drainage Commission, 168
Manitoba Escarpment, 159–60, 163, 164–65, 172
Manning, Richard, 200
maps, 144, 286n16
Maracle, Lee, 264
Markham, Clements, 148
Mart (character), 181–83
material interaction: and creativity, 109–113
materialism, 5, 6, 23, 25, 29, 56, 62, 70, 73, 77, 107n35, 119, 200, 202, 244, 256, 257, 266, 268, 270, 279, 285n11; of ethical subject formation, 53
McKay, Bernard, 309
McKay, Don: comparison with Wagoner, 129–30; description of nature by, 127, 134; language and terminology used by, 128; life and career of, 134n3; on limits of language, 128–29; poems of, 6, 118, 126, 127, 193, 283; on purpose of poetry, 123; value of accuracy, 131; vocabulary and poetic language of, 127, 132, 201–2
McKay, Ian, 303, 307, 312, 317
McLuhan, Marshall, 65, 68, 71, 320n1

McNabb, Mary Ann, 15, 21
media of art objects, 109–10, 329
Melville, Herman, 90
Men There Were Then, The (White), 179
Métis culture, 27, 28
Midgley, Mary, 125
Mill Creek Bridge: debris and junk under, 241, 244, 249; fears of riding under, 246; graffiti on, 247; history of building, 243; homelessness and, 251; impact on environment, 249; as marker, 246; overview of, 6, 241; in political and social implications, 255; public space and, 249–50; reading, 241–42; as symbol of capitalism, 249; weight capacity of, 247
minoritization. *See* becoming-minor
Mitchell, Alanna, 261
Mitmann, Greg, 161
modernism, 293–94
Monbiot, George, 11n8
Montana, 19
Morantz, Alan, 273
Morris, Julie (character), 183, 185–86
Morton, Peter, 298
Morton, Timothy, 120, 121, 135n13, 290, 291
Morwood, William, 153n2
mountaineering, history of, 153n4
Mount Brown: discovery of, 141; Douglas' travel to, 149–50, 152; in fiction, 149; on maps, 143*fig*, 145*fig*; measurement of, 146, 148, 156n37; photograph of, 151*fig*; position of bolt and cairn, 157n52; Collie's search for, 147
Mount Hooker, 141, 143*fig*, 145*fig*, 146, 148
Mount Kitchener. *See* Peak Douglas
Mount Lefroy, 147
Mount McKinley, 152
Mount Robson, 145
Murphy, Michelle, 161
Murray, John, 142, 144, 150, 153
museumification, 51, 54, 61n29
museums, 58, 82
Musqueam Creek, 263
Musqueam Ecosystem Conservation Society, 263
Musqueam, 45, 46

Nadeau, Sylvie, 258n32
Nancy, Jean-Luc, 211, 212, 218, 228
Narayan, R. K., 279
Nash, Roderick, 180
nations and nation states, 212–14, 215–16, 220, 221, 267, 277
Native Plant Society of Saskatchewan, 17
Natural Alien, The (Evernden), 67
natural history, 49, 54, 196
naturalist clubs, 48, 49
nature: benefits of conservation of, 54; *vs.* capital, 214; culture and, 52–53, 55; definitions of, 71–72; health of, 194–95, 199, 205; human relationships with, 6, 7, 9, 48, 311; knowledge and, 7, 39, 161; poetic representation of, 129, 226, 233, 234; *vs.* science, 48; technological progress and, 226; writers and, 49
"Nature, History and Poetry" (Auden), 124
Nature Vancouver, 48
Nelles, H. V., 303
nest, metaphor of, 74–75
New, W. H.: communication with, 286n12; ecological thinking of, 277; literary career of, 285n9; poetic language of, 277–80, 283, 284, 286n17; on wilderness myth, 285n4; works of, 273, 274–75, 282
Niagara Escarpment, 159–60, 163–64, 172
Nietzsche, Friedrich, 303
North Saskatchewan River Valley, 251
nuclear power, 222
Nyhart, Lynn, 50

ocean, 261, 266
O'Connell, Beatrice, 137n65
"Oh Lovely Rock" (Jeffers), 130
Oil! (Sinclair), 216
oil exploration: accidents during, 214, 216, 222; caribou migration and, 87–88; ethics of, 219–20; poetic representation of, 227–28; water pollution and, 265. *See also* bitumen industry; petroculture
Okanagan, 33, 34
Oliver, Frank, 254
Oliver, Joe, 9, 12n12

Olympic Peninsula, 117, 127, 131
"On a Mountainside" (Wagoner), 118, 129
Ontario, 159, 161, 166–67
onto-ethology, 105
ontology, 65, 66–68, 70, 75, 90, 91
Open: Man and Animal, The (Agamben), 242
oral culture, 39
Organic Machine, The (White), 161
Osborne, Lyndal, 6, 329
outside, metaphor of, 267–68

Palliser, John, 148
Papaschase First Nation, 254, 255
Parfitt, Ben, 177, 188, 189
Park, Ondine, 5
Parr, Joy, 161
Parry, William, 155n10
pastahowin (blasphemous act), 28–29
Peak Douglas, 148, 156n33
Pearson, Lester, 221
pesticides, 202–3, 208n53, 208n54
Peterson, Roger Tory, 195
petroculture, 220, 329
Phillips, Dana, 126, 290
photography, 26, 80, 82–83, 303
Pierre, Keom, 317
Pierre, Suzanne, 317
Piper, Liza, 161
pitseed goosefoot *(Chenopodium berlandieri)*, 246
place: dynamic nature of, 277; in ecocriticism, 137n80; geography and ontology of, 273–74, 275, 276; perception of, 198, 298; poetic representation of, 133, 195, 297; sense of belonging to, 10, 274, 275. *See also* Elsewhere; Here; There
Plato, 67
poetics, 266–67, 268
poetry: characteristics of, 123, 124, 133; in comparison to science, 125–26; figuration in, 126; as form of knowledge, 123, 124; genre of long poem, 286n17; power of, 134; representation of environment in, 118, 123, 126–31, 133–34, 227, 236–37; scholars on, 123–24; vocabulary and language in, 132

Pollan, Michael, 17
Poole, Mike, 305, 306
Porcupine Caribou Herd, 87, 88, 102
postcolonial, 119, 137n80, 276, 277, 282, 287n20
Pound, Ezra, 293
Practical Ecocriticism: Literature, Biology, and the Environment (Love), 119
prairies, 4, 16, 18–21. *See also* land; grassland
Pratt, E. J., 291–92
Pritchard, Allan, 181
Pushing the Limits (Scott), 151

Rae, Jen, 11
Raglon, Rebecca, 5
Raibmon, Paige, 311
rainforest, loss of, 16
Rajala, Richard, 184, 189, 190note
reading: environment, 242; as form of listening, 245; graffiti, 246; horizontal, 244, 255; paranoid, 80; practice of, 242–43; spatial way of, 242, 328; storybook, 278–79; theriomorphic, 243
reality: death of, 303; denaturing, 303, 320n2; documentary films and separation from, 313–14; escape from, 317–18; masking of, 308, 319–20; photography and, 303–4
Reckoning, The, 304, 305–6, 308, 317, 319
Refuge (Williams), 193, 195–96, 198, 205
Regional District of the Central Okanagan, 35–36
resourcism, 67, 68
restoration: advocacy of, 59; aims of, 55–56; approaches to, 57; authenticity and forgery in, 58; characteristics of, 52, 56; *vs.* conservation, 59; design, 55; ecology, 54; historic sites and, 62n55; natural history and, 54
return, metaphor of, 284
Richards, I. A., 125
Richards, Thomas, 155n15
Richardson, John, 154n10
Ricou, Laurie, 117
Riffaterre, Michael, 123
Robbins, David, 153n4
Robinson, Harry, 35
Robinson, Zac, 4, 7, 11, 142*fig*, 151*fig*

Robisch, S. K., 121–22
Rocky mountains: descriptions of, 148; exploration of, 141, 145, 147, 149, 156n24; heights of, 149, 155n13, 157n37; as literary subject, 294; maps of, 143, 144, 145*fig*, 148; measurement of peaks, 155n15, 157n37; in poetry, 289–90, 296; tourism in, 145–46; weather in, 150
Roethke, Theodore, 137n65
Ronell, Avital, 268
Rowley, Mari-Lou, 227
Rueckert, William, 134n4

sacrifice, 219, 221, 227
Saletan, William, 222
salmon, 234, 235, 260, 262, 263, 264
Saltfleet Township, 163, 164
Sandford, Robert, 152
Sandilands, Catriona, 5
Saner, Reg, 205
sanitation, 24–25
Sarbach, Peter, 147
Schäffer, Mary T. S., 296
Schama, Simon, 7
Schindler, David, 213
sci-arts collaborative projects, 72–75
science, 95, 119–20, 121, 125–26
Science and Poetry (Richards), 125
Scott, Chic, 151
Scott, J. R., 170
Sea Sick: The Global Ocean in Crisis (Mitchell), 261
Sedgwick, Eve, 80
self-sacrifice. *See* death
Sellers, Christopher, 161
Selters, Andy, 151
Semchuk, Sandra, 69
sensations, 94–95, 106n10
Serres, Michel, 268
Service, Robert, 318
Seton, Ernest Thompson, 49
shame, 252
Shields, Robert, 5
Shiva, Vandana, 260, 270
Shukin, Nicole, 223
Significance of the Frontier in American History, The (Turner), 180
Silverstone, Peter, 213

Simpson, Aemilius, 149, 156n37
Simpson, Sir George, 149
Sinclair, Upton, 216
Siporin, Ona, 195
Siwash (character), 274, 275
Siwash Rock, 285n8
Smith, A. J. M., 277
Smith, Conrad, 166–67
Smith, Damaris, 166–67
Smith, E. D. (Ernest D'Israeli), 159, 167, 168, 170
Smith, Michael V., 40
Snow, C. P., 125
Snyder, Gary, 38, 124, 132
Sokalski, Mitch, 47
Solnit, Rebecca, 268
SongBird Oratorio, 73–74
SongBird Project, 72–75, 76
space, 279, 280, 281
Sparrow, Willard, 263
Spirit Lives in the Mind, The, 30
Stanley Park (Taylor), 274
sternwheeler SS *Minto*, 312, 313*fig*
Stewart, Christine, 6, 10, 328
Stewart, D. F., 169, 170
stone: etymological associations of, 127, 128; poetic representation of, 129
Stoney Nakoda, 146
Strike/Slip (McKay), 118, 127, 128, 133
Stutchbury, Bridget, 200
sublime, idea of, 82–83
Sugars, Cynthia, 277, 282
Sullivan, J. G., 168–69
Sullivan Report, 169, 170
Syncrude Canada, 213–14, 225
Szabo-Jones, Lisa, 1, 3, 9, 11, 49
Szeman, Imre, 214
Szerszynski, Bronislaw, 55

Tackle Climate Change – Use Wood (Ward), 177, 178
Tallmadge, John, 195
tar sands, origin of name, 271n9
Taxonomia photo series, 81, 83
Taylor, Alan, 161, 171, 275
Taylor, John, 129
Taylor, Timothy, 44, 48, 54, 274
There: geographical location of, 279; overview, 276; poetic representation

of, 279–81; transformation of Here to, 282; travel between Here and, 281. *See also* Elsewhere; Here
This Elusive Land: Women and the Canadian Environment, 5
Thompson, David, 157n37
Thorington, James Monroe, 149, 150, 151, 152, 157n39
thrumming, 96–97
tidalectics, 269
Timber (Haig-Brown), 183–86, 189, 192n36
time, perception of, 233, 245, 255
Tomah, Robert, 317
tombs, cultural significance of, 216–17
Topophilia (Tuan), 198
Tranströmer, Tomas, 284
Traveling Light (Wagoner), 129
Treaty Eight (1899), 77n8
Treaty Six (1876), 254, 255, 258n41
Truth of Ecology, The (Phillips), 126
Tse Keh Nay (also Tsay Keh Nay, Tsay Keh Dene, Tall Grass Indians), 8, 305, 309, 312, 312*fig*, 315, 319, 323n71, 324n72, 352
Tuan, Yi-fu, 198
tundra, 92, 95
Turner, Frederick Jackson, 180
Turner, Ted, 18
Turning Timber into Dollars (Hak), 181
Turton, Perry, 218
Tyrell, Joseph B., 157n37

Uexküll, Jakob von, 242, 253, 256
uncertainty, 162, 171, 277. *See also* certainty
underbridge: as abstraction, 252; art, 248; debris and junk, 241, 249, 251; dust, 256; ecology of, 255; environment, 241; experience of reading, 246–47, 256, 328; fears of, 246; life, 248; as metaphor of border, 256–57
Underwood Log (New), 274–77, 282, 283, 285, 285n9, 286n12
Underwood typewriter, 280, 286n17
United Food and Commercial Workers (UFCW) strike, 247
University of British Columbia (UBC), 47
U.S. Environmental Protection Act (EPA), 26
Utah, 3, 195

Valenčius, Conevery Bolton, 161, 171
Vancouver (B.C.), 74, 185, 262, 263, 264, 266
Vancouver Island, 127
Vancouver Natural History Society, 47, 48, 73
Vancouver Sun, 314
Venne, Sharon, 255
Vermeulen, Pieter, 211, 212, 215, 228

W.A.C. Bennett Dam, 304, 307, 317, 321n11, 321n12
Wagoner, David: characteristics of poetry of, 6, 126, 131; comparison with McKay, 129–30; language and literary forms used by, 125, 130, 132; life and career of, 134n3, 137n65; poems of, 117, 118; representation of nature by, 129
Walcott, Derek, 279
Waldo (B.C.), 308, 309
Walt Whitman Bathing (Wagoner), 129
Ward, Roxane, 177
Warshall, Peter, 267
Washington (D.C.), 102
waste: artistic representation of, 26; cultural values and, 24, 31; greed and, 29–30; human relationship to, 23–24; idea of, 23; ideology and, 24; industrial, 26; public perception of, 25; sanitation and, 25; separation between self and, 25; in traditional cultures, 27, 28
Waste Land (Hanson), 26
waste management facilities, 3. *See also* Ituna (Sask.) Landfill
wastewest, 24, 25–26
water: ecological crisis and, 264; methods of cleaning, 265; pollution, 264–65; power of, 117; sacred nature of, 262; tracking flow of, 265–66; wild stream protection, 263
Waterton, Claire, 55
Ways to the Sky (Selters), 151
Weisman, Alan, 264
Wenkchemna (Whyte), 289, 291, 297
West, 3–4, 8
wetlands, 193–206; bogs, 53, 63n68; marsh, 46, 196, 197, 198; swamp, 315. *See also* landscapes
What the Crow Said (Kroetsch), 325

Wheeler, Arthur O., 150, 151
Where Is Here? (Morantz), 273
White, Howard, 179
White, Richard, 161, 180
Whitehead, Alfred North, 125
Whiteman, Maria, 6, 79*fig*, 81*fig*, 83*fig*, 84*fig*
Whyte, Jon, 289–93, 294, 295, 296, 297, 298–99
Wilderness and the American Mind (Nash), 180
wilderness myth, 273, 285n4
Wilkin, L. C., 159
Williams, Terry Tempest: death of mother, 197; education of, 206n9; family and ancestors, 206n5; life and experience of, 194–95; on nuclear weapons testing, 205; testimony of, 208n59; on understanding nature, 206; works of, 194
Williston, Ray, 307
Williston Lake Reservoir, 306, 307, 316*fig*, 317
Wilson, J. Tuzo, 294
Wilson, Tom, 148
Winterson, Jeanette, 71
Wisakaychak stories, 29–31
Wisconsinan glacier, 251
Wittgenstein, Ludwig, 220

Wonders, Karen, 44, 45, 50, 54–55, 57, 58
Wong, Rita, 8, 10, 131, 132, 328
Woodall, Christopher W., 176
Wood Bison Gateway, 224*fig*
Wood Buffalo National Park, 266
Woodcock, Brian, 47
Woodhaven Eco Art Project, 35, 38, 40, 41, 329
Woodhaven Nature Conservancy, 33, 34–35, 37, 40
Woodsmen of the West (Grainger), 181–83, 189
workers, killed and injured, 217*fig*, 218
world: human relationship with, 67, 76; metaphor of transformation of, 235–36
World Network of Biosphere Reserves, 171–72
World Without Us, The (Weisman), 264
Wright, Stephen, 37–38, 41
writing, act of, 280, 281, 287n20
Wyile, Herb, 287n20

Young, Geoffrey Winthrop, 147
Yuggoth (Pluto), 313, 323n49
Yukon, 3, 8, 87, 104

Zhu, Kai, 176
Zwicker, Heather, 253

Books in the Environmental Humanities Series
Published by Wilfrid Laurier University Press

Animal Subjects: An Ethical Reader in a Posthuman World ▪ Jodey Castricano, editor ▪ 2008 ▪ ISBN 978-0-88920-512-3

Open Wide a Wilderness: Canadian Nature Poems ▪ Nancy Holmes, editor ▪ 2009 ▪ ISBN 978-1-55458-033-0

Technonatures: Environments, Technologies, Spaces, and Places in the Twenty-first Century ▪ Damian F. White and Chris Wilbert, editors ▪ 2009 ▪ ISBN 978-1-55458-150-4

Writing in Dust: Reading the Prairie Environmentally ▪ Jenny Kerber ▪ 2010 ▪ ISBN 978-1-55458-218-1

Ecologies of Affect: Placing Nostalgia, Desire, and Hope ▪ Tonya K. Davidson, Ondine Park, and Rob Shields, editors ▪ 2011 ▪ ISBN 978-1-55458-258-7

Ornithologies of Desire: Ecocritical Essays, Avian Poetics, and Don McKay ▪ Travis V. Mason ▪ 2013 ▪ ISBN 978-1-55458-630-1

Ecologies of the Moving Image: Cinema, Affect, Nature ▪ Adrian J. Ivakhiv ▪ 2013 ▪ ISBN 978-1-55458-905-0

Avatar and Nature Spirituality ▪ Bron Taylor, editor ▪ 2013 ▪ ISBN 978-1-55458-843-5

Moving Environments: Affect, Emotion, Ecology, and Film ▪ Alexa Weik von Mossner ▪ 2014 ▪ ISBN 978-1-77112-002-9

Found in Alberta: Environmental Themes for the Anthropocene ▪ Robert Boschman and Mario Trono, editors ▪ 2014 ▪ ISBN 978-1-55458-959-3

Sustaining the West: Cultural Responses to Canadian Environments ▪ Liza Piper and Lisa Szabo-Jones, editors ▪ 2015 ▪ ISBN 978-1-55458-923-4